THE
EQUATIONS
WORLD

BORIS PRITSKER

Dover Publications, Inc.
Mineola, New York

Bibliographical Note

The Equations World is a new work, first published by
Dover Publications, Inc., in 2019.

International Standard Book Number
ISBN-13: 978-0-486-83280-7
ISBN-10: 0-486-83280-5

Manufactured in the United States by LSC Communications
83280501   2019
www.doverpublications.com

Dedication

"The Equations World" is dedicated to people very close to me, who I love and admire and who had great influence on my life.

To my teachers, former colleagues and close friends at Kiev high school #49; Yarovaya Faina Romanovna, for her friendship and love, encouragement and constant support; and in memory of the late Elgort Leonid Isaakovich and Anisimova Lyudmila Veniominovna.

To my college professor and math scientific advisor at Kiev State Pedagogical University Mikhalin Gennadi Aleksandrovich, who inspired, supported, and guided my first steps in the profession.

To my loving wife Irina, for her understanding, inspiration, sense of humor, and her patience towards my recent book-writing activities.

With great love to my sons Aleksandr and Bryan, whose futures I believe are bright and boundless. Special gratitude goes to Bryan for his help in editing several chapters in the book.

In memory of my beloved parents Polina and Samuil, whose unconditional love, support, and faith in me always helped me in the past and will stay with me for the rest of my life.

Contents

Preface

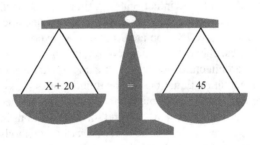

"The weight of a cake equals one kilogram and half of the cake's weight. What is its weight?" my dad asked me with a smile, expecting a quick response. I was eight or nine years old back then and loved those cute logical puzzles he gave me from time to time. Without hesitation I answered, "It's easy. The cake's weight is 1.5 kilograms." (I was born and grew up in Kiev, Ukraine, where metric system measurements were used). "Wrong. Think again." It took me a while to finally realize the trick and solve the problem with his help. I have to admit, I admire my dad's patience because every time he explained it to me I still was not convinced that my original solution was wrong. "OK. Let's try this. Assume you put this cake on one side of a weighing scale and you place half of the same cake and a 1 kilogram weight on the other scale. The weighing scale must be in balance. Then just assume you cut the cake in half and take from each scale half of it. Obviously, the scales will still be in balance. On the left-hand side there is half of a cake and there is a one kilogram weight on the other side. So, what does it mean?" This time I gave the correct response—since half of a cake weighs one kilogram, then the whole cake should weigh 2 kilograms. WOW!!!

That problem is still one of my favorite illustrations of the brilliance of the use of equations in our lives. A weighing scale, balance, or seesaw is often presented as an analogy to an equation. Indeed, if we denote the cake's weight by x, then it's easy to set up an obvious equation: $x = \frac{1}{2}x + 1$, which will result in $x = 2$. It was one of the first algebra lessons in my life. It was a very convincing example of the importance of equations in problems' solutions.

A few years ago I came across one interesting fact about Yuri Gagarin's first human flight into outer space. Scientists had no idea how the human body and brain would react to unusual space conditions. Their biggest concern was psychological endurance. They were afraid that the astronaut might lose self-control and become insane and destroy the operational panel. As a precaution a special defense system was created to not allow the astronaut to suppress the automatic operation to manual. In order to switch the space ship's pilot program from auto to manual in case of an emergency, Yuri Gagarin would have to open a sealed envelope with a math problem inside. Only after solving the problem would he be able to gain the code for unblocking the operational system from auto to manual. It's hard to imagine a better example of the importance of mathematics in human life!

"What science can there be more noble, more excellent, more useful for men, more admirably high and demonstrative, than this of mathematics," said one of the founding fathers of the United States, Benjamin Franklin. Mathematics is the queen of all technical sciences. In today's world, mathematical principles and mathematical modeling are incorporated into almost every aspect of any scientific research. Physics, chemistry, biotechnology, engineering, robotics, statistics, computer science, architecture, geodesy, you name it, cannot survive without the application of mathematics. Equations are the foundation of the great field of mathematics. They are the lifeblood of mathematics, science, and technology. In one way or another you come across setting up and solving equations in many problems.

This book is about equations. It is about various methods and techniques applied in equations' solutions. From simple problems involving linear equations to difficult and challenging math contest problems, we will explore the amazing world of variables combined in statements that help make complicated things easy and elegant, and reveal deep and beautiful patterns and regulations. As a matter of fact, equations serve as a universal language for all people no matter what nationality or origin they are.

People invented Esperanto as an international auxiliary language to help communicate in various matters. But there is no need for Esperanto when mathematicians talk with each other using math symbols and equations.

Babylonian mathematicians could solve problems involving calculation of areas and perimeters of rectangles as early as 2000 BC. Mathematicians of Greece,

Egypt, India, and China used geometric methods to solve linear and quadratic equations back in 1650 BC. The introduction of the Cartesian coordinate system by the great French mathematician and philosopher René Descartes (1596–1650) provided the link between Euclidean geometry and algebra. The studies of François Viète (1540–1603), Sir Isaac Newton (1642–1726), Gottfried Wilhelm Leibniz (1646–1716), Abraham de Moivre (1667–1754), Leonhard Euler (1707–1783), Joseph-Louis Lagrange (1736–1813), Pierre Laurent Wantzel (1814–1848), Carl Friedrich Gauss (1777–1855), Augustin-Louis Cauchy (1789–1857), Niels Henrik Abel (1802–1829), Évariste Galois (1811–1832), and other prominent mathematicians significantly revolutionized and developed algebraic methods that are broadly used today.

The major topics in the book are: linear equations and systems of linear equations, quadratic equations, Viète's formulas, application of quadratic function properties in solution of the equations with parameters, cubic and quartic equations, Cardano's formula, Ferrari's method, and other techniques for the solutions of cubic and quartic equations, division of polynomials, Little Bezout's Theorem and its applications, irrational equations, exponential and logarithmic equations, trigonometric equations and their applications, conditions for constructability of regular polygons and how they relate to cyclotomic equations, Diophantine and other multivariable equations, systems of nonlinear equations with two or more variables, search for integer solutions with the help of linear homogeneous recurrence relations properties, the Golden Ratio, and the Fibonacci sequence properties. Along with going over some basic fundamental definitions and theorems, we offer a set of non-routine problems that stimulate readers to explore unfamiliar or little-known aspects of mathematics. The main idea is to present an insight into the great field of mathematics by showing diverse problem-solving methods and techniques. The book furnishes a bridge across the mathematical disciplines of algebra, geometry, number theory, and trigonometry. It exposes various equations and systems of equations and covers a wide spectrum of techniques and algorithms for their solutions and applications in more than 280 problems. We restrict our studies mostly to the field of real numbers and try to stay in the boundaries of the high school curriculum. Discussing the history and development of the equations' study is impossible without mentioning complex numbers. The introduction and the use of complex numbers in solving algebraic equations was a great revolution. We will be referring to these numbers in several chapters. However, we feel their properties represent more advanced topics. The detailed discussions of complex numbers, therefore, will be omitted, and while solving equations, we will be seeking real, not imaginary, solutions, unless indicated otherwise.

Additional exercises, varying widely in difficulty, are included at the end of almost every chapter. The book also provides short historical tours, and traces the ideas and techniques for attacking equations developed by the prominent mathematicians through the centuries. It describes not only the discovery of the equations' solutions, but the foundations of mathematical thought as well. We believe that readers should gain a deeper understanding of mathematical concepts in their historical context.

Math is not boring and difficult to understand, as many perceive it. It can be accessible, inspiring, and above all amazingly creative. That feeling of joy that makes your heart tingle when you solve a complicated problem is incomparable to any other feeling in the world! The book demonstrates equations' double role in a problem-solving process. First of all, we try to demystify math by concentrating on useful and efficient hints, tricks, equation solving methods, and techniques demonstrating that problem solving is fun and enjoyable. The book provides important principles and broad strategies for solving equations. Secondly, equations and systems of equations often serve as important tools for expressing the data in a problem. Readers will discover how the analysis of problems and utilization of equations involve fitting parts together to solve a puzzle. We truly hope readers will enjoy the journey into the fascinating *Equations World*.

Chapter 1

Linear Equations

An equation is a statement that claims that two quantities are equal. In mathematics the term "equation" means equality with one or more variables or unknowns. "Variable" comes from the Latin translation "capable of changing." Our ancestors in ancient Babylonia and Egypt stated and solved problems using equations, but not in the same way it is done today. They did not identify any formulas or algorithms for solving equations, but everything was "solved" in verbal descriptions and instructions. Representing known and unknown quantities by letters (variables) was first done by the French attorney François Viète (1540–1603). It was his brilliant idea to work with letters as if they are numbers and obtain the result in which the numbers are substituted at the very end to get the answer. Though not a professional mathematician by education, Viète made quite remarkable contributions to the field of mathematics. He was the first to introduce notations for problem solutions and to formalize mathematics, which allowed for a simplification of algebraic manipulations. Without his accomplishments, it would not be possible to work easily with equations. His approach to use the "variable" as a supplemental tool for solving different problems was a huge step forward in the development of contemporary algebra. He created a new language that allows us to express ideas, logic, and problems in an organized and systematic way.

Viète suggested three stages in solving a problem: First, summarize the problem in the form of an equation. Denote the known quantities by constants-consonants (Viète was the first to use the term *coefficients* to define the numbers in an equation) and unknown by variables-vowels. In 1637 the great French mathematician and philosopher René Descartes (1596–1650) suggested using x, y, z

(the last letters in the alphabet) as variables and a, b, c (the first letters) as constants. This convention is still commonly used today.

Second step—analyze the equation and solve it. At that stage we assess the characteristic of the problem, or *porisma*, as he called it.

Final step—*exegetical analysis*. Find the numerical or geometrical solution of the initial problem.

As a matter of fact, this algorithm is the best description of the whole problem-solving process. Analyze which variable to introduce; decide how to use the conditions of a problem to set up an equation; solve it. The introduction of variables basically translates a problem from a regular language into mathematical terms, providing the ability to obtain a result and then translate it back into a specific outcome.

Let's demonstrate this on the following problem:

Find three consecutive natural numbers such that the square of the middle number is greater by 1 than the product of the other two numbers.

Solution. Denoting by x the smaller of the numbers, we express the following numbers as $x + 1$ and $x + 2$. According to the given conditions, $(x + 1)^2 = x(x + 2) + 1$. This statement can be rewritten as $x^2 + 2x + 1 = x^2 + 2x + 1$ and it clearly holds for any value of x. Thus, we conclude that any three consecutive natural numbers will satisfy the problem's statement! The problem was solved by merely translating its conditions into algebra terms.

By a mathematical equation we will understand a mathematical sentence that states that two expressions name the same number. Thus $5x - 30 = 0$ is an equation that is true when x is 6 and false for any other value of x. To "solve" an equation means to find all the solutions (roots) or prove they do not exist. When solving a polynomial equation (equation of the form $P = Q$, where P and Q are polynomials with coefficients from the same number field), it is important to always specify a set of numbers in which the solutions are allowed. In this book we will be interested mostly in real solutions, unless it is specified otherwise.

The simplest of all polynomial equations are the linear equations, which have the form $ax + b = 0$, where a and b are constants ($a \neq 0$) and x is unknown-variable. The equation we considered above, $5x - 30 = 0$, is an example of a linear equation that has just one root.

Many types of problems may be solved by changing the verbal statements into one or more equations. We will start our journey by examining a few of the most commonly encountered "family related" word problems which are easily solvable by linear equations. Under "family related" we mean the problems where the solutions are based on the exact same principles and can be applied to any problem in that family in the same way.

Age Problems

Problem 1. Eight years from now Elly will be twice the age she was 6 years ago. What is her present age?

Solution. The key to the solution of this problem is to use Elly's age now as a variable x. Then 6 years ago she was $x - 6$ years old and 8 years from now she will be $x + 8$ years old. Based on the conditions of the problem, in 8 years she will be twice as old as she was 6 years ago. Therefore,

$$2(x - 6) = x + 8,$$
$$2x - 12 = x + 8,$$
$$x = 20.$$

Elly is 20 years old now.

Problem 2. Mr. Hughes, who is 36 years of age, has a daughter who is 6 years old. In how many years will the father be 4 times as old as his daughter?

Solution. Let x be the number of years until Mr. Hughes is 4 times as old as his daughter. At that time, he will be $36 + x$ years old. His daughter will be $6 + x$ years old. Since in that year he will be 4 times as old as his daughter, it can be represented by the equation

$$36 + x = 4(6 + x),$$
$$36 + x = 24 + 4x,$$
$$3x = 12, \text{ from which } x = 4.$$

In 4 years Mr. Hughes will be 40 years old, his daughter will be 10 years old, and his age that year is 4 times that of her age.

Motion Problems

The formula
$$Speed \times Time = Distance$$
is a major tool in the solutions of a family of problems.

Problem 3. Gary left home at city A at 8 am and was traveling at 40 miles per hour to city B. His brother Mark, who lives in city B, left his home at 9 am and was traveling along the same route to city A with the speed 60 miles per hour. At what time will they meet, if the distance between cities A and B is 240 miles?

Solution. Let x be the number of hours Gary traveled till they met. Mark was traveling 1 hour less, because he left his home 1 hour later that morning. Then Gary covered the distance of $40x$ miles and Mark covered the distance of $60(x - 1)$ miles. Since they traveled in the direction facing each other, the sum of the distances covered by both of them till the point of meeting must equal the distance between the cities, 240 miles. Thus,

$$40x + 60(x - 1) = 240,$$
$$40x + 60x - 60 = 240,$$

$$100x = 300,$$
$$x = 3.$$

Gary traveled 3 hours. Since he left home at 8 am, the brothers met at 11 am.

Problem 4. Two automobiles travel in the same direction with speeds of 40 miles per hour and 50 miles per hour, respectively. How many hours after they are alongside of each other will they be 18 miles apart?

Solution. Let x be the desired number of hours. Then the first automobile will cover the distance of $40x$ miles, and the second automobile will cover the distance of $50x$ miles. The distance between them is 18 miles. Thus,

$$50x - 40x = 18,$$
$$x = 1.8.$$

They will be traveling 1.8 hours or 1 hour and 48 minutes to be at the distance of 18 miles apart from each other.

Problem 5. Two people go down along a moving escalator in a subway. The first person counted 40 steps during his run, the second person counted 60 steps. One of them moves twice as fast as the other. How many steps would they need to go down to reach the bottom of the escalator if it is not in motion (what is its length)?

Solution. They are moving with the escalator, whereby the first person takes less number of steps but more steps come out meanwhile as he is slow. Assume the first person moves with a speed x steps per minute. Then the second person moves with a speed $2x$ steps per minute. The first person spent $\frac{40}{x}$ minutes for his walk and the second person spent $\frac{60}{2x} = \frac{30}{x}$ minutes for his walk. Denote by y steps per minute the speed of the moving escalator. The escalator moved down $(y \cdot \frac{40}{x})$ steps during the first person's run and its length is $(40 + \frac{40y}{x})$ steps. The escalator moved down $(y \cdot \frac{30}{x})$ steps during the second person's run and its length can be expressed as $(60 + \frac{30y}{x})$ steps. Therefore, we obtain the equation $40 + \frac{40y}{x} = 60 + \frac{30y}{x}$. Making simple modifications we get $\frac{10y}{x} = 20$, from which $y = 2x$. Plugging this expression for y in either of the expressions for the escalator's length (in the first one, for example) we get $40 + \frac{40y}{x} = 40 + \frac{40 \cdot 2x}{x} = 40 + 80 = 120$ steps.

Answer: the escalator's length is 120 steps.

An interesting variation of the "motion problems" is the family of so-called clock problems.

Clock Problems

Problem 6. At 3 pm a clock's hands are perpendicular to each other. At what time after 3 pm will they be perpendicular again?

Solution. At 3 pm the clock's hands are perpendicular to each other, because the central angle they form is $1/4$ of the full circle of $360°$, or exactly $90°$. As the hands move, the angle they form will change. The problem is about the speed covered by each hand and the distance between them (in minutes) after specific time passed. In this case the distance each hand travels is measured by the minute markers of the clock. It means that our goal is to express somehow the distance on the circumference covered by each hand until they form the central angle of $90°$ again, which is equivalent to 15 minutes time distance between them. Comparing the speeds of minute and hour hands, we know that the minute hand moves 12 times as fast as the hour hand. Rephrasing the question we may ask: At what time after 3 pm will the difference in distances covered by minute and hour hands equal to 15 minutes? Assume it will happen in x minutes. Then the distance covered by the minute hand is x and the distance covered by the hour hand is $\frac{1}{12}x$. We get the equation $x - \frac{1}{12}x = 15$. It follows $\frac{11}{12}x = 15$, from which $x = 16\frac{4}{11}$ minutes. Therefore, the clock hands will become perpendicular the first time after 3 pm at 3:16$\frac{4}{11}$ pm.

Similar logic can be applied to many problems-siblings of the clock family relatives.

Problem 7. At 12 o'clock the minute and the hour hands overlap. Can you specify each time when it happens again?

Solution. Let's find the very first time when the hands will overlap after 12 noon. Since the hour hand travels 12 times slower than the minute hand, and the minute hand moves in front of the hour hand, they will not overlap during the first hour. As we get to 1 pm they switch the positions and now the hour hand stands in front of the minute hand. So, now the minute hand has to catch up with the hour hand. The distance between them is $\frac{1}{12}$ of the full circle, which is equivalent to 5 minutes time frame. Assume they will meet in x minutes. Comparing the distances covered by each clock hand till meeting point, we get $x - \frac{1}{12}x = 5$, from which $x = 5\frac{5}{11}$ minutes.

We conclude that the first overlap happens at 1:5$\frac{5}{11}$ pm. Obviously, the next time they will overlap is in 1 hour and 5$\frac{5}{11}$ minutes, or at 2:10$\frac{10}{11}$ pm. By adding every time 1 hour and 5$\frac{5}{11}$ minutes to the time at meeting point we can calculate the time of all the next remaining 9 overlaps (we leave that exercise to the reader). In total there will be 11 meetings, the last at 10:54$\frac{6}{11}$ pm.

Work Problems

The idea behind all of the "work related" problems is to express the job done in one unit of time and then translate the conditions of a problem into an equation connecting those job performances of all people or devices involved.

Problem 8. Alex can paint a house in 8 days, and his brother Bryan can do the same job in 12 days. How long would it take them to finish the job if they work together?

Solution. Since Alex can complete the job in 8 days, $\frac{1}{8}$ of the full project will be finished by him in 1 day. Similarly, Bryan would finish $\frac{1}{12}$ of the job in 1 day, also working alone. Let's assume they will finish the project in x days if they work together. Then Alex will perform $\frac{x}{8}$ portion of the job and Bryan will perform $\frac{x}{12}$ portion of the job. Adding those quantities together gives the full job. Or expressing it in algebra terms,

$$\frac{x}{8} + \frac{x}{12} = 1$$
$$\frac{3x + 2x}{24} = 1,$$

$5x = 24$, from which $x = 4.8$. The brothers will be able to finish the job in 4.8 days when working together.

Problem 9. Three pipes are used to fill a pool with water. One pipe alone can fill the pool in 9 hours. The second pipe can fill it in 6 hours. The third pipe can fill it in 3 hours. How long will it take to fill the pool if all three pipes are used simultaneously?

Solution. We will apply the same logic as in the problem above. Assume it will take x hours to fill the pool when three pipes are opened at the same time. It follows that $\frac{x}{9}$ portion of the pool is filled by the first pipe, $\frac{x}{6}$ portion of the pool is filled by the second pipe, and $\frac{x}{3}$ portion of the pool is filled by the third pipe. Hence, we get an equation

$$\frac{x}{9} + \frac{x}{6} + \frac{x}{3} = 1,$$
$$\frac{2x + 3x + 6x}{18} = 1,$$
$$11x = 18,$$
$$x = 1\frac{7}{11}.$$

It will take $1\frac{7}{11}$ of an hour to fill the pool.

Mixture and Solutions Problems

Mixture and solutions problems are most often faced in chemistry or in food-related recipes when one needs to find a specific proportion in a desired mixture. The main idea is to utilize the following principle in setting up an equation:

$$\frac{\text{Quantity of substance dissolved}}{\text{Total quantity of solution}}$$
$$= \text{Fractional part of solution containing dissolved substance.}$$

Problem 10. How many pounds of pure salt must be added to 40 pounds of a 3 percent solution of salt and water to increase it to a 20 percent solution?

Solution. The total amount, in pounds, of a saline solution can be expressed as the pounds of each type of saline solution multiplied by the percent of the saline solution. We have 40 pounds of a mixture of water and salt, in which salt has the weight of $40 \cdot 3\% = 1.2$ pounds. Adding to 1.2 pounds of salt another x pounds gives 20% of salt concentration in the new mixture. The ratio of the new amount of salt $(x + 1.2)$ pounds to the total mixture of $(x + 40)$ pounds must equal 20%, which can be represented by the equation

$$\frac{x + 1.2}{x + 40} = 0.2,$$
$$0.2x + 8 = x + 1.2,$$
$$0.8x = 6.8,$$
$$x = 8.5.$$

You need to add 8.5 pounds of salt in order to get 20% salt in the new mixture.

Problem 11. A cafeteria sells peanuts for \$2 per pound and unsalted cashews for \$8 per pound. They decided to sell the mixture of both items instead of separate products. How many pounds of peanuts should be mixed with 10 pounds of unsalted cashews to get a mixture that sells at \$5 per pound?

Solution. Let x be the number of pounds of peanuts to be mixed. So, the total weight of the mixture will be $(10 + x)$ pounds. The cost of the mixture, in dollars, is $(10 + x) \cdot 5$. On the other hand, the same cost, in dollars, is calculated using the expression $2x + 8 \cdot 10$. Thus, we have an equation

$$(10 + x) \cdot 5 = 2x + 8 \cdot 10.$$

Simplifying and solving the equation gives

$$50 + 5x = 2x + 80,$$
$$3x = 30,$$
$$x = 10.$$

Answer: 10 pounds of peanuts should be used in the new mixture.

Income – Expenses – Tax-Related Problems

This family of problems includes problems involving net income, discounted prices, and tax-related calculations. The main idea is to properly identify the variable selection and apply the formulas for net income and expenses or specific tax calculations to set up an equation expressing a problem's conditions.

Problem 12. Lawn and Landscaping Inc. collected $10,700 from its customers during the month of May. Assuming the company charges a 7% sales tax, a bookkeeper calculated $749.00 of sales tax to be paid to the State Department of Taxation. Is it a correct amount to be remitted?

Solution. It is important to understand that the total cash collections of $10,700 consists of gross income and sales tax charged to customers. Denoting dollar gross sales receipts by x gives $0.07x$ dollars of sales tax collected. Then $x + 0.07x = 10,700$ or equivalently $1.07x = 10,700$. Solving this equation yields $x = 10,000$, which means that the gross income collected for the month of May was $10,000 and the sales tax payable has to be $10,000 \cdot 7\% = \$700$, not $749, which most likely was calculated by a sloppy bookkeeper as $10,700 \cdot 7\% = \$749$. His sales tax amount is overstated by $49.

Problem 13. The partnership paid to partner A $50,000 compensation as a guaranteed payment for the year. It was decided to make a pension contribution to a SEP IRA account on his behalf and treat it as an additional guaranteed payment. The maximum SEP IRA allowable for the year is 20% of the net self-employment earnings. What is the maximum SEP IRA contribution that can be made on partner A's behalf, if his net self-employment earnings for the year equal the difference between the guaranteed payment and half of the self-employment tax at 14.13% (92.35% times 15.3%) rate?

Solution. Assume the maximum SEP IRA contribution to be made is x dollars. Thus, the total guaranteed payment earned is $(50,000 + x)$ dollars. The self-employment tax equals $(50,000 + x) \cdot 0.1413$, half of which is respectively $(50,000 + x) \cdot 0.1413 \cdot 0.5 = (50,000 + x) \cdot 0.07065$. Since the maximum SEP IRA has to equal 20% of the difference between guaranteed payment and half of the self-employment tax, we get the equation

$$x = 0.2 \cdot ((50,000 + x) - (50,000 + x) \cdot 0.07065),$$
$$x = 0.2 \cdot (50,000 + x) \cdot 0.92935,$$
$$x = 0.18587(50,000 + x),$$
$$x - 0.18587x = 9,293.5, \text{ or equivalently } 0.81413x = 9,293.5, \text{ from which}$$
$$x = 11,415.25.$$

The maximum SEP IRA contribution to be made is $11,415.25.

Problem 14. Given p is the price per unit sold in dollars, v is the variable cost per unit sold in dollars, and c is the total fixed cost. Find the number of units to be sold to break even.

Solution. At the break-even point the revenues of the business are equal its total costs. Denoting by x the number of units sold to break even, the total revenue equals px and the total cost is $vx + c$. We have an equation $vx + c = px$, solving which gives $x = \frac{c}{p-v}$.

Digits-Numbers or Integer Problems

Dealing with these problems often it's not enough to use one variable to solve the problem. You may need to introduce two or more variables and link them together in the equation or system of equations.

Problem 15. Is it possible to express the number 1990 as the sum of several (two or more) consecutive natural numbers?

Solution. Let us consider n consecutive natural numbers the smallest of which is x: $x, x+1, x+2, \ldots, x+(n-3), x+(n-2), x+(n-1)$. So, the question is if it is possible to find x that for a specific number n, $n > 2$, the sum of the consecutive n numbers starting from x equals 1990. Adding two numbers equidistant from both ends we notice that their sum will always be the same number, $2x + (n-1)$. Indeed,

$$x + x + (n-1) = 2x + (n-1),$$
$$x + 1 + x + (n-2) = 2x + (n-1),$$
$$x + 2 + x + (n-3) = 2x + (n-1), \ldots.$$

Since the number of these sums is $\frac{n}{2}$, the sum of the consecutive numbers can be expressed as $\frac{(2x+(n-1))\cdot n}{2}$ and we obtain the equation $\frac{(2x+(n-1))\cdot n}{2} = 1990$ or $(2x + (n-1)) \cdot n = 3980$. Let's show that if one of the numbers $(2x + (n-1))$ and n is even then the other one has to be odd. Indeed, if n is even, that is, $n = 2m$, then $2x + (n-1) = 2x + 2m - 1 = 2(x+m) - 1$ is an odd number. If n is odd, that is, $n = 2m + 1$, then $2x + (n-1) = 2x + 2m = 2(x+m)$ is an even number. Therefore, to solve the problem we need to express 3980 as the product of an even and an odd number. It can be done in one of the following ways:

$$3980 = 4 \cdot 995 = 5 \cdot 796 = 20 \cdot 199 = 199 \cdot 20 = 796 \cdot 5 = 995 \cdot 4.$$

Consider the first choice, $3980 = 4 \cdot 995$. Then $n = 4$ and $2x + (n-1) = 995$. Therefore, $2x + (4-1) = 995$. Solving this equation for x gives $x = 496$ and we see that in this case the solution of the problem is the sum of the four numbers starting from 496: $496 + 497 + 498 + 499 = 1990$. We leave it for readers to verify that in the second case it will be the sum of five consecutive numbers starting from 396; in the third case, the sum of twenty consecutive numbers starting from 90; and the fourth, fifth, and sixth cases give the negative solutions, which do not satisfy the conditions of the problem.

Problem 16. The sum of the digits of a two-digit number equals 12. If you add 36 to that number you will get a result, which will be the two-digit number written with the same digits but in a reverse order. Find the original number.

Solution. In this case we need to introduce two variables, x as the first digit and y as the second digit in the original two-digit number. Therefore, $x + y = 12$. Since we are dealing with a two-digit number, it may be written as $10x + y$. The two-digit number written with the same digits but in a reverse order can be expressed as $x + 10y$. Translating the second condition of the problem into algebra gives the equation

$$10x + y + 36 = x + 10y, \text{ or } y - x = 4.$$

We get the system of two simultaneous equations (both must be true at the same time):

$$x + y = 12,$$
$$y - x = 4,$$

It's a simple system, which can be solved either by solving the top equation for one variable in terms of the other one, and then substituting it into the bottom equation; or just by adding those two equations and eliminating variable x. The second option looks easier and we will get $2y = 16$, from which $y = 8$. Substituting 8 for y in $x + y = 12$ gives $x + 8 = 12$, or $x = 4$. Thus, the original number was 48.

Other–Various

The above families of related problems can be extended by readers in their own way. I strongly believe that if you can distinguish a specific problem as belonging to a family, it would significantly simplify the solution. You just need to apply the methods commonly used for that family. There are a lot of problems though that are hard to consider as siblings related to the same family-group. So, let's agree that in this section we will refer to a few interesting problems, not separated into some specific family.

Problem 17. (*Quantum*, May/June 1994, Étienne Bézout's problem.) According to a contract, a worker is to be paid 48 francs for each day worked and is to give up 12 francs for each day not worked. After 30 days the worker is owed nothing. How many days did the worker work during these 30 days?

Solution. Assume he worked x days. Then he did not work $(30 - x)$ days. He was paid $48x$ francs and he gave up $12 \cdot (30 - x)$ francs. It follows

$$48x - 12 \cdot (30 - x) = 0,$$
$$48x - 360 + 12x = 0,$$
$$60x = 360,$$
$$x = 6.$$

He worked 6 days out of the 30 days.

Problem 18. There were 22 people at Bryan's birthday party. Veronica danced with 7 boys, Nicole with 8 boys, Elly with 9 boys and so forth all the way up to Irina, who danced with every boy present at the party. How many boys attended Bryan's birthday party?

Solution. This problem looks very confusing at first glance. How can we link the number of boys dancing with various girls with their attendance at the party? Let's assume there were x girls at the party. Starting with Veronica, who was dancing with 7 boys, every following girl was dancing with an increasing by 1 number of boys. Irina was the only girl who danced with every boy, so she was dancing with $(x + 6)$ boys.

$$\text{First Veronica danced with} \quad 6 + 1 \text{ boys}$$
$$\text{The second girl danced with} \; 6 + 2 \text{ boys}$$
$$\text{The third girl danced with} \quad 6 + 3 \text{ boys}$$
$$\dotfill$$
$$x\text{th Irina danced with} \qquad 6 + x \text{ boys}$$

Hence, the total number of people at this party is expressed by the equation $x + (x + 6) = 22$. It follows that $2x = 16$, resulting in $x = 8$, which means there were 8 girls at the party. Therefore, $22 - 8 = 14$ boys were in attendance.

   This problem is an excellent example of how elegant and efficient a solution becomes after the problem is translated into an equation. We got a simple equation, which in the beginning was not obvious at all. Indeed, as soon as you realize that the best selection of the variable is not the number of boys asked for in the problem, but rather the number of girls, then the rest is easy.

Problem 19. After a father passed away, his estate was inherited by his sons and allocated among them according to his will as following: the oldest son got $1,000 and $\frac{1}{6}$ of the remainder, the second son got $2,000 and $\frac{1}{6}$ of the remaining estate, the third son got $3,000 and $\frac{1}{6}$ of the remainder, and so forth up to the last son who received the rest of the estate remaining after each brother's cut was taken. It turned out that each brother received an equal share of the estate. What was the estate's worth and how many brothers were there in the family?

Solution. Since each brother inherited the same amount, let's denote it by x dollars. The oldest son received $1,000 and $\frac{1}{6}$ of the remainder of the estate. It implies that the whole estate's worth was $1,000 + 6(x - 1,000) = 6x - 5,000$ dollars. The second brother inherited $2,000 + \frac{1}{6}(6x - 5,000 - x - 2,000) = \frac{5x + 5,000}{6}$ dollars. We know that the inherited amounts are equal, which can be expressed in the equation $x = \frac{5x + 5,000}{6}$. Solving the equation gives $6x = 5x + 5,000$. Hence, $x = 5,000$. So, each brother inherited $5,000 from the estate. Therefore, the estate's original worth was $6 \cdot 5,000 - 5,000 = 25,000$ dollars. It follows there were $25,000 : 5,000 = 5$ brothers in this family. Let's verify that the third, fourth, and fifth brother received $5,000 as well. Indeed, the third brother's share is $3{,}000 + \frac{1}{6}(25{,}000 - 10{,}000 - 3{,}000) = 5{,}000$, the fourth brother's share is $4{,}000 + \frac{1}{6}(25{,}000 - 15{,}000 - 4{,}000) = 5{,}000$, and the fifth brother's remainder equals $25{,}000 - 4 \cdot 5{,}000 = 5{,}000$.

Before we begin our discussion of systems of linear equations and the methods of their solution, it has to be noted that occasionally (as we will show in the next problems) you can get the final result even without actually solving the system. In some problems one might end up setting up the Diophantine linear equation with a few variables, which will be covered in detail in later chapters. If such an equation is solvable, it might have an indefinite number of solutions. Depending on the conditions of a problem and the question asked, one could be able to draw important conclusions by analyzing the findings after manipulations and simplifications of the equations and get the solutions satisfying the given restrictions on the variables.

Problem 20. A math professor offered the following problem to his students about his age: "In 1992 my age was exactly the sum of the digits of the year I was born in. What year was I born in?"

Solution. This problem belongs to the "Integer Problems" family. If we denote by a and b ($0 \le a \le 9, 0 \le b \le 9$) the last digits of the year in which the professor was born, we write it as year $19ab$. Therefore, in 1992 his age must be $1 + 9 + a + b = 10 + a + b$ years. On the other hand, the same number equals the difference between 1992 and the year in which he was born:

$$1992 - 19ab = 92 - 10a - b.$$

It follows that

$$10 + a + b = 92 - 10a - b, \text{ or equivalently } 11a + 2b = 82.$$

Recalling the restrictions on a and b ($0 \le a \le 9, 0 \le b \le 9$), we can analyze the possible outcomes and eventually arrive at the desired result. If $a > 8$, then $11a + 2b > 88$, which contradicts $11a + 2b = 82$. If $a \le 5$, then even for the maximum possible $b = 9$, $11a + 2b < 55 + 18 = 73 < 82$. Again, we got a contradiction with the equality $11a + 2b = 82$. Thus, the only possible values a can have are 6 or 7. If $a = 7$, then solving the equation $11 \cdot 7 + 2b = 82$ we get $b = 2.5$, which is impossible; a and b have to be integers. Therefore, the only remaining option for a is $a = 6$. Solving the equation $11 \cdot 6 + 2b = 82$ we obtain $b = 8$ and conclude that the professor was born in 1968.

Problem 21. Two cars left simultaneously from city A and city B. They were driving in a direction facing each other with different speeds. At some point of time they met. After meeting, the first car spent 4 hours for the rest of its trip and the second car spent 9 hours for the rest of its trip. How much time did each car spend for the entire trip?

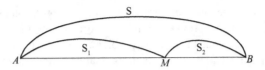

Solution. This is typical "Motion Problems" relative. Let's denote the distance between A and B by S, the distance covered by the first car up to the meeting point M by S_1, and the distance covered by the second car up to the meeting point M by S_2. Assume the speed of the first car was v_1 and the speed of the second car was v_2. Finally, since they drove the same time till they met at point M, if we denote this time by t, then the first car spent for the full trip $t_1 = t + 4$ hours and the second car spent for the full trip $t_2 = t + 9$ hours. Obviously, then

$$t_2 - t_1 = 5. \qquad (*)$$

Now we are ready to set up a few equations expressing the conditions of the problem in algebra terms: $S_1 = v_1 t$, as the distance covered by the first car before the meeting point or $S_1 = 9v_2$, as the distance covered by the second car after the meeting point. Therefore,

$$v_1 t = 9v_2, \text{ or } t = 9(v_2/v_1). \qquad (1)$$

$S_2 = v_2 t$, as the distance covered by the second car before the meeting point or $S_2 = 4v_1$, as the distance covered by the first car after the meeting point. Therefore,

$$v_2 t = 4v_1, \text{ or } t = 4(v_1/v_2). \qquad (2)$$

Comparing equations (1) and (2) leads to $9(v_2/v_1) = 4(v_1/v_2)$, from which

$$(v_2/v_1)^2 = 4/9 \text{ and } v_2/v_1 = 2/3. \qquad (3)$$

On the one hand, the distance between cities is $S = v_1 t_1$ (if you look at it from the first driver's perspective); on the other hand, the same distance $S = v_2 t_2$ (from the second driver's perspective). Thus, $v_1 t_1 = v_2 t_2$, and therefore, $v_2/v_1 = t_1/t_2$. Substituting this into the ratio v_2/v_1 from (3) gives $t_1/t_2 = 2/3$, from which $t_1 = (2/3)t_2$. Recalling expression (*) that $t_2 - t_1 = 5$, we get $t_2 - (2/3)t_2 = 5$. It follows that $(1/3)t_2 = 5$, from which $t_2 = 15$ hours, and then $t_1 = 10$ hours. The first car spent 10 hours and the second car spent 15 hours for the entire trip.

When you face a problem similar to this one, it can seem as if there are some missing attributes; you do not have enough information to solve it. Do not be afraid to introduce as many variables as you need to translate it into algebra. It will help to establish relationships not seen before through equations. This cute problem is a great example of the significance of equations in the problem-solving process. By manipulating a few equations, we were able to get rid of all the variables other than the desired one representing the time asked for in the problem.

As we saw in the very first problem of this chapter, sometimes you don't even need to solve the equation to get to the desired result. Equations serve as an essential tool for expressing the data in a problem. It suffices to introduce one or few variables, set up some equation or equations, analyze them, and make conclusions from them. The following problems illustrate vivid examples.

Problem 22. Irina lost 10% of her weight during the spring, gained 20% during the summer, lost 5% during the autumn, and lost another 5% during the winter. Did she gain or lose weight by the year end?

Solution. This is a classic example of a problem in which you are not asked to find a specific quantity, but rather the change in it. Is it possible to solve it without introducing variables? Sure it is. However, the introduction of a variable and translation of the question into algebra greatly simplifies the route to the final result.

Assume her weight at the beginning of the year was x pounds. Therefore, by the end of the spring it becomes $x - 0.1x = 0.9x$ pounds. Then she gained 20% or $0.9x \cdot 0.2 = 0.18x$ pounds, so by the end of the summer her weight was $0.9x + 0.18x = 1.08x$ pounds. Losing 5% of her weight during the autumn resulted in $1.08x - 1.08x \cdot 0.05 = 1.026x$ pounds by the beginning of the winter. Finally, she lost another 5% of her weight during the winter months. Hence, her weight by the year end was $1.026x - 1.026x \cdot 0.05 = 0.9747x$ pounds. She should be proud of herself because she eventually lost weight by the year end; to be exact, she lost $100\% - 97.47\% = 2.53\%$ of her original weight.

Problem 23. A problem given by a teacher was solved only by a few students in a class. It turned out that the number of boys who solved the problem equals the number of girls who did not solve it. Compare the number of students who solved the problem with the number of girls in the class.

Solution. Introducing several variables, as indicated in the table below, allows obtaining the algebraic interpretation of the given conditions.

Boys		Girls	
Solved problem	Did not solve	Solved problem	Did not solve
x	y	z	x

We see that the number of students who solved the problem is $(x + z)$; the number of girls in the class is $(x + z)$ as well. Therefore, these two numbers are equal. Would it be that easy to answer the problem's question without translating the problem into algebra terms?

Many geometrical problems, as the one illustrated below, have pure algebraic solutions applying the equations expressing the given conditions.

Problem 24. A circle is inscribed in the triangle ABC. The point of its tangency with side AB is D such that $AC \cdot BC = 2AD \cdot DB$. Prove that ACB is a right triangle ($\angle C = 90°$).

Solution. By the properties of a circle inscribed in a triangle, the segments from each vertex to the points of tangency on the respective sides have the same length. Let's introduce three variables x, y, and z: $AD = AN = x$, $CN = CM = y$, and $BD = BM = z$. We know that in a right triangle the Pythagorean Theorem holds true and conversely, if in a triangle the sum of squares of two sides equals to the square of the third side, it has to be a right triangle with the biggest side being its hypotenuse. Therefore, if we manage to prove that $AB^2 = AC^2 + BC^2$, then angle C has to be a right angle.

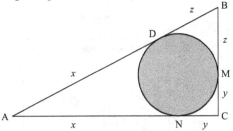

First, notice that by the conditions of the problem $AC \cdot BC = 2AD \cdot DB$ or $(x + y)(z + y) = 2xz$. Simplifying the last equality gives

$$xz + yz + xy + y^2 = 2xz, \text{ or equivalently, } xy + yz + y^2 - xz = 0. \quad (1)$$

Now find the difference $AB^2 - (AC^2 + BC^2) = (x + z)^2 - (x + y)^2 - (y + z)^2 = x^2 + 2xz + z^2 - x^2 - 2xy - y^2 - y^2 - 2yz - z^2 = 2(xz - xy - yz - y^2)$. Substituting (1) into the last expression yields $AB^2 - (AC^2 + BC^2) = 0$, or $AB^2 = AC^2 + BC^2$ which implies that the Pythagorean Theorem holds true for triangle ABC and thus angle C must be a right angle.

The equation of a straight line in a Cartesian coordinate system is $y = ax + b$, where y and x are variables and a and b are constant numbers. Coefficient a is called the slope, and coefficient b represents the y-intercept. To solve the linear equation graphically means to find the point of intersection of the respective straight line with the X-axis.

In order to graphically solve, for instance, the equation $2x + 3 = 7$, we need to find the abscissa of the point of intersection of the graph of the function $y = 2x - 4$ with the X-axis. The solution of this equation is $x = 2$. See the figure below.

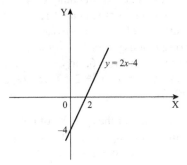

If you come across a problem whose solution requires a system of linear equations, then the graphical solution of such a system in the XY-plane will be the point of intersection of the straight lines representing the graphs of the equations in the system. For the system of two linear equations there will be one solution, if the lines intersect; no solutions, if the lines are parallel; and indefinite number of solutions, if the lines coincide.

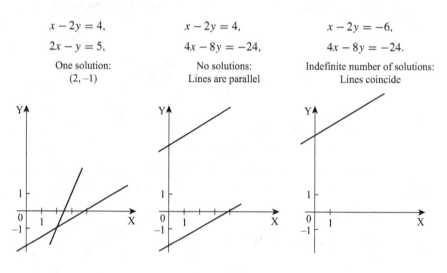

$$x - 2y = 4,$$
$$2x - y = 5,$$

One solution:
$(2, -1)$

$$x - 2y = 4,$$
$$4x - 8y = -24,$$

No solutions:
Lines are parallel

$$x - 2y = -6,$$
$$4x - 8y = -24.$$

Indefinite number of solutions:
Lines coincide

Problem 25. Solve the system of two linear equations

$$x + 3y = 11,$$
$$3x + y = 9.$$

There are a few classic methods for the analytical solution of the system of linear equations. We can multiply the first equation by -3 and then add it to the second equation, eliminating x. It yields the equation $-8y = -24$, from which $y = 3$. After substituting the value of y in either of the given equations we find $x = 2$. Another way to solve the system is to solve it for one of the variables and then substitute it into the other equation. Both of these methods are well covered in the school curriculum. There exists one more technique, applicable in the specific cases when the left sides of two equations are symmetric in each variable (the left sides will not change by permutations of the variables) as in the system suggested in this problem. Adding the two equations yields $4x + 4y = 20$, or $x + y = 5$. Subtracting this from the first equation gives $2y = 6$, or $y = 3$. Subtracting $x + y = 5$ from the second equation gives $2x = 4$, or $x = 2$. Therefore, the solution to the given system of equations is $(2, 3)$.

   To conclude our discussion of the systems of two linear equations with two variables, let's examine the following age-related word problem as an example of a more complicated challenge solvable through a system of two linear equations.

Problem 26. Steven is twice as old now as Robert was in a year when Steven was as old as Robert is old now. In a year when Robert will be as old as Steven is old now, the sum of their ages will equal 63. How old is each of them now?

Solution. The table below summarizes the problem's conditions translated into algebra.

	Age now	Past age	Future age
Steven	y	x	$(y+(y-x))$
Robert	x	$(x-(y-x))$	y

Indeed, assume Steven's age now is y and Robert's age now is x. Steven was as old as Robert is old now $(y-x)$ years ago. So, Robert's age in that year was $(x-(y-x))$. Knowing that Steven's age now, y, is twice the number $(x-(y-x))$, Robert's age in a year when Steven's age was the same as Robert's age now, we arrive at a first equation

$$2(x-(y-x)) = y.$$

$(y-x)$ years will pass till the year when Robert becomes as old as Steven is old now, y years old. Steven will be $(y+(y-x))$ years old in that year. According to the problem's conditions, the sum of their ages has to be 63. It allows us to set up a second equation

$$(y+(y-x))+y = 63.$$

To find the guys' ages we have to solve the system of the following equations

$$2(x-(y-x)) = y,$$
$$(y+(y-x))+y = 63.$$

Making simplifications, we get

$$4x-3y = 0,$$
$$3y-x = 63.$$

Adding the equations gives $3x = 63$, from which $x = 21$. Substituting this value in either equation yields $y = 28$.

Answer: Steven is 28 years old and Robert is 21 years old now.

In discussing systems of three linear equations we need to recall that the graph of a linear equation with three variables is a plane in 3-dimensional space. Therefore, to get the solutions graphically, we would need to consider the possible intersections of three planes. One solution exists when the planes intersect at one point. No solutions exist if there are no points in common with all three planes. When planes intersect at one straight line or planes coincide there are indefinitely many solutions (if they have one line in common, then every point on that line is the solution of a system; if the planes coincide, then every point on the plane is the solution of a system). In order to solve a system of three or more linear equations analytically, you can step by step eliminate the variables in the equations and get to an equation with just one variable. After finding its root, you substitute it back in the original equations and find the other variables. This method is called *Gaussian Elimination* after "the king of all mathematicians" or *Princeps mathematicorum*, the great German scholar Johann Carl Friedrich Gauss (1777–1855). We will come across his name a few more times during our journey through the Equations World.

One of the greatest mathematicians who ever lived, Carl Gauss made tremendous contributions in algebra, geometry, number theory, physics, astronomy, mechanics, geodesy, and optics. It's a well-known fact that Gauss was joking about his childhood when he said that he learned how to count before learning how to speak.

He was the one who proved the fundamental theorem of algebra, which we will talk about in later chapters. Gauss made significant contributions in the development of the theory of infinite series and their convergence criteria. He contributed to the development of the theory of differential equations, and modular arithmetic; he was the first to solve the equation $x^{17} - 1 = 0$ and to demonstrate the construction of the regular 17-gon with straightedge and compass. His list of achievements in the field of mathematics is almost endless. He was way ahead of his time. From his correspondence prior to 1829 it was clear that he was playing with the idea of impossibility of proving the Euclidean postulate about parallel lines. He even indicated that he was very close to accepting the idea of the existence of non-Euclidean geometry before it was independently published by Janos Bolyai (1802–1860) and Nikolai Lobachevsky (1792–1856).

Being a perfectionist, Gauss always refused to publish his ideas or theories that were not finished and fully proved. He kept those ideas and problems in his private notes and diaries. Even though there was no formal proof found of his being in full possession of non-Euclidean geometry, probably he was just afraid to publish his findings because of controversy. Some historians believe that he viewed non-Euclidean geometry as too revolutionary to be revealed at that time. Going back to the topic of this chapter and speaking about the Gaussian Elimination method, it has to be noted that, ironically, "Gaussian Elimination" is not quite a fair name because this method of solving linear equations first appeared in Chinese mathematical texts as early as approximately 150 BCE. Sir Isaac Newton (1642–1726) developed the detailed description of the method, which was included in algebra textbooks during his lifetime. Gauss in 1810 devised a notation for symmetric elimination that was adopted in the 19th century. The algorithm we are going to discuss below was named as the Gaussian Elimination method only in the second half of the 20th century probably as the result of recognition of one of its last contributors.

The idea behind the Gaussian Elimination method of solving the system of linear equations is to identify the variable, which is the easiest to get rid of in the system by performing arithmetic operations with equations. Then the first step is repeated for the second variable and so forth, till you get the solution of last equation for the last remaining variable. There is no strict rule in what order to eliminate the variables. It has to be decided based on each of the system equation's conditions.

Problem 27. Solve the system of linear equations

$$x - y + z = 5,$$
$$x + 3y - 2z = -9,$$
$$2x + y - z = -2.$$

Solution. It does not matter which variable to select first for the elimination in this case. Let's do it for x. We will multiply the first equation by -1 and add it to the second equation; then we will multiply the first equation by -2 and add it to the third equation.

$$x - y + z = 5,$$
$$4y - 3z = -14,$$
$$3y - 3z = -12.$$

Subtracting the third equation from the second equation gives $y = -2$.

$$x - y + z = 5,$$
$$4y - 3z = -14,$$
$$y = -2.$$

After determining y, we can substitute its value in the second equation and find z, $-8 - 3z = -14$, from which $z = 2$.

The final step is to substitute the values of y and z into the first equation and find x

$$x + 2 + 2 = 5,$$
$$x = 1.$$

The solution to the system is $(x, y, z) = (1, -2, 2)$.

Problem 28. Solve the system of linear equations

$$x + 2y + z = 14,$$
$$2x + y + z = 12,$$
$$x + y + 2z = 18.$$

Solution. If you look carefully at this system, you should notice that it is an example of a symmetric system. The solution might be simplified by application of the technique described in problem 25. Instead of utilizing Gaussian Elimination, we will add all three equations to get $4x + 4y + 4z = 44$, or $x + y + z = 11$. The next few steps are very similar. We will subtract this equation in turn from the first, the second, and the third equation to find $y = 3$, $x = 1$, $z = 7$.

Answer: $(1, 3, 7)$.

Exercises

Problem 29. The exact dates of birth and death of the famous ancient Greek mathematician Diophantus are unknown. However, a legend states that the following epitaph is written on his grave:

> Here lays Diophantus, the wonders behold.
> Through art algebraic, the stone tells how old:
> God gave him his boyhood one-sixth of his life,
> One-twelfth more as youth while whiskers grew rife;
> And then yet one-seventh ere marriage begun;
> In five years there came a bouncing new son.
> Alas, the dear child of master and sage
> After attaining half the measure of his father's life
> chill fate took him. After consoling his fate by the
> science of numbers for four years, he ended his life.

How old was he when he died?

Problem 30. Find two natural numbers, the difference and ratio of which is the same natural number.

Problem 31. Alex left home at 7 am and was driving with the speed 50 m/h. Bryan was sleeping until 7.30 am and left home at 8 am. At what speed would he need to drive in order to catch Alex by noon?

Problem 32. (*Isaac Newton's problem from "Arithmetica Universalis."*) A certain merchant increases his estate yearly by a third part, deducting 100 pounds, which he spends yearly in his family; and after three years he finds his estate doubled. What was he worth?

Problem 33. Solve the system of linear equations

$$x + y - z = 0,$$
$$x - y + z = 2,$$
$$-x + y + z = 4.$$

Problem 34. A man changed a $1 bill and received 14 coins in nickels and dimes. How many nickels did he get?

Problem 35. Water is poured into a swimming pool from 12 tubes. Normally it takes 9 hours to fill the pool. All tubes were opened simultaneously, but in 4 hours some new tubes were opened, so the pool was filled with water 2 hours earlier than planned. How many new tubes were opened?

Problem 36. Ten years ago a father was 20 years older than his son. Now he is twice as old as his son. What is the age of each of them now?

Chapter 2

Quadratic Equations

Any equation written as $ax^2 + bx + c = 0$, in which x is an unknown variable and a $(a \neq 0)$, b, and c are constant coefficients, is called a quadratic equation. The history of quadratic equations goes as far back as 2000 BC. To follow the major steps in the discovery and development of the solutions of quadratic equations, we have to first mention our ancestors in Babylonia and China who found the method of completing the square and used it to solve problems involving area calculations. Later the prominent Greek mathematicians Pythagoras (around 500 BC) and Euclid (around 300 BC) used a strictly geometrical method of solving quadratic equations. Pythagoras considered only rational solutions, but Euclid extended the allowed roots to irrational numbers. In 628 AD the Indian mathematician Brahmagupta discovered an almost modern solution of quadratics and provided the first explicit description of the formula for the general solution of a quadratic equation. Next, the Persian mathematician Muhammad ibn Musa al-Khwarizmi (780–850) gave a classification of the different types of quadratics and a rule for solving each type of equation. He was the first who solved quadratic equations algebraically. Many historians credit his work for the beginning of algebra as a separate mathematical discipline, moving away from the traditional Greek concept of mathematics, which was purely geometrical. The word "algebra" came from one of the words, "al-jabr" in the title of his book *al-Kitāb al-mukhtasar fi hisāb al-jabr wal-muqābala* (*The Compendious Book on Calculation by Completion and Balancing*) published around 830 CE. It is interesting to note that the word "algorithm" was originated as the Latin interpretation of "al-Khwarizmi";

we use it today to describe a method for solving a problem by following specific steps. His work was further advanced and developed by Jewish astronomer, philosopher, and mathematician Abraham bar Hiyya Ha-Nasi (Savasorda) (1070–1136 or 1145), who gave the complete solution to the quadratic equation in his book *Treatise on Measurement and Calculation*, published in 1145. In the western world, the general solutions of the quadratic equation were first introduced by the Flemish mathematician and engineer Simon Stevin (1548–1620) in his book *Arithmetic*, published in 1594. Finally, in 1637 the great French mathematician René Descartes published his work *La Géometrié* containing the quadratic formula we use today.

If $ax^2 + bx + c$ factors into the product of two linear factors, solving $ax^2 + bx + c = 0$ amounts to solving two linear equations. Assuming you are able to find numbers p, q, m, and n such that $ax^2 + bx + c = (px + q)(mx + n)$, then the solutions of the original equation will be $x = -\frac{q}{p}$ and $x = -\frac{n}{m}$. The factorization is conveniently achieved by completing the square. For example, let's consider the equation $x^2 + 4x - 5 = 0$. Using the well-known formulas for a square, $a^2 + 2ab + b^2 = (a + b)^2$ and for the difference of two squares, $a^2 - b^2 = (a - b)(a + b)$, we obtain

$$(x^2 + 4x + 4) - 4 - 5 = 0,$$
$$(x + 2)^2 - 9 = 0,$$
$$(x + 2)^2 - 3^2 = 0,$$
$$(x + 2 - 3)(x + 2 + 3) = 0,$$
$$(x - 1)(x + 5) = 0,$$
$$x - 1 = 0 \text{ or } x + 5 = 0, \text{ from which}$$
$$x = 1, x = -5.$$

The equation has two solutions 1 and −5.

Performing the simple manipulations with any constant coefficients a, b, and c in the same manner as it was done above, one would be able to factor $ax^2 + bx + c$ as

$$ax^2 + bx + c = a\left(x - \frac{-b - \sqrt{b^2 - 4ac}}{2a}\right)\left(x - \frac{-b + \sqrt{b^2 - 4ac}}{2a}\right).$$

Denoting $D = \sqrt{b^2 - 4ac}$ gives the general formula for the solutions of a quadratic equation as

$$x = \frac{-b \pm \sqrt{D}}{2a}.$$

If the second coefficient is an even number, the formula may be simplified into

$$x = \frac{-\frac{b}{2} \pm \sqrt{\frac{D}{4}}}{a}.$$

D in the above formula is called the *discriminant*, and a, b, and c represent respectively the first, the second, and the third coefficients in the quadratic equation.

Let's examine another approach, proposed by François Viète, for solving equation $x^2 + px + q = 0$ with the leading coefficient reduced to 1. He did it by using the substitution $x = y + z$.

$$(y+z)^2 + p(y+z) + q = 0,$$
$$y^2 + 2yz + z^2 + py + pz + q = 0,$$
$$y^2 + y(2z + p) + z^2 + pz + q = 0.$$

Since z can be any number, assume that the coefficient of y to the first power equals 0. Then $2z + p = 0$, from which $z = -\frac{p}{2}$. Thus,

$$z^2 + pz + q = \left(-\frac{p}{2}\right)^2 - \frac{p^2}{2} + q = \frac{p^2}{4} - \frac{p^2}{2} + q = q - \frac{p^2}{4}.$$

Accordingly, the original equation can be rewritten as

$$y^2 + q - \frac{p^2}{4} = 0.$$

Solving this equation we obtain

$$y = \sqrt{\frac{p^2}{4} - q}, \quad \text{or} \quad y = -\sqrt{\frac{p^2}{4} - q}.$$

Therefore,

$$x = z + y = -\frac{p}{2} \pm \sqrt{\frac{p^2}{4} - q}.$$

It's not difficult to notice in the above solutions that the roots x_1 and x_2 of the quadratic equation must satisfy

$$x_1 + x_2 = -p,$$
$$x_1 x_2 = q.$$

These formulas were first introduced and named after Francois Viète, who is often recognized as the father of the new algebra. His great contributions in algebra, trigonometry, geometry, and astronomy are astonishing. In addition to great achievements in the development of symbolic algebra briefly mentioned in chapter 1, he also presented methods for solving equations of second, third, and fourth degree, managed to calculate π to 10 places using a polygon of 393,216 sides, created two books of trigonometric tables with the amazing precision of 10^{-8}, and wrote numerous works on trigonometry, algebra, and geometry, most famous of which is *In Artem Analyticien Isagoge*. Being an attorney by education, he served as a privy councilor to the French kings Henry III and Henry IV. The times in which he lived were full of adventures and political conflicts between France and its neighbors, Spain being one of them. Viète took part in these conflicts, managing to decode the secret Spanish cipher used in the correspondence between Spanish agents and supporters in France. It was so unbelievable that Spanish inquisitors accused him of magic and evil powers and sentenced him to death. Since Viète was not within their reach, that sentence was not executed. As a matter of fact, his work leading to deciphering the enemy's secret codes laid down the foundation for cryptography. But his name is most often associated with the formulas for the link between the roots and coefficients of a quadratic polynomial, Viète's formulas. The above formulas derived for the quadratic equation with the absent first coefficient ($a = 1$) are easily extended to general formulas for an equation $ax^2 + bx + c = 0$ as the following:

$$x_1 + x_2 = -\frac{b}{a}, \tag{1}$$

$$x_1 x_2 = \frac{c}{a}. \tag{2}$$

As it was proved in high school, Viète's theorem works both ways.

Direct Statement: If x_1 and x_2 are the roots of a quadratic equation $ax^2 + bx + c = 0$, then formulas (1) and (2) hold true.

Converse Statement: The numbers x_1 and x_2 are the roots of a quadratic equation $x^2 - (x_1 + x_2)x + x_1 x_2 = 0$.

These formulas are very useful not just for figuring out the roots of a quadratic equation in many cases when they are easily identifiable by their sum and product, but in many other problems containing quadratic equations as well. The formulas are also helpful in the approximation of the roots of a quadratic function, especially when one root is much smaller than the other one. It is interesting to note that the quadratic equation whose leading coefficient is 1 ($a = 1$) and whose roots are the conjugate numbers $m + n\sqrt{t}$ and $m - n\sqrt{t}$, where m, n, and t are integers, has integer coefficients. Indeed, according to Viète's formulas, to find the second and the third coefficients in the quadratic equation, we need to find the sum and the product of the roots:

$$m + n\sqrt{t} + m - n\sqrt{t} = 2m,$$
$$(m + n\sqrt{t})(m - n\sqrt{t}) = m^2 - n^2 t.$$

Thus, the quadratic equation can be written as $x^2 - 2mx + m^2 - n^2 t = 0$. Since m, n, and t are integers, it follows the equation has integer coefficients.

Let's now turn to practical application of the general formula for the solutions of a quadratic equation and Viète's formulas in problem solving.

Problem 1. Solve the equation $x^2 + 6x = 91$.

Solution. The geometrical solution for the positive root of this equation was given by Italian mathematician Gerolamo Cardano (1501–1576) whose math achievements will be discussed in the next chapter. Consider two squares, one with side x and another one with side $x + 3$. Calculating the area of the big square, we see that $x^2 + 2 \cdot 3x + 9 = x^2 + 6x + 9 = (x + 3)^2$. It is given that $x^2 + 6x = 91$. Comparing these two equalities yields $(x + 3)^2 = 91 + 9 = 100$, from which $x + 3 = \pm 10$ and respectively $x = 7$ or $x = -13$ (not a satisfactory solution in Cardano's case).

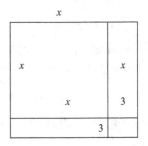

Solving the equation $x^2 + 6x - 91 = 0$ by the general formula gives the same solutions.

$$D = 6^2 + 4 \cdot 91 = 400.$$

$$x = \frac{-6 \pm \sqrt{400}}{2}, \quad x_1 = 7, \quad x_2 = -13.$$

The third alternative solution is to use Viète's formulas: $x_1 + x_2 = -6$ and $x_1 \cdot x_2 = -91$. Two numbers satisfying those conditions are $x_1 = 7$ and $x_2 = -13$.

Problem 2. Given the equation $x^2 + ax + b = 0$. Without calculating its roots write another quadratic equation such that its roots equal the squares of the roots of the first equation.

Solution. If x_1 and x_2 are the roots of the given equation, then $x_1 + x_2 = -a$ and $x_1 x_2 = b$. Squaring both sides of each equality, we obtain

$$x_1^2 + 2x_1 x_2 + x_2^2 = a^2,$$
$$x_1^2 \cdot x_2^2 = b^2.$$

Rewriting these equalities to express the sum and product of the squares of the roots, we get

$$x_1^2 + x_2^2 = a^2 - 2x_1 x_2 = a^2 - 2b,$$
$$(x_1 \cdot x_2)^2 = b^2.$$

Denoting the roots of the second equation by y_1 and y_2, we know that according to the conditions of the problem, $y_1 = x_1^2$ and $y_2 = x_2^2$. Thus, the following equalities have to hold true

$$y_1 + y_2 = x_1^2 + x_2^2 = a^2 - 2b,$$
$$y_1 \cdot y_2 = x_1^2 \cdot x_2^2 = b^2.$$

The last step is to set up the quadratic equation using Viète's formulas:
$$y^2 - (a^2 - 2b)y + b^2 = 0.$$

Problem 3. Find the sum of the cubes of the roots of the quadratic equation $x^2 - 5x - 6 = 0$ without solving the equation.

Solution. By Viète's formulas, $x_1 + x_2 = 5$ and $x_1 x_2 = -6$. It follows that

$$(x_1 + x_2)^3 = 5^3 = 125. \tag{1}$$

On the other hand,

$$(x_1 + x_2)^3 = x_1^3 + 3x_1^2 x_2 + 3x_1 x_2^2 + x_2^3 = x_1^3 + 3x_1 x_2 (x_1 + x_2) + x_2^3.$$

After substituting the values of the sum and the product of the roots from Viète's formulas in the above equality, we obtain that

$$(x_1 + x_2)^3 = x_1^3 + 3 \cdot (-6) \cdot 5 + x_2^3 = x_1^3 + x_2^3 - 90. \tag{2}$$

Combining equalities (1) and (2) yields

$$125 = x_1^3 + x_2^3 - 90.$$

Therefore, $x_1^3 + x_2^3 = 125 + 90 = 215$.

Having now reviewed Viète's formulas of the relationship of the coefficients of a quadratic polynomial with the sum and product of its roots, it has to be noted that François Viète extended his discovery to the similar properties for a general polynomial of degree $n > 2$, $P(x) = a_n x^n + a_{n-1} x^{n-1} + \cdots + a_1 x + a_0$ ($a_n \neq 0$) with the roots $x_1, x_2, \ldots, x_n$ (while Viète considered only the positive roots; another French mathematician Albert Girard (1595–1632) expanded the formulas for any real roots):

$$x_1 + x_2 + \cdots + x_n = -\frac{a_{n-1}}{a_n},$$

$$(x_1 x_2 + x_1 x_3 + \cdots + x_1 x_n) + (x_2 x_3 + x_2 x_4 + \cdots + x_2 x_n) + \cdots + x_{n-1} x_n = \frac{a_{n-2}}{a_n},$$

$$\ldots$$

$$x_1 \cdot x_2 \cdot \ldots \cdot x_n = (-1)^n \frac{a_0}{a_n}.$$

In later chapters we will come across these formulas for cubic and other polynomials.

Quadratic equations are very important in our journey into the Equations World. I would say they serve as a backbone of the foundation for many methods and techniques offered throughout the book. We will demonstrate how to reduce the power, simplify, and transform many complicated equations, solutions of which will eventually arrive at a solution of a standard quadratic equation. Alternatively, sometimes during the solution of a problem we will perform the opposite task, finding a quadratic trinomial whose roots are given numbers. That technique is a very effective tool in many difficult problems as will be demonstrated later on.

As we agree to restrict the roots of any equation in our book to the field of real numbers, we can analyze how many real solutions of a quadratic equation exist depending on the discriminant value. The number $D = b^2 - 4ac$ is called the *discriminant*, because its value determines the nature of the solutions. Since $x = \frac{-b \pm \sqrt{D}}{2a}$, when $D > 0$, the equation has two distinct real solutions; when $D = 0$, there is only one real solution (two coinciding solutions); and when $D < 0$, no real solutions exist.

Problem 4. Prove that at least one of the equations

$$(a^2 + b^2)x^2 + 2acx + c^2 - b^2 = 0 \quad \text{and} \quad 2ay^2 + 2cy + b = 0,$$

where a, b, c are the coefficients; x and y are the variables, has real solutions.

Solution. Let's find the discriminant of each quadratic equation:

$$\begin{aligned} D_1 &= 4a^2c^2 - 4(a^2 + b^2)(c^2 - b^2) = 4a^2c^2 - 4a^2c^2 - 4b^2c^2 + 4a^2b^2 + 4b^4 \\ &= -4b^2c^2 + 4a^2b^2 + 4b^4 = 4b^2(b^2 + a^2 - c^2). \\ D_2 &= 4c^2 - 8ab. \end{aligned}$$

If we assume that the second equation has no real solutions, then $D_2 < 0$, or equivalently,

$$4c^2 - 8ab < 0, \text{ from which } c^2 < 2ab. \tag{1}$$

Using (1) in the evaluation of D_1 gives

$$D_1 = 4b^2(b^2 + a^2 - c^2) > 4b^2(b^2 + a^2 - 2ab) = 4b^2(b - a)^2 \geq 0$$

(the product of two nonnegative numbers must be nonnegative). Therefore, making an assumption that $D_2 < 0$, we conclude that the first equation must have solutions. It follows that at least one of the equations has roots. If, on the contrary, $D_2 \geq 0$, then the second equation must have solutions and there is no need to analyze the first equation. Again, at least one of the equations has roots. So, in either case, the problem's statement is proved valid.

Problem 5. Find all values of k for which quadratic trinomial

$$x^2 + 2(k - 9)x + (k^2 + 3k + 4)$$

becomes a perfect square.

Solution. A quadratic trinomial is a perfect square when it has two coinciding roots, which occurs when its discriminant equals 0. Let us find the discriminant D.

$$\begin{aligned} D &= 4(k - 9)^2 - 4(k^2 + 3k + 4) = 4(k^2 - 18k + 81 - k^2 - 3k - 4) \\ &= 4(77 - 21k). \end{aligned}$$

Therefore, $D = 0$ when $4(77 - 21k) = 0$, from which $k = \frac{11}{3}$.

Problem 6. For what value of p does the ratio of the roots of the equation $x^2 + px - 16 = 0$ equal -4?

Solution. Let $\frac{x_1}{x_2} = -4$, then $x_1 = -4x_2$, where x_1 and x_2 are the roots of the given equation. Since $x_1 + x_2 = -p$ and $x_1 \cdot x_2 = -16$, then $-4x_2 + x_2 = -p$, or equivalently, $-3x_2 = -p$, from which $p = 3x_2$ and $-4x_2 \cdot x_2 = -16$, or equivalently, $4x_2^2 = 16$, from which $x_2 = \pm 2$. Substituting these values of x_2 into $p = 3x_2$, we get $p = \pm 6$.

You can't study quadratic equations without mentioning the properties of a quadratic function. As we will demonstrate in the following discussion, those properties are crucial in the solution of many complicated problems, especially problems with so-called parameters.

A single-variable quadratic function has the form $f(x) = ax^2 + bx + c$, for $a \neq 0$. To find the zeros or roots of that function means to solve the quadratic equation $ax^2 + bx + c = 0$. Graphically those solutions will be the abscissas of the points of intersection of the parabola (graph of a quadratic function) with the X-axis. So, it is important to understand the location and configuration of the graph of a quadratic function depending on the values of the first coefficient a and the discriminant D. Let us consider all the possible scenarios for a and D.

Case 1.
$a > 0$
$D > 0$

Two points of intersection.
Two roots.

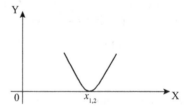

Case 2.
$a > 0$
$D = 0$

The parabola is tangent to the X-axis.
One root.

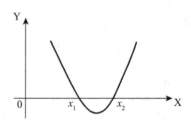

Case 3.
$a > 0$
$D < 0$

The parabola is above the X-axis.
No points of intersection.
No roots exist.

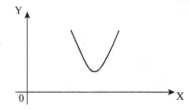

Case 4.
$a < 0$
$D > 0$

Two points of intersection.
Two roots.

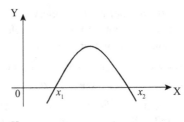

Case 5.
$a < 0$
$D = 0$

The parabola is tangent to the X-axis.
One root.

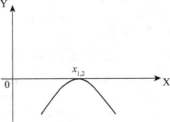

Case 6.
$a < 0$
$D < 0$

The parabola is below the X-axis.
No points of intersection.
No roots exist.

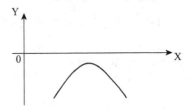

Problem 7. Find all the values of the coefficient a, such that the graph of the function $f(x) = x^2 + ax + 25$ is tangent to the X-axis.

Solution. In this problem we are dealing with Case 2. The parabola will be tangent to the X-axis if the discriminant D of the quadratic equation $x^2 + ax + 25 = 0$ equals to 0. Since $D = a^2 - 100$, then $a^2 - 100 = 0$, which yields $a = \pm 10$.

Problem 8. Find the minimum integer value of the parameter t such that the inequality $(t^2 - 1)x^2 + 2(t - 1)x + 1 > 0$ holds true for any real value of x.

Solution. Don't forget, t is some parameter or constant number, x is the variable. Our goal is to find a specific value of the parameter t such that the given quadratic polynomial has positive values for any real x. That is possible under the conditions of Case 3 when $a > 0$ and $D < 0$. So, the following inequalities must hold true at the same time:

$$t^2 - 1 > 0,$$
$$4(t - 1)^2 - 4(t^2 - 1) < 0.$$

Solving this system of inequalities leads to

$(t - 1)(t + 1) > 0,$

$t^2 - 2t + 1 - t^2 + 1 < 0,$ or equivalently

$(t - 1)(t + 1) > 0,$

$t > 1.$

We see that in order for t to satisfy both of the above inequalities, it has to belong to the set of numbers from 1 (not including 1) to $+\infty$, $]1, +\infty[$. The minimum integer value from that segment is 2.

Problem 9. Prove that for any real numbers $a_1, a_2, \ldots, a_n$ the following inequality is always true: $(a_1 + a_2 + \cdots + a_n)^2 \leq n(a_1^2 + a_2^2 + \cdots + a_n^2)$.

Solution. This problem has several different solutions. We will demonstrate an elegant and efficient way to solve it by introducing an auxiliary quadratic function, the properties of which makes the final outcome very simple and obvious. Let us consider the function

$$f(x) = (a_1x + 1)^2 + (a_2x + 1)^2 + \cdots + (a_nx + 1)^2.$$

First, notice that since $f(x)$ is a sum of squares, and the square of a real number is always nonnegative, then $f(x) \geq 0$ for any real number x. Secondly, this function is a quadratic function. Indeed, after opening the parenthesis and making a few simplifications, combining like terms, and regroupings, we get

$$f(x) = (a_1^2 + a_2^2 + \cdots + a_n^2)x^2 + 2(a_1 + a_2 + \cdots + a_n)x + n.$$

The coefficients are $a = a_1^2 + a_2^2 + \cdots + a_n^2 > 0$, $b = a_1 + a_2 + \cdots + a_n$, and $c = n$. Since we observed that $f(x) \geq 0$ for any real x, then the discriminant D for this quadratic function must be less than or equal to 0 (Case 3):

$$D = 4(a_1 + a_2 + \cdots + a_n)^2 - 4n(a_1^2 + a_2^2 + \cdots + a_n^2) \leq 0,$$

which yields the desired result:

$$(a_1 + a_2 + \cdots + a_n)^2 \leq n(a_1^2 + a_2^2 + \cdots + a_n^2).$$

The method of auxiliary elements is a very useful and powerful tool. It is applied frequently throughout the book. The main idea is to introduce some new item, variable, or function in order to clarify the entire picture by making use of the properties of a new element. It helps simplify the process of finding ties and connections that were hidden before. This great technique is equally efficient for geometry problems as well, which I addressed in a separate chapter in my book *Geometrical Kaleidoscope*, Dover Publications, 2017.

Let's extend our analysis of quadratic function properties further by demonstrating its applications in the solution of problems when you have to find the parameters depending on the restrictions on the roots of a quadratic function. The graph of a quadratic function, called a parabola, is a symmetric curve with its axis of symmetry passing through its vertex. For the given function $f(x) = ax^2 + bx + c$, it can be rewritten (by completing the square, for example) as

$$f(x) = a\left(x + \frac{b}{2a}\right)^2 - \frac{(b^2 - 4ac)}{4a}.$$

It follows that the vertex of the parabola has the coordinates $(-\frac{b}{2a}, -\frac{D}{4a})$. Depending on the value of the first coefficient a in the equation of a quadratic function,

the function has a maximum value at its vertex (when $a < 0$) or a minimum value
(when $a > 0$). Now assume that the quadratic function has two roots x_1 and x_2
(its discriminant has to be a positive number, $D > 0$), and m and n are some
real numbers. There are four possible configurations among the roots and given
parameters:

(i) $m < x_1 \le x_2$

(ii) $x_1 \le x_2 < m$

(iii) $x_1 < m < x_2$

(iv) $m < x_1 \le x_2 < n$.

Let's examine each case on the graph.

Case 1.

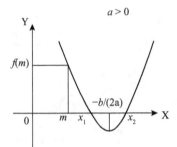
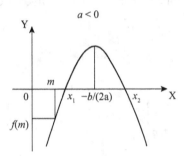

The following inequalities are true for Case 1:

$$a \cdot f(m) > 0,$$
$$m < -b/(2a),$$
$$D \ge 0.$$

Case 2.

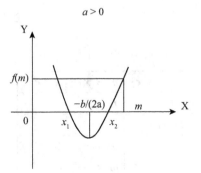
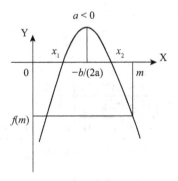

The following inequalities are true for Case 2:

$$a \cdot f(m) > 0,$$
$$-b/(2a) < m,$$
$$D \geq 0.$$

Case 3.

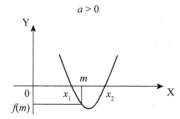

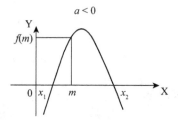

The following inequalities are true for Case 3:

$$a \cdot f(m) < 0,$$
$$D > 0.$$

Case 4.

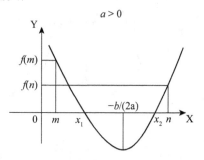

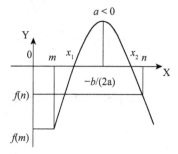

The following inequalities are true for Case 4:

$$a \cdot f(m) > 0,$$
$$a \cdot f(n) > 0,$$
$$m < -b/(2a) < n,$$
$$D \geq 0.$$

For convenience we organize the above results in a small table:

$m < x_1 \leq x_2$	$x_1 \leq x_2 < m$	$x_1 < m < x_2$	$m < x_1 \leq x_2 < n$
$D \geq 0$	$D \geq 0$	$D > 0$	$D \geq 0$
$a \cdot f(m) > 0$	$a \cdot f(m) > 0$	$a \cdot f(m) < 0$	$a \cdot f(m) > 0$
$m < -\frac{b}{2a}$	$m > -\frac{b}{2a}$		$a \cdot f(n) > 0$
			$m < -\frac{b}{2a} < n$

The established properties give a very useful technique in solving the following problems.

Problem 10. Find the values of the parameter d for which both roots of the equation $x^2 - (d+1)x + d + 4 = 0$ will be negative numbers.

Solution. Since both roots of the equation have to be negative, the Case 2 conditions are satisfied, $x_1 \le x_2 < m$ for $m = 0$. So, $x_1 \le x_2 < 0$. Solve the following system of linear inequalities:

$$D \ge 0,$$
$$1 \cdot f(0) > 0,$$
$$0 > \frac{(d+1)}{2}.$$

Since $D = (d+1)^2 - 4(d+4)$ and $f(0) = d+4$, the system is modified to the following:

$$(d+1)^2 - 4(d+4) \ge 0,$$
$$d+4 > 0,$$
$$d < -1.$$

Note that

$$(d+1)^2 - 4(d+4) = d^2 + 2d + 1 - 4d - 16 = d^2 - 2d - 15 = (d+3)(d-5).$$

In order to factor the quadratic polynomial, we had to find its roots. It is easily done by Viète's formulas. Indeed, $d_1 + d_2 = 2$ and $d_1 d_2 = -15$, then obviously, $d_1 = -3$ and $d_2 = 5$. So, finally we get the system

$$(d+3)(d-5) \ge 0,$$
$$d > -4,$$
$$d < -1.$$

Solving this system yields the common values of the parameter d, which must satisfy each condition at the same time. The inequality $-4 < d \le -3$, derived from combining the inequalities $d > -4, d < -1, d \le -3$, and $d \ge 5$, represents the set of all possible values of d.

Answer: $-4 < d \le -3$.

Problem 11. For what values of d are both roots of the equation

$$x^2 - 6dx + 2 - 2d + 9d^2 = 0 \text{ greater than 3?}$$

Solution. In this problem the conditions of Case 1 are satisfied:

$$D \geq 0,$$
$$1 \cdot f(3) > 0,$$
$$3 < -\frac{-6d}{2}.$$

Calculating D and $f(3)$ we get

$$36d^2 - 4(2 - 2d + 9d^2) \geq 0,$$
$$9 - 18d + 2 - 2d + 9d^2 > 0,$$
$$d > 1.$$

After simplifications of the first and second inequalities the system can be rewritten as

$$d \geq 1,$$
$$9d^2 - 20d + 11 > 0,$$
$$d > 1.$$

Let's find the roots of the quadratic polynomial in the second inequality. $D = 400 - 396 = 4$. Then $d_1 = 1$, and $d_2 = \frac{11}{9} = 1\frac{2}{9}$. Therefore,

$$9d^2 - 20d + 11 = 9(d - 1)\left(d - 1\frac{2}{9}\right) > 0.$$

The system converts to

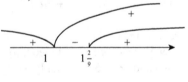

$$d \geq 1,$$
$$(d - 1)\left(d - 1\frac{2}{9}\right) > 0,$$
$$d > 1.$$

The common solutions must satisfy the condition $1\frac{2}{9} < d < +\infty$.

Problem 12. Find the values of the parameter k for which the equation $5x^2 + 2kx + 5 = 0$ has two equal positive roots.

Solution. As in the previous problem, the conditions of Case 1 are satisfied. However, in this problem the roots are set to be equal, therefore

$$D = 0,$$
$$5f(0) > 0,$$
$$0 < -\frac{2k}{10}.$$

Solving the above system gives

$$k^2 - 25 = 0,$$
$$25 > 0,$$
$$k < 0.$$

Solving the first equation in the system gives $k = \pm 5$. Since $k < 0$ in the last inequality, only -5 satisfies the system. Therefore, when $k = -5$ both roots are equal positive numbers.

Problem 13. Prove that for any real numbers a and b, if $4b + a = 1$, then $a^2 + 4b^2 \geq \frac{1}{5}$.

Solution. Expressing from the given equality a in terms of b gives $a = 1 - 4b$. It follows $a^2 + 4b^2 = (1 - 4b)^2 + 4b^2 = 1 - 8b + 16b^2 + 4b^2 = 20b^2 - 8b + 1$. Now introduce the auxiliary function $f(x) = 20x^2 - 8x + 1$. Since the first coefficient is positive, $20 > 0$, the graph of this quadratic function is a parabola that opens upward. The discriminant is negative, $D = 64 - 80 = -16 < 0$. Hence, the parabola is located above the X-axis and has no points of intersection with the X-axis. The abscissa of its vertex is $x_0 = \frac{8}{40} = \frac{1}{5}$. The minimum value of a quadratic function appears as a constant in the vertex form of its equation, $f(\frac{1}{5}) = 20 \cdot \frac{1}{25} - 8 \cdot \frac{1}{5} + 1 = \frac{1}{5}$. Therefore, $20x^2 - 8x + 1 \geq \frac{1}{5}$ for any real x. Recalling that $20b^2 - 8b + 1 = a^2 + 4b^2$, we arrive at the desired result, $a^2 + 4b^2 \geq \frac{1}{5}$.

Problem 14. Real numbers x, y, and a are such that the following two equalities are true at the same time:

$$x + y = a - 1,$$
$$xy = a^2 - 7a + 14.$$

For what values of the parameter a does the sum $(x^2 + y^2)$ have a maximum value?

Solution. Using the first of the given equalities, let's express the sum $(x^2 + y^2)$ in terms of a:

$$x^2 + y^2 = (x + y)^2 - 2xy = a^2 - 2a + 1 - 2a^2 + 14a - 28$$
$$= -a^2 + 12a - 27 = -((a - 6)^2 - 9) = 9 - (a - 6)^2.$$

At this point one may conclude that the maximum value of the last expression is attained when the difference $(a - 6)$ has a minimum value, which is obtained when $a = 6$. As good as this looks, unfortunately it is not correct. Substituting $a = 6$ into each of the given equalities yields

$$x + y = 5,$$
$$xy = 8.$$

The above system of equations does not have any solutions in real numbers, which should not be hard for readers to verify. So, we have to find some other way of evaluating a. Using Viète's formulas introduce an auxiliary quadratic equation for the variable z with the coefficients from the given equalities:

$$z^2 - (a-1)z + (a^2 - 7a + 14) = 0.$$

This equation will have roots when its discriminant expressed in terms of a is a nonnegative number,

$$D = (a-1)^2 - 4(a^2 - 7a + 14) = a^2 - 2a + 1 - 4a^2 + 28a - 56$$
$$= -3a^2 + 26a - 55 \geq 0.$$

Let's now solve the inequality $-3a^2 + 26a - 55 \geq 0$, or equivalently, $3a^2 - 26a + 55 \leq 0$. In order to factor the left-hand side, we need to find the roots of the quadratic trinomial.

$$D = 676 - 660 = 16.$$
$$a = \frac{26 \pm 4}{6}.$$

Hence, $a_1 = 5$, $a_2 = \frac{11}{3}$ and respectively the inequality can be rewritten as

$$3(a-5)\left(a - 3\frac{2}{3}\right) \leq 0.$$

The inequality $3\frac{2}{3} \leq a \leq 5$, derived from combining the inequalities $a \geq 3\frac{2}{3}$ and $a \leq 5$, represents the set of all satisfactory values of a. In other words, the desired value of a is to be determined from the segment $\left[3\frac{2}{3}, 5\right]$. Recalling that $x^2 + y^2 = 9 - (a-6)^2$, we see that the maximum value of the sum $(x^2 + y^2)$ is obtained when a gets the closest value to 6 from all the numbers in the segment. Therefore, 5 is the required result.

Answer: $(x^2 + y^2)$ has the maximum value when $a = 5$.

It is worth emphasizing here the nonstandard technique of utilizing an auxiliary quadratic equation constructed from Viète's formulas. It was a real savior in the clarification of the problem's conditions, helping to apply them in the most efficient and effective way.

The discussion of quadratic equations and quadratic functions' properties cannot be completed without mentioning geometrical links and connections. An interesting and vivid geometrical interpretation of the solutions of a quadratic equation is presented by a Carlyle Circle (after Thomas Carlyle (1795–1881)).

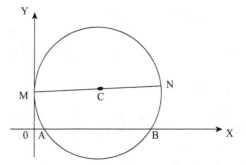

We assume readers are familiar with the basic definitions, properties, and equations related to the applications of the Cartesian coordinate system in a two-dimensional plane. Cartesian coordinates are very helpful in establishing links between geometry and algebra, between geometrical shapes and their presentation by Cartesian equations. We are about to see one such enlightening interpretation below.

Let's consider the circle with diameter MN in a Cartesian coordinate system, where $M(0, 1)$ and $N(s, p)$. The circle is tangent to the Y-axis at point M and has two points of intersection with the X-axis at $A(x_1, 0)$ and $B(x_2, 0)$. If C is the center of this circle, then the coordinates of C can be found as the coordinates of the midpoint of the segment MN:

$$x_c = \frac{s+0}{2} = \frac{s}{2}, \quad y_c = \frac{p+1}{2}.$$

The length of MN is determined as

$$\sqrt{(x_N - x_M)^2 + (y_N - y_M)^2} = \sqrt{(s - 0)^2 + (p - 1)^2} = \sqrt{s^2 + (p - 1)^2}.$$

The radius of this circle r equals half the length of the segment MN:

$$r = \frac{1}{2}\sqrt{s^2 + (p - 1)^2}.$$

The general equation of any circle with center $C(x_c, y_c)$ and radius r in a Cartesian coordinate system is $(x - x_c)^2 + (y - y_c)^2 = r^2$. It follows that the equation of our circle can be written as

$$\left(x - \frac{s}{2}\right)^2 + \left(y - \frac{p+1}{2}\right)^2 = \frac{1}{4}(s^2 + (p - 1)^2).$$

The points of intersection of the circle with the X-axis must have coordinates that satisfy the equation of the circle. Therefore, since the ordinate of each of those points is 0, their abscissas may be found from

$$\left(x - \frac{s}{2}\right)^2 + \left(\frac{p+1}{2}\right)^2 = \frac{s^2}{4} + \frac{(p-1)^2}{4}.$$

Squaring and combining like terms gives

$$x^2 - 2x \cdot \frac{s}{2} + \left(\frac{s}{2}\right)^2 + \frac{p^2 + 2p + 1}{4} - \frac{s^2}{4} - \frac{p^2 - 2p + 1}{4} = 0,$$

$$x^2 - sx + \frac{s^2}{4} - \frac{s^2}{4} + \frac{4p}{4} = 0.$$

After a final simplification we get to the equation $x^2 - sx + p = 0$, solutions of which are the abscissas of points A and B. So, for a quadratic equation $x^2 - sx + p = 0$ the circle in the coordinate plane having the line segment joining the points $M(0, 1)$ and $N(s, p)$ as a diameter defines the solutions of that equation as abscissas of its points of intersection with the X-axis. If the Carlyle Circle is tangent to the X-axis, there is only one real solution of the quadratic equation; if the circle has no points of intersection with the X-axis, there are no real solutions.

We encourage readers to do further research of the utilization of Carlyle circles or the Carlyle algorithm in developing compass-straightedge constructions of regular polygons, particularly the pentagon, the 17-gon, the 257-gon, and the 65537-gon.

It will be a great exercise in getting more prepared for our coming discussion of the conditions for constructability of regular polygons with a compass and straightedge. We will talk about Carl Gauss's discovery, Évariste Galois's theory, and the link between analytic geometry and algebra, clarifying the fact that constructible lengths must come from base lengths by the solution of some sequence of quadratic equations.

Exercises

Problem 15. The coefficients in a quadratic equation are replaced with asterisks:

$$*x^2 + *x + * = 0.$$

Imagine two people playing the following game: The first player names three numbers. The second one writes them at will replacing the asterisks. Can the first player ensure that the resulting equation has distinct rational roots no matter how the second one arranges the coefficients?

This problem was offered by A. Berzins in the currently defunct magazine *Quantum*, May/June 1994 issue.

Problem 16. Given the equation $x^2 - x - q = 0$. For what values of the parameter q does the sum of the cubes of the equation roots equal 19?

Problem 17. Given the equation $ax^2 + bx + c = 0$, in which the coefficients satisfy the condition $2b^2 - 9ac = 0$. Prove that the ratio of the roots of the equation equals $\frac{1}{2}$.

Problem 18. Prove that for any real values of the parameter p and $a \neq 0$ the following equation has roots

$$\frac{1}{x+p} + \frac{1}{x-p} = \frac{1}{a}.$$

Problem 19. The lengths of the legs of a right triangle are the roots of the quadratic equation $ax^2 + bx + c = 0$. Find the hypotenuse of that triangle without solving the equation.

Problem 20. Find the values of the parameter a for which the equation $ax^2 - 2x + 3 = 0$ has only one real solution.

Problem 21. Find the value of the parameter p, such that both roots of the equation $x^2 + 2(p+1)x + 9p - 5 = 0$ are negative numbers.

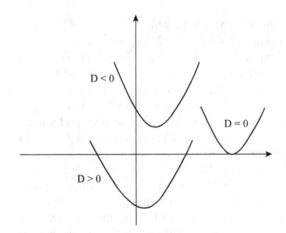

Chapter 3

Cubic and Quartic Equations

It's great to have the explicit formula $x = \frac{-b \pm \sqrt{b^2 - 4ac}}{2a}$ for the solutions of a quadratic equation $ax^2 + bx + c = 0$ ($a \neq 0$), which was covered in the previous chapter. What about third and fourth degree polynomial equations? How do you find their solutions? Is it possible to get similar formulas for them? The history of mathematics tells us that cubic equations were known to mathematicians in the ancient world. However, prior to the 11th century no classification of cubic equations existed, until the great Persian poet, philosopher, astronomer, and mathematician Omar Khayyam (1048–1131) made huge contributions in developing the theory of cubic equations. Omar Khayyam, once called by another medieval scholar Al-Zamakhshari (1075–1143) "the philosopher of the world," is better known in the western world for his quatrains, four-line philosophical poems. His brilliant poetry has not only survived through centuries but remains as popular as ever in modern society. His mathematical genius influenced his writing style and the way he conveyed his sharp and insightful thoughts. Here is one of his great "formulas" to live a life:

> To wisely live your life, you need to know much.
> The main two rules absorb for the beginning.
> You better off to starve, than eat whatever,
> And better be alone, than with whoever.

Omar Khayyam's contributions in mechanics, geometry, astronomy, mineralogy, and especially in algebra are invaluable. Many scholars consider his *Treatise on Demonstration of Problems of Algebra* as one of the most influential treatises on algebra written before modern times. He gave a classification of cubic equations with general geometric solutions found with ruler and compass, equations solvable by means of intersecting conic sections, which he stated cannot be solved using compass and ruler constructions, and equations that involve the inverse of the unknown. He observed an interesting geometric solution of the cubic equation $x^3 + 2x - 2x^2 - 2 = 0$. He transformed it into the equation $x^3 + 2x = 2x^2 + 2$ and then managed to find the roots as abscissas of the points of intersection of a hyperbola and a circle, represented respectively on the left-hand side and right-hand side of the equation. As a matter of fact, a similar approach might be used for the geometrical solution of any cubic equation. Despite his great success studying cubic equations, Omar Khayyam did not manage to provide a general formula similar to the one used for the solution of quadratic equations.

Such success was achieved through the works of the Italian mathematicians Scipione del Ferro (1465–1526), Niccolò Tartaglia (1500–1557), and Gerolamo Cardano (1501–1576). The story of discovery of the general formula for the solution of a cubic equation is one of the most controversial and fascinating in the history of mathematics. Even today the question of the origin of the formula cannot be answered unambiguously. The very first person who managed to solve the equation $x^3 + ax = b$ was an Italian professor of mathematics from Bologna, Scipione del Ferro. He conveyed his method to his son-in-law Aniballe della Nave and to his student Antonia Maria Fior. It is believed that no publications of del Ferro's discovery were available to the general public. During a math competition with Fior, a Venetian mathematician Niccolò Tartaglia discovered his own method of solving a cubic equation, which he preserved as a big secret and did not share with anybody else. From history we know that Gerolamo Cardano eventually convinced Tartaglia to reveal the secret to him. Tartaglia insisted, however, on not allowing the method to be published without his permission. He never published his results by himself. Cardano spent many years putting together his own research, del Ferro's work (Cardano and his student Ferrari were allowed by del Ferro's son-in-law to work with the late mathematician's papers), and Tartaglia's results. In 1545 he published everything he knew about cubic equations' solutions in his book *Ars Magna*. Many historians still believe that Tartaglia had not been given enough credit for his discovery. Ironically, even though Gerolamo Cardano, a prominent physician, who studied philosophy, astrology, and mathematics in his spare time, was not the one who actually proved it, the formula for the solutions of cubic equation bears his name. There are many publications about this very interesting chapter in mathematical history. I would recommend one such great article by S. Gindikin "The controversial origins of Cardano's formula" published in the May/June 1995 issue of the currently defunct magazine *Quantum*.

We will demonstrate Cardano's classical method of solving cubic equations. It provides an explicit formula for the roots of any cubic equation in addition to the algebraic techniques such as the factorization by grouping. Any cubic equation $a_0 x^3 + a_1 x^2 + a_2 x + a_3 = 0$ ($a_0 \neq 0$) can be "depressed" to the simpler form of $y^3 + py + q = 0$ and then after a few substitutions and simplifications its solutions may be derived by making use of Viète's formulas. Let's now follow this process step by step. First, since $a_0 \neq 0$, dividing both sides of the equation by a_0 and then introducing a new variable $x = y - \frac{a_1}{3a_0}$ gives

$$\left(y - \frac{a_1}{3a_0}\right)^3 + \frac{a_1}{a_0}\left(y - \frac{a_1}{3a_0}\right)^2 + \frac{a_2}{a_0}\left(y - \frac{a_1}{3a_0}\right) + \frac{a_3}{a_0} = 0.$$

After expanding parentheses and combining like terms, we get the equation $y^3 + py + q = 0$, where p and q are some real numbers—coefficients expressed in terms of the coefficients of the original equation. Now we introduce two more new variables u and v linked by the condition $y = u + v$. It follows

$$(u + v)^3 + p(u + v) + q = 0,$$
$$u^3 + v^3 + 3uv(u + v) + p(u + v) + q = 0,$$
$$u^3 + v^3 + (u + v)(3uv + p) + q = 0.$$

At this point imposing another condition on the new variables, $uv = -\frac{p}{3}$, and substituting it into the above equation yields

$$u^3 + v^3 + (u + v)\left(3 \cdot \left(-\frac{p}{3}\right) + p\right) + q = 0,$$
$$u^3 + v^3 + (u + v) \cdot 0 + q = 0,$$
$$u^3 + v^3 + q = 0,$$
$$u^3 + v^3 = -q. \qquad (*)$$

Recalling that $uv = -\frac{p}{3}$ leads to $u^3 v^3 = -\frac{p^3}{27}$. The last equality along with equality $(*)$ allows us to observe that according to Viète's formula, u^3 and v^3 are the roots of the quadratic equation $z^2 + qz - \frac{p^3}{27} = 0$. By the general formula then $z = -\frac{q}{2} \pm \sqrt{\frac{q^2}{4} + \frac{p^3}{27}}$ and $z_1 = u^3$, $z_2 = v^3$.
Going back to the variable $y = u + v$, we finally obtain that

$$y = \sqrt[3]{-\frac{q}{2} + \sqrt{\frac{q^2}{4} + \frac{p^3}{27}}} + \sqrt[3]{-\frac{q}{2} - \sqrt{\frac{q^2}{4} + \frac{p^3}{27}}}.$$

This formula for the solutions of a cubic equation is called *Cardano's formula*. For the general cubic equation of the form $a_0 x^3 + a_1 x^2 + a_2 x + a_3 = 0$ ($a_0 \neq 0$) the number and types of roots is determined by the discriminant of the cubic equation,

$$D = 18a_0 a_1 a_2 a_3 - 4a_1^3 a_3 + a_1^2 a_2^2 - 4a_0 a_2^3 - 27a_0^2 a_3^2.$$

If $D > 0$, then there are three distinct real roots. If $D = 0$, then there are either three equal real roots (triple root) or two equal roots (double root) and a simple root distinct from them. All roots are real. If $D < 0$, then there is only one real root. The other two roots are complex conjugate numbers and are beyond the scope of our coverage in the book.

Cardano's formula has the drawback that it may bring square roots of negative numbers into calculations in intermediate steps, even when those numbers do not appear in the problem or its answer. It might be the case for a cubic equation with real coefficients and three real roots. When it happens, ultimately the square roots of negative numbers would cancel out during the computation, but still they come into play and one should be able to operate with them. We cannot perform those calculations without additional discussion of complex numbers. Since complex numbers are a more advanced topic, as we agreed to stay in the field of real numbers, we will avoid discussing specific problems for Cardano's formula application. When you encounter a cubic equation, the objective is to find one real root, which will always exist, and then the other two real roots can be found by polynomial division and the quadratic formula. From a practical standpoint, if you manage to find an integer solution, then there is no need for you to make calculations using Cardano's formula. In a cubic equation with integer coefficients, the rational root test is a great tool to search for a possible solution. We are not going to cover it here. Since this method is applicable for an equation of any power greater than 2, we will demonstrate it in detail in the next chapter. Some cubic equations are easily solved by completing the cube. Another unusual and surprising technique for solving cubic equations linked to trigonometry will be covered in the chapter devoted to trigonometric equations. Meanwhile, let's show a few examples of a practical approach to solving a cubic equation by factoring a cubic polynomial. The idea is to factor the cubic polynomial into linear and quadratic polynomial factors and then solve two simple equations.

Problem 1. Solve the equation

$$x^3 - 2x - 4 = 0.$$

Solution. This is the good example of a cubic equation where one of the roots is easily identifiable. It's not hard to see that one of the roots is 2. Indeed, $2^3 - 4 - 4 = 0$. The next step is to factor the cubic polynomial. One of the factors has to be $(x - 2)$. Adding and subtracting $2x^2$ and regrouping the terms gives

$$x^3 - 2x^2 + 2x^2 - 2x - 4 = 0,$$
$$x^2(x - 2) + 2(x^2 - x - 2) = 0.$$

Using Viète's formulas, $x^2 - x - 2$ has roots -1 and 2 and is factored as $(x + 1)(x - 2)$. Hence, we can rewrite the equation as

$$x^2(x - 2) + 2(x + 1)(x - 2) = 0,$$
$$(x - 2)(x^2 + 2x + 2) = 0.$$

By the zero-product property, set each factor on the left-hand side of the last equation equal to 0 and solve each resulting equation for x.

$$x - 2 = 0 \quad \text{or} \quad x^2 + 2x + 2 = 0.$$

From the first equation we get the root $x = 2$. The discriminant of the second quadratic equation is negative: $D = 4 - 8 = -4 < 0$. Therefore, it has no real solutions. The original equation has one real root 2.

Problem 2. Solve the equation

$$x^3 - 6x^2 + 11x - 6 = 0.$$

Solution. This equation is easily solvable by completing the cube with a further factorization. Using the formula $a^3 - 3a^2b + 3ab^2 - b^3 = (a - b)^3$, the equation can be rewritten as

$$(x^3 - 6x^2 + 12x - 8) - x + 2 = 0,$$
$$(x - 2)^3 - (x - 2) = 0,$$
$$(x - 2)((x - 2)^2 - 1) = 0;$$

using the formula for the difference of squares gives

$$(x - 2)(x - 2 + 1)(x - 2 - 1) = 0,$$
$$(x - 2)(x - 1)(x - 3) = 0,$$

from which

$$x = 2, \quad x = 1, \quad \text{or} \quad x = 3.$$

The original equation has three real roots 1, 2, and 3.

Notice that the formula $a^3 + 3a^2b + 3ab^2 + b^3 = (a + b)^3$ is often used in a similar way, as it is demonstrated in the problem below.

Problem 3. Solve the equation

$$x^3 - 3x^2 - 3x - 1 = 0.$$

Solution. Observing that $x^3 + 3x^2 + 3x + 1 = (x + 1)^3$, we can rewrite the equation as

$$2x^3 - x^3 - 3x^2 - 3x - 1 = 0$$

with the further transition to

$$2x^3 - (x^3 + 3x^2 + 3x + 1) = 0,$$
$$2x^3 = (x + 1)^3 \text{ or } \sqrt[3]{2}x = x + 1.$$

From this it follows that $x = \frac{1}{\sqrt[3]{2} - 1}$.

Answer: $\frac{1}{\sqrt[3]{2} - 1}$.

It is worth noting that depending on the specific problems, as will be illustrated in later chapters, both formulas may be rewritten and utilized as

$$(a-b)^3 + 3ab(a-b) = a^3 - b^3,$$
$$(a+b)^3 - 3ab(a+b) = a^3 + b^3.$$

Cubic equations as well as quadratic equations often are very helpful in solving many non-routine problems involving the evaluation of numerical expressions or inequalities. Here is one such example.

Problem 4. Prove that

$$\underbrace{\sqrt[3]{6+\sqrt[3]{6+\cdots+\sqrt[3]{6}}}}_{n \text{ cube roots}} + \underbrace{\sqrt{6+\sqrt{6+\cdots+\sqrt{6}}}}_{m \text{ square roots}} < 5.$$

Solution. An elegant solution can be derived by introducing an auxiliary element—some number $\alpha > 0$ such that $6 + \alpha = \alpha^3$. After n repetitive steps it allows us to make an assessment of the first addend:

$$\sqrt[3]{6+\sqrt[3]{6+\cdots+\sqrt[3]{6}}} < \sqrt[3]{6+\sqrt[3]{6+\cdots+\sqrt[3]{6+\alpha}}} = \sqrt[3]{6+\alpha} = \alpha.$$

To be specific in our valuation, we now need to find α by solving the equation $6 + \alpha = \alpha^3$. Applying the formula for the difference of cubes, we get

$$\alpha^3 - 8 + 2 - \alpha = 0,$$
$$(\alpha^3 - 8) - (\alpha - 2) = 0,$$
$$(\alpha - 2)(\alpha^2 + 2\alpha + 4) - (\alpha - 2) = 0,$$
$$(\alpha - 2)(\alpha^2 + 2\alpha + 3) = 0,$$

$\alpha = 2$ or $\alpha^2 + 2\alpha + 3 = 0$. The quadratic equation has no roots, because its discriminant is a negative number, $D = 4 - 12 = -8$. We get that the only solution is $\alpha = 2$ and respectively

$$\sqrt[3]{6+\sqrt[3]{6+\cdots+\sqrt[3]{6}}} < 2. \tag{1}$$

In the same fashion, we introduce an auxiliary number $\beta > 0$ such that $6 + \beta = \beta^2$. Hence, after m steps we get

$$\sqrt{6+\sqrt{6+\cdots+\sqrt{6}}} < \sqrt{6+\sqrt{6+\cdots+\sqrt{6+\beta}}} = \sqrt{6+\beta} = \beta. \tag{2}$$

To find β we need to solve the equation $6 + \beta = \beta^2$ and select its positive root. $\beta^2 - \beta - 6 = 0$, from which $\beta = 3$ or $\beta = -2$. For $\beta = 3$ the inequality (2) holds true. It follows from inequalities (1) and (2) that

$$\sqrt[3]{6+\sqrt[3]{6+\cdots+\sqrt[3]{6}}}+\sqrt{6+\sqrt{6+\cdots+\sqrt{6}}} < 2+3 = 5,$$

as was to be proved.

The equation $a_4x^4 + a_3x^3 + a_2x^2 + a_1x + a_0$ $(a_4 \neq 0)$ is called a quartic equation. These equations are very much involved in today's life. They are often encountered in the solution of many problems in computer-aided manufacturing, software design, computer graphics, computational geometry, and even optics.

A quartic equation, similar to a quadratic and cubic equation, has a general formula for its solution. The Italian mathematician Lodovico Ferrari (1522–1565), one of the best students of Gerolamo Cordano, is considered to be the first to solve the quartic equation, as he did so in 1540. The general case solution and formula are pretty complicated and cumbersome. To give you an idea about its complexity, just look at the discriminant calculation through the equation's coefficients:

$$\begin{aligned} D = {} & 256a_4^3a_0^3 - 192a_4^2a_3a_1a_0^2 - 128a_4^2a_2^2a_0^2 + 144a_4^2a_2a_1^2a_0 - 27a_4^2a_1^4 \\ & + 144a_4a_3^2a_2a_0^2 - 6a_4a_3^2a_1^2a_0 - 80a_4a_3a_2^2a_1a_0 + 18a_4a_3a_2a_1^3 + 16a_4a_2^4a_0 \\ & - 4a_4a_2^3a_1^2 - 27a_3^4a_0^2 + 18a_3^3a_2a_1a_0 - 4a_3^3a_1^3 - 4a_3^2a_2^3a_0 + a_3^2a_2^2a_1^2. \end{aligned}$$

We leave it to readers to investigate the determination of the formulas for the solutions of a quartic equation in a general case. You may try to derive it yourself or do some research. It's not worth wasting time here because the formula is too unwieldy for general use. Instead, in the following discussion, we will analyze a few commonly employed practical methods and techniques simplifying solutions of quartic equations.

1   Ferrari's method

We will start with Ferrari's classical method. His method relies on reducing the quartic equation to quadratic equations by completing a square. We are sure readers are well familiar with the formulas for the square of the sum and the difference of two numbers:

$$(a \pm b)^2 = a^2 \pm 2ab + b^2.$$

In algebra we often use a technique of recasting mathematical objects in one way or another. You may find many such transformations in this book. One method consists of adding and subtracting the same term in expression, preserving the total but making it helpful to rearrange the way you need it. For example, if you already have in some expression a^2 and b^2 and your goal is to get to a perfect square, you add and subtract $2ab$. Then you just group the three terms together to complete a perfect square and work with the rest of the expression depending on the goal you want to achieve. You already witnessed its application during the solution of problem 1.

Working with a quartic equation, Ferrari's goal was to reduce it to quadratic equations by factoring the quartic polynomial. His idea was to get the difference of two squares and then factor it by using the formula $x^2 - y^2 = (x - y)(x + y)$. Consider the equation

$$a_4 x^4 + a_3 x^3 + a_2 x^2 + a_1 x + a_0 = 0. \quad (a_4 \neq 0),$$

we can always get to an equivalent equation with the first coefficient equal to 1 by dividing both sides by $a_4 \neq 0$.

$$x^4 + ax^3 + bx^2 + cx + d = 0.$$

Applying the suggested technique for adding and subtracting the same number, in this case $\frac{a^2}{4} x^2$, we rewrite the equation as

$$x^4 + 2 \cdot \frac{a}{2} x^3 + \frac{a^2}{4} x^2 - \frac{a^2}{4} x^2 + bx^2 + cx + d = 0,$$

$$\left(x^2 + \frac{a}{2} x \right)^2 + \left(b - \frac{a^2}{4} \right) x^2 + cx + d = 0.$$

If the coefficients a, b, c, and d are such that $\left(\frac{a^2}{4} - b \right) x^2 - cx - d$ is a perfect square, then the solution is simplified by using the difference of squares formula to factor the left side and getting to the solution of two quadratic equations.

If you are not so lucky and the last expression is not a perfect square, then you need to do some extra work. Introducing a new variable y and adding and subtracting the same term

$$2y \left(x^2 + \frac{a}{2} x \right) + y^2,$$

leads to

$$\left(x^2 + \frac{a}{2} x + y \right)^2 - \left(\left(2y + \frac{a^2}{4} - b \right) x^2 + (ay - c)x + (y^2 - d) \right) = 0.$$

The expression

$$P(x) = \left(2y + \frac{a^2}{4} - b \right) x^2 + (ay - c)x + (y^2 - d)$$

is to be looked at as a quadratic polynomial for x. It will become a perfect square when the first coefficient is a nonnegative number and its discriminant equals to 0,

$$D = (ay - c)^2 - 4 \left(2y + \frac{a^2}{4} - b \right) (y^2 - d) = 0.$$

The last equation is a cubic equation in y; it is called the Ferrari resolvent for $P(x)$ and it is solvable, for example, by Cardano's formula. As you get its real roots, you then substitute them back for y and should be able to get the factorization of the original quartic equation. Hence, the solution of the original equation will be reduced to the solution of two quadratic equations. Instead of completing the general case solution (we invite readers to do it by themselves), we will demonstrate Ferrari's method application for the specific examples.

Problem 5. Solve the equation

$$x^4 - 4x^3 + 12x - 9 = 0.$$

Solution. First, we will add and subtract $4x^2$ to complete a square on the left-hand side and see if there is a need for introducing a new variable:

$$x^4 - 4x^3 + 4x^2 - 4x^2 + 12x - 9 = 0,$$
$$(x^4 - 4x^3 + 4x^2) - (4x^2 - 12x + 9) = 0,$$
$$(x^2 - 2x)^2 - (4x^2 - 12x + 9) = 0,$$

In this case we just got lucky. Noticing that $4x^2 - 12x + 9$ is a perfect square as well, $4x^2 - 12x + 9 = (2x - 3)^2$, the next steps will not require us solving for a cubic equation and we can proceed directly to factoring:

$$(x^2 - 2x)^2 - (2x - 3)^2 = 0,$$
$$((x^2 - 2x) - (2x - 3))((x^2 - 2x) + (2x - 3)) = 0,$$
$$(x^2 - 4x + 3)(x^2 - 3) = 0,$$
$$x^2 - 4x + 3 = 0, \text{ or } x^2 - 3 = 0.$$

Solving the first equation by Viète's formulas gives the roots 1 and 3. Solving the second equation gives the roots $\pm\sqrt{3}$.

Answer: $1, 3, \pm\sqrt{3}$.

Problem 6. Solve the equation

$$x^4 - 10x^2 - 8x + 5 = 0.$$

Solution. First, rewrite the equation as

$$x^4 = 10x^2 + 8x - 5.$$

In accordance with Ferrari's method, introducing the new variable y and adding and subtracting the same expression $2yx^2 + y^2$ on each side of the last equation gives

$$x^4 + 2yx^2 + y^2 = 10x^2 + 8x - 5 + 2yx^2 + y^2,$$
$$(x^2 + y)^2 = (10 + 2y)x^2 + 8x + y^2 - 5.$$

Let's now consider the conditions when the right-hand side becomes a perfect square. The right-hand side is the quadratic trinomial in x. It becomes a perfect square when its discriminant equals 0 and the first coefficient is a positive number, which means y has to satisfy the conditions $10 + 2y > 0$, that is, $y > -5$.

$$D = 64 - 4(10 + 2y)(y^2 - 5) = -8y^3 - 40y^2 + 40y + 264 = 0,$$

or equivalently

$$y^3 + 5y^2 - 5y - 33 = 0.$$

Looking at the factors of -33, we see that one of the rational roots of the above equation is -3. We leave it to readers to verify if that equation has other solutions. You may get back to it after studying the next chapter for the division of polynomials or try to factor the cubic polynomial by some other method and investigate its factors.

Substituting this value for y, we obtain:

$$(x^2 - 3)^2 = 4x^2 + 8x + 4,$$
$$(x^2 - 3)^2 - (2x + 2)^2 = 0,$$
$$(x^2 - 3 - 2x - 2)(x^2 - 3 + 2x + 2) = 0,$$
$$x^2 - 2x - 5 = 0 \text{ or } x^2 + 2x - 1 = 0.$$

Solving the first quadratic equation gives

$$D = 4 + 20 = 24,$$
$$x = \frac{2 \pm \sqrt{24}}{2}, \quad x = 1 \pm \sqrt{6}.$$

Solving the second quadratic equation gives

$$D = 4 + 4 = 8,$$
$$x = \frac{-2 \pm \sqrt{8}}{2}, \quad x = -1 \pm \sqrt{2}.$$

Answer: $1 \pm \sqrt{6}, -1 \pm \sqrt{2}$.

Even though Ferrari's method might be looked at as a universal tool for solving any quartic equation, from a practical standpoint, in specific cases, you would be better off applying various short cuts and useful techniques. Let's go over the most commonly encountered situations below.

2 Biquadratic equations

Biquadratic equations are the quartic equations in which the coefficients $a_3 = 0$ and $a_1 = 0$, $a_4 x^4 + a_2 x^2 + a_0 = 0$. The easiest way to solve these equations is to use substitution $x^2 = y$ and get to the solution of a standard quadratic equation.

Problem 7. Solve the equation

$$x^4 - 13x^2 + 36 = 0.$$

Solution. Let $x^2 = y$, then

$$y^2 - 13y + 36 = 0,$$
$$D = 169 - 144 = 25,$$
$$y = \frac{13 \pm \sqrt{25}}{2},$$
$$y_1 = 9, \text{ or } y_2 = 4.$$

So, $x^2 = 9$ or $x^2 = 4$, and we conclude that the roots of the original equation are $x = \pm 3$ or $x = \pm 2$.

3 Equations containing reciprocal quantities

The key here will be substituting one of the reciprocal terms containing the variable x by another variable y. Then the other reciprocal term will equal $\frac{1}{y}$. The original equation will be transformed into a quadratic equation.

Problem 8. Solve the equation

$$\left(\frac{x}{x+1}\right)^2 + \left(\frac{x+1}{x}\right)^2 = \frac{17}{4}, \quad x \neq 0, x \neq -1.$$

Solution. Let's make the substitution $y = \left(\frac{x}{x+1}\right)^2$. Obviously, being a square, y has to be a nonnegative number, $y \geq 0$. It implies that we will need to eliminate negative roots (if any) of the equation for y. We have

$$y + \frac{1}{y} = \frac{17}{4},$$
$$4y^2 - 17y + 4 = 0,$$
$$D = 289 - 64 = 225.$$
$$y = \frac{17 \pm \sqrt{225}}{8},$$
$$y = 4, \text{ or } y = \frac{1}{4}.$$

Going back to our substitution for x, we get two equations, each of which leads to the solution of another two quadratic equations:

$$\left(\frac{x}{x+1}\right)^2 = 4 \text{ or } \left(\frac{x}{x+1}\right)^2 = \frac{1}{4}.$$
$$\frac{x}{x+1} = 2 \text{ or } \frac{x}{x+1} = -2; \quad \frac{x}{x+1} = \frac{1}{2} \text{ or } \frac{x}{x+1} = -\frac{1}{2}.$$

Solve each of these equations.

The first equation becomes $2x + 2 = x$, from which $x = -2$.

The second equation becomes $-2x - 2 = x$, from which $x = -\dfrac{2}{3}$.

The third equation becomes $x + 1 = 2x$, from which $x = 1$.

The last equation becomes $-x - 1 = 2x$, from which $x = -\dfrac{1}{3}$.

Answer: $-2, -\frac{2}{3}, 1, -\frac{1}{3}$.

4 Equations of the type $(x + a)(x + b)(x + c)(x + d) = m$, where $m \neq 0$

These equations are reduced to quadratic equations if $a + b = c + d$, or $a + c = b + d$, or $b + c = a + d$.

Problem 9. Solve the equation

$$(x + 2)(x - 3)(x + 1)(x + 6) = -96.$$

Solution. In this case $a = 2$, $b = -3$, $c = 1$, $d = 6$. Notice that $a + c = b + d$. Therefore, we rewrite the equation as

$$(x + 2)(x + 1)(x - 3)(x + 6) = -96,$$
$$(x^2 + 3x + 2)(x^2 + 3x - 18) = -96.$$

At this point it should not be hard to see how to introduce the new variable y, $y = x^2 + 3x$. The new equation to solve is $(y + 2)(y - 18) = -96$.

$$y^2 - 16y + 60 = 0,$$
$$D = 256 - 240 = 16.$$
$$y = \frac{16 \pm \sqrt{16}}{2},$$

$y = 10$, or $y = 6$. Then $x^2 + 3x = 10$ or $x^2 + 3x = 6$.

Solve each of the above equations for x.

$$x^2 + 3x = 10,$$
$$x^2 + 3x - 10 = 0,$$
$$D = 9 + 40 = 49,$$
$$x = \frac{-3 \pm \sqrt{49}}{2},$$
$$x_1 = 2, \quad x_2 = -5.$$

Second equation: $x^2 + 3x = 6$, or

$$x^2 + 3x - 6 = 0,$$
$$D = 9 + 24 = 33,$$
$$x = \frac{-3 \pm \sqrt{33}}{2},$$
$$x_3 = \frac{-3 + \sqrt{33}}{2}, \quad x_4 = \frac{-3 - \sqrt{33}}{2}.$$

Answer: 2, -5, $\frac{-3+\sqrt{33}}{2}$, $\frac{-3-\sqrt{33}}{2}$.

5   Equations of the type $\frac{ax}{px^2+mx+q} + \frac{bx}{px^2+nx+q} = c, \ (c \neq 0)$

If the constant $c \neq 0$, then, obviously, the equation will not have the solution $x = 0$. Keeping in mind that $x \neq 0$, dividing by x the numerator and denominator in each fraction on the left-hand side of the equation, yields

$$\frac{a}{px + m + \frac{q}{x}} + \frac{b}{px + n + \frac{q}{x}} = c.$$

The next step is to introduce the new variable $y = px + \frac{q}{x}$, which leads to the equation

$$\frac{a}{y + m} + \frac{b}{y + n} = c.$$

Finding the common denominator on the left-hand side and making a few simplifications yields the standard quadratic equation

$$cy^2 + (c(m + n) - (a + b))y + (cmn - an - bm) = 0.$$

Before proceeding to demonstrate the solution with a specific example, it has to be noted that it is very important to always keep in mind the domain of an equation during every step in the solution process. As you introduce the new variable y, you must observe the domain of this new equation, y can be any real number, but $y \neq -m$, $y \neq -n$.

Problem 10. Solve the equation

$$\frac{2x}{2x^2 - 5x + 3} + \frac{13x}{2x^2 + x + 3} = 6.$$

Solution. First, we need to determine the domain of the equation and exclude the values of x for which the denominator of each fraction equals 0:

$$2x^2 - 5x + 3 \neq 0 \quad \text{and} \quad 2x^2 + x + 3 \neq 0.$$

Basically, we have to solve two equations and exclude their roots (if any) from the solutions of the original equation.

$$2x^2 - 5x + 3 = 0,$$
$$D = 25 - 24 = 1,$$
$$x = \frac{5 \pm 1}{4}.$$

Therefore,

$$x_1 = \frac{3}{2}, \quad x_2 = 1.$$
$$2x^2 + x + 3 = 0,$$
$$D = 1 - 24 = -23 < 0,$$

which means that this equation has no real solutions.

From the above analysis it follows that the domain of the original equation includes all real numbers, except $\frac{3}{2}$ and 1.

Now let's proceed to the solution of the equation. Since $x \neq 0$, dividing the numerator and denominator in each fraction by x, the equation transforms to

$$\frac{2}{2x - 5 + \frac{3}{x}} + \frac{13}{2x + 1 + \frac{3}{x}} = 6.$$

Substituting $y = 2x + \frac{3}{x}$ yields the new equation

$$\frac{2}{y - 5} + \frac{13}{y + 1} = 6.$$

Observe that the domain of this new equation consists of all real numbers except 5 and -1, $y \neq 5$ and $y \neq -1$.

$$\frac{2(y + 1) + 13(y - 5)}{(y - 5)(y + 1)} = 6,$$
$$2y + 2 + 13y - 65 = 6(y^2 - 4y - 5),$$
$$6y^2 - 39y + 33 = 0,$$
$$D = 1521 - 792 = 729,$$
$$y = \frac{39 \pm \sqrt{729}}{12},$$
$$y_1 = 5.5, \quad y_2 = 1.$$

Substituting these values back into $y = 2x + \frac{3}{x}$ gives two equations for x:

$$2x + \frac{3}{x} = 5.5 \quad \text{or} \quad 2x + \frac{3}{x} = 1.$$

Solving the first equation gives

$$2x + \frac{3}{x} = 5.5,$$
$$2x^2 - 5.5x + 3 = 0,$$
$$D = 30.25 - 24 = 6.25,$$
$$x = \frac{5.5 \pm \sqrt{6.25}}{4},$$
$$x_1 = 2, \quad x_2 = \frac{3}{4}.$$

Solving the second equation gives

$$2x + \frac{3}{x} = 1,$$
$$2x^2 - x + 3 = 0.$$
$$D = 1 - 24 = -23 < 0.$$

There are no real solutions.

Since each of the two roots that we found belongs to the domain of the original equation, we conclude that the equation is solved and has two solutions 2 and $\frac{3}{4}$.

Answer: 2, $\frac{3}{4}$.

6   Equations of the type $(x + a)^4 + (x + b)^4 = c \ (c > 0)$

The solution of these equations is simplified by the substitution

$$x = y - \frac{a+b}{2}.$$

Problem 11. Solve the equation

$$(x + 6)^4 + (x + 4)^4 = 82.$$

Solution. As indicated above, the substitution to use is

$$x = y - \frac{6+4}{2} = y - 5.$$

It follows that

$$(y + 1)^4 + (y - 1)^4 = 82.$$

We assume that the readers are familiar with the use of Pascal's triangle in finding the coefficients in a binomial expansion and can easily get from the original equation to the following:

$$y^4 + 4y^3 + 6y^2 + 4y + 1 + y^4 - 4y^3 + 6y^2 - 4y + 1 = 82,$$
$$2y^4 + 12y^2 - 80 = 0,$$
$$y^4 + 6y^2 - 40 = 0.$$

The last equation is a biquadratic equation and is solvable by another substitution

$$z = y^2, \quad z \geq 0.$$
$$z^2 + 6z - 40 = 0,$$
$$D = 36 + 160 = 196,$$
$$z = \frac{-6 \pm \sqrt{196}}{2}$$

so $z_1 = 4$, or $z_2 = -10$, which should be rejected because satisfactory roots should be nonnegative, $z \geq 0$. Therefore $y^2 = 4$. It follows that $y = \pm 2$.

Finally, we get to our first substitution, $x = y - 5$. Substituting the found values of y, we find $x = 2 - 5 = -3$ or $x = -2 - 5 = -7$.

Answer: $-3, -7$.

7 Equations of the type $\frac{1}{(x+a)^2} + \frac{1}{(x+b)^2} = c, (c > 0)$

The domain of the equation is the set of all real numbers except $x = -a$ and $x = -b$. The solution of these equations is simplified by making use of the substitution $y = \frac{1}{(x+a)(x+b)}$. However, in this case some preliminary work has to be done before introducing the new variable. The goal is to complete a perfect square on the left-hand side of the equation. In order to do that, we subtract the same expression $\frac{2}{(x+a)(x+b)}$ from both sides of the equation.

$$\frac{1}{(x+a)^2} - \frac{2}{(x+a)(x+b)} + \frac{1}{(x+b)^2} = c - \frac{2}{(x+a)(x+b)},$$

$$\left(\frac{1}{x+a} - \frac{1}{x+b}\right)^2 = c - \frac{2}{(x+a)(x+b)},$$

$$\left(\frac{b-a}{(x+a)(x+b)}\right)^2 = c - \frac{2}{(x+a)(x+b)},$$

$$(b-a)^2 \cdot \left(\frac{1}{(x+a)(x+b)}\right)^2 = c - \frac{2}{(x+a)(x+b)},$$

Now it's clear that the substitution $y = \frac{1}{(x+a)(x+b)}$ turns the original equation into a quadratic equation easily solvable by regular methods.

Problem 12. Solve the equation

$$\frac{1}{(x-3)^2} + \frac{1}{(x-4)^2} = 3.$$

Solution. As it was explained above, we will subtract $\frac{2}{(x-3)(x-4)}$ from both sides of the equation:

$$\frac{1}{(x-3)^2} - \frac{2}{(x-3)(x-4)} + \frac{1}{(x-4)^2} = 3 - \frac{2}{(x-3)(x-4)},$$

$$\left(\frac{1}{x-3} - \frac{1}{x-4}\right)^2 = 3 - \frac{2}{(x-3)(x-4)},$$

$$\left(\frac{-1}{(x-3)(x-4)}\right)^2 = 3 - \frac{2}{(x-3)(x-4)},$$

$$\left(\frac{1}{(x-3)(x-4)}\right)^2 = 3 - \frac{2}{(x-3)(x-4)}.$$

Introducing $y = \frac{1}{(x-3)(x-4)}$ leads to $y^2 = 3 - 2y$, or equivalently, $y^2 + 2y - 3 = 0$. By Viète's formulas, $y_1 = 1$, $y_2 = -3$. Returning to the substitution for x gives

$$\frac{1}{(x-3)(x-4)} = 1 \quad \text{or} \quad \frac{1}{(x-3)(x-4)} = -3.$$

Modifying the first equation, we get

$$x^2 - 7x + 11 = 0,$$
$$D = 49 - 44 = 5,$$
$$x = \frac{7 \pm \sqrt{5}}{2}.$$

Modifying the second equation, we get

$$3x^2 - 21x + 37 = 0,$$
$$D = 441 - 444 = -3.$$

Since $D < 0$, the second equation has no real solutions. Therefore, the original equation has two real roots $\frac{7+\sqrt{5}}{2}$ and $\frac{7-\sqrt{5}}{2}$.

Answer: $\frac{7+\sqrt{5}}{2}$, $\frac{7-\sqrt{5}}{2}$.

8 Completing the square

Sometimes you should be able to convert an equation into a simpler equation using the methods outlined above by regrouping the terms and completing the square.

Problem 13. Solve the equation

$$x^4 + 6x^3 + 5x^2 - 12x + 3 = 0.$$

Solution. The idea is trying to separate a perfect square in such a way that the original equation could be converted to a standard quadratic equation after the selection of a new variable. Selecting it should be pretty obvious. Let's add and subtract $9x^2$ on the left-hand side of the equation:

$$(x^4 + 6x^3 + 9x^2) - 9x^2 + 5x^2 - 12x + 3 = 0,$$
$$(x^4 + 6x^3 + 9x^2) - 4x^2 - 12x + 3 = 0,$$
$$(x^2 + 3x)^2 - 4(x^2 + 3x) + 3 = 0.$$

Substitution to make is $y = x^2 + 3x$. The new equation is easily solvable as a standard quadratic equation $y^2 - 4y + 3 = 0$. By Viète's formulas its solutions are $y_1 = 3$ and $y_2 = 1$. It follows that

$$x^2 + 3x = 3 \text{ or } x^2 + 3x = 1.$$

Solve each of these two equations for x.
    Solving the first equation gives

$$x^2 + 3x - 3 = 0,$$
$$D = 9 + 12 = 21,$$
$$x = \frac{-3 \pm \sqrt{21}}{2}.$$

Solving the second equation gives

$$x^2 + 3x - 1 = 0,$$
$$D = 9 + 4 = 13,$$
$$x = \frac{-3 \pm \sqrt{13}}{2}.$$

Answer: $\frac{-3 \pm \sqrt{21}}{2}$, $\frac{-3 \pm \sqrt{13}}{2}$.

Problem 14. Solve the equation

$$(x^2 - 2x + 2)^2 - 2(x^2 - 2x + 2) + 2 = x.$$

Solution. We can rewrite the equation as

$$((x^2 - 2x + 2)^2 - 2(x^2 - 2x + 2) + 1) + 1 = x.$$

Completing the square on the left-hand side gives

$$((x^2 - 2x + 2) - 1)^2 + 1 = x,$$
$$(x^2 - 2x + 1)^2 - x + 1 = 0,$$
$$((x - 1)^2)^2 - (x - 1) = 0,$$
$$(x - 1)^4 - (x - 1) = 0,$$
$$(x - 1)((x - 1)^3 - 1) = 0,$$
$$x - 1 = 0 \text{ or } (x - 1)^3 = 1.$$
$$x = 1 \text{ or } x = 2.$$

Alternatively, this problem has an unorthodox solution, which represents a very interesting technique of introducing an auxiliary function. In this case such a function is going to be

$$f(x) = x^2 - 2x + 2.$$

Therefore,

$$f(f(x)) = (x^2 - 2x + 2)^2 - 2(x^2 - 2x + 2) + 2.$$

It follows from the conditions of the problem that $f(f(x)) = x$. The original equation can be rewritten as

$$(f(x) - 1)^2 + 1 = x.$$

Clearly, since the left-hand side is greater than or equal to 1,

$$\underbrace{(f(x) - 1)^2}_{\geq 0} + 1 \geq 1,$$

then the right-hand side has to be greater than or equal to 1 as well, $x \geq 1$. The quadratic function $f(x) = x^2 - 2x + 2$ is increasing for all $x \geq 1$. Indeed, for all x such that $x \geq 1$,

$$f(x) - f(1) = x^2 - 2x + 2 - (1 - 2 + 2)$$
$$= x^2 - 2x + 2 - 1 = x^2 - 2x + 1 = (x - 1)^2 \geq 0,$$

which means $f(x) \geq f(1)$. It implies that the following equations

$$f(f(x)) = x \tag{1}$$
$$\text{and} \quad f(x) = x \tag{2}$$

are equivalent. Indeed, let x_0 be a solution of (2). Then $f(f(x_0)) = f(x_0) = x_0$, so it is a solution of (1) as well. Assume now that x_0 is a solution of (1). It follows $f(f(x_0)) = x_0$. Let $x_0 \neq f(x_0)$. Hence, it is either $x_0 > f(x_0)$ or $x_0 < f(x_0)$. Since f increases for the selected values of $x > 1$, then in the first case,

$f(x_0) > f(f(x_0)) = x_0$, or in the second case, $f(x_0) < f(f(x_0)) = x_0$, which contradicts our assumption. Thus, any solution of the equation $f(x) = x$ will also be a solution of the equation $f(f(x)) = x$. Therefore, the solution of the original equation is reduced to the solution of the equation $x^2 - 2x + 2 = x$, or equivalently,

$$x^2 - 3x + 2 = 0.$$

By Vièle's formulas the solutions are $x_1 = 1$, $x_2 = 2$.

Answer: 1, 2.

Even though completing the square may look easier in solving this particular equation, I strongly recommend remembering the auxiliary function method. By adding this method to your arsenal, you get a powerful tool for solving algebraic equations (any equations, not necessarily quartic or cubic). It might significantly simplify the solution when other methods fail.

9 Reciprocal equations of the type $ax^4 + bx^3 + cx^2 + bx + a = 0, \ (a \neq 0)$

These equations are distinguished from other equations by the equal coefficients symmetrical with respect to the middle term. Since the coefficient $a \neq 0$, then, obviously, x can't equal to 0 and we can divide both sides of the equation by x^2. It will be transformed into the equation

$$ax^2 + bx + c + \frac{b}{x} + \frac{a}{x^2} = 0.$$

Regrouping the terms in the last equation allows us to introduce a new variable. The solution of the original equation would be reduced to the solution of a quadratic equation.

$$\left(ax^2 + \frac{a}{x^2}\right) + \left(bx + \frac{b}{x}\right) + c = 0,$$

$$a\left(x^2 + \frac{1}{x^2}\right) + b\left(x + \frac{1}{x}\right) + c = 0.$$

Substituting $x + \frac{1}{x} = y$ gives $\left(x + \frac{1}{x}\right)^2 = y^2$ or $x^2 + \frac{1}{x^2} + 2 = y^2$, from which $x^2 + \frac{1}{x^2} = y^2 - 2$. Therefore, we can rewrite the original equation as $a(y^2 - 2) + by + c = 0$, or $ay^2 + by + (c - 2a) = 0$, which is a standard quadratic equation.

The same technique can be extended to a reciprocal equation of any even power $2n$. Dividing both sides of the equation by x^{2n} and grouping the symmetrical terms it is simplified through the introduction of a new variable and then expanded to an equation of a reduced power.

Problem 15. Solve the equation

$$2x^4 + 3x^3 - 16x^2 + 3x + 2 = 0.$$

Solution. Divide both sides of the equation by x^2 ($x \neq 0$):

$$2x^2 + 3x - 16 + \frac{3}{x} + \frac{2}{x^2} = 0,$$

Regroup the terms as

$$\left(2x^2 + \frac{2}{x^2}\right) + \left(3x + \frac{3}{x}\right) - 16 = 0,$$

$$2\left(x^2 + \frac{1}{x^2}\right) + 3\left(x + \frac{1}{x}\right) - 16 = 0.$$

Substituting $x + \frac{1}{x} = y$ gives $2(y^2 - 2) + 3y - 16 = 0$ or equivalently,

$$2y^2 + 3y - 20 = 0,$$
$$D = 9 + 160 = 169,$$
$$y = \frac{-3 \pm \sqrt{169}}{4},$$

therefore, $y_1 = 2.5$, $y_2 = -4$.

Going back to the substitution for x gives $x + \frac{1}{x} = 2.5$ or $x + \frac{1}{x} = -4$. Solving the first equation gives

$$x + \frac{1}{x} = 2.5,$$
$$x^2 - 2.5x + 1 = 0,$$
$$D = 6.25 - 4 = 2.25,$$
$$x = \frac{2.5 \pm \sqrt{2.25}}{2}.$$

It follows that $x_1 = 2$, $x_2 = 0.5$.

Solving the second equation gives

$$x + \frac{1}{x} = -4,$$
$$x^2 + 4x + 1 = 0,$$
$$D = 16 - 4 = 12,$$
$$x = \frac{-4 \pm \sqrt{12}}{2}.$$

It follows that $x_3 = -2 - \sqrt{3}$, $x_4 = -2 + \sqrt{3}$.

Answer: $2, 0.5, -2 - \sqrt{3}, -2 + \sqrt{3}$.

We restricted our investigation to a few types of the most commonly encountered cubic and quartic equations. In the next chapter we extend our study of equations to those of degree greater than four, demonstrating various power reduction methods leading to lower degree equations, including cubic and quartic equations. The techniques exhibited here will be broadly applied in their solutions. We invite readers to investigate further on their own other methods and techniques for cubic and quartic equations solutions. Meanwhile, the problems below can sharpen your skills in the methods reviewed in this chapter.

Exercises

Problem 16. Solve the equation $\frac{x^2+x-5}{x} + \frac{3x}{x^2+x-5} + 4 = 0$.

Problem 17. Solve the equation $(x+3)(x-1)(x+2)(x+6) = -20$.

Problem 18. Solve the equation $\frac{4x}{4x^2-8x+7} + \frac{3x}{4x^2-10x+7} = 1$.

Problem 19. Solve the equation $(x+3)^4 + (x+5)^4 = 16$.

Problem 20. Solve the equation $6x^4 - 13x^3 + 12x^2 - 13x + 6 = 0$.

Problem 21. Evaluate $\sqrt[3]{9+\sqrt{80}} + \sqrt[3]{9-\sqrt{80}}$.

Problem 22. Verify that the number $x = \sqrt[3]{4+\sqrt{80}} - \sqrt[3]{\sqrt{80}-4}$ is one of the roots of the equation $x^3 + 12x - 8 = 0$.

Chapter 4

Division of Polynomials. Little Bézout's Theorem and its Consequences

The word "Algebra" is translated from Arabic "*al-jabr*" as "reunion of broken parts." As a matter of fact, it explains what is achieved by setting up and solving equations. Is it always possible to "reunite the broken parts" or solve an equation of any power and express its roots in terms of radicals (through four basic arithmetic operations and nth roots)? That question was raised by prominent mathematicians throughout the centuries. As simple as it may look, it actually was very complicated to get general results that were sufficient to give an answer. In previous chapters we went through the formulas and their applications for quadratic, cubic, and quartic equations and saw that the most significant results have been achieved only in the 15th–17th centuries. In the theory of equations new advances in the 18th century were brought through the work of many outstanding mathematicians including Leonhard Euler (1707–1783), Joseph-Louis Lagrange (1736–1813), Carl Friedrich Gauss (1777–1855), Étienne Bézout (1730–1783), and Alexandre-Théophile Vandermonde (1735–1796). However, up until the beginning of the 19th century, there was no definite conclusion regarding the solvability in radicals of equations with a degree greater than four. The Italian mathematician Paolo Ruffini (1765–1822) was the first to be credited for the proof that fifth-degree or higher order equations generally cannot be solved in radicals. However, he provided only an incomplete proof in 1799. The major breakthrough in solving equations of higher powers should be attributed to the independent work of two young geniuses; the Norwegian mathematician Neils Henrik Abel (1802–1829) and the French mathematician Évariste Galois (1811–1832). It is absolutely unbelievable how much they managed to achieve in their short and tragic lives—Abel died in poverty suffering from tuberculosis at the age of 26 and Galois's life was cut short in a duel when he was not even 21 years old. Abel completed Ruffini's proof in 1823. Since then the theorem is well-known as the Abel-Ruffini Theorem and it states that *there is no general algebraic solution (solution in radicals) to polynomial equations of degree five or higher*. It's important to emphasize the words "general solutions" in the theorem. It does not say that there are no solutions at all, but it does state that there is no general solution in radicals to all the equations of degree five or higher. Abel completely changed the approach to equations from just attempts to find solutions to demand a proof of the solutions' existence for certain types of quintic and higher degrees' equations.

There is no Nobel Prize award in mathematics. For some reason known only to him, Alfred Nobel did not include mathematics in the list of the scientific achievements for the most prestigious award in today's world. The award that is equivalent to the Nobel Prize award in mathematics is named after Neils Abel and is called the Abel Prize. Along with the Fields Prize (awarded to mathematicians younger than 40 years of age), it is regarded as the highest honor a mathematician can receive. What else could you add to characterize Neils Henrik Abel's contributions to the field of mathematics?

The same theorem was independently proved in 1846 by Évariste Galois. He was able to introduce a new approach in modern algebra which was named after him as Galois Theory. He gave an answer to the question why there is no general formula for the roots of a fifth or higher degree polynomial equation in terms of its coefficients using the algebraic operations of addition, subtraction, division, multiplication, and application of radicals. In addition to this, he explained why it would be possible to solve equations of degree four or lower in a specific manner. Amazingly, his theory also explains the conditions for the constructability of regular polygons by means of compass and straightedge, providing a connecting link between classical geometrical issues and algebraic solutions of polynomial equations. The night before his fatal duel Galois wrote a letter to his friend Auguste Chevalier, which years later revolutionized modern algebra and created a foundation of a group theory that basically became the foundation of abstract algebra. We will not examine the basics of his theory. It is way beyond the scope of this book. Ambitious readers, however, may wish to get acquainted with his findings in greater detail and explore further on their own. One such great source I can refer to is Jean-Pierre Tignol's book *Galois' Theory of Algebraic Equations*, World Scientific, 2011.

Even though we now know (thanks to Abel's and Galois's work) that there are no general algebraic solutions to polynomial equations of degree five or higher, some of those equations are solvable in radicals. In this chapter we will focus on a few classic practical and useful techniques for solving such equations. As it was promised in the previous chapter, we will get back to cubic and quartic equations and demonstrate the application of the methods to them as well.

In 1798 Carl Gauss proved the fundamental theorem of algebra, which states that any single-variable polynomial with complex coefficients has at least one root and the number of roots of a non-zero polynomial over the field of complex numbers is equal to the degree of the polynomial, counting each root with its multiplicity. Staying in the field of real numbers only, the theorem asserts that any polynomial with real coefficients can be factored into a product of polynomials of the first or second degree. If you manage to factor a polynomial of degree three or greater, you would be facing new polynomials—factors with lower degrees, thus bringing the problem of finding the roots to some solvable path. Sometimes by repeating that process until you get to the quadratic equation, you may be able to solve the original equation and find all of its roots. To be able to do that, you first need to identify some possible roots. In his book *Disquisitiones Arithmeticae*, Carl Gauss proved a few important properties in number theory, Gauss's Lemma being one of them. It has a special case known as the *Rational Root Theorem*, which answers the question of how to identify a polynomial's possible rational roots.

Theorem. *For any polynomial equation with integer coefficients*

$$a_n x^n + a_{n-1} x^{n-1} + \cdots + a_1 x + a_0 = 0$$

(when $a_n \neq 0$, $a_0 \neq 0$), each rational solution can be found as $x = \frac{p}{q}$ (p and q are integers with the greatest common devisor of 1). p is an integer factor of the coefficient a_0 and q is an integer factor of the coefficient a_n.

Proof. Substituting $x = \frac{p}{q}$ into the equation and multiplying both sides by q^n give

$$a_n p^n + a_{n-1} p^{n-1} q + \cdots + a_1 p q^{n-1} + a_0 q^n = 0.$$

Then

$$a_n p^n + a_{n-1} p^{n-1} q + \cdots + a_1 p q^{n-1} = -a_0 q^n,$$

and factoring p out, we get

$$p(a_n p^{n-1} + a_{n-1} p^{n-2} q + \cdots + a_1 q^{n-1}) = -a_0 q^n.$$

Obviously, the left side of the last equality is some composite number with one of the factors being p. Therefore, the right side of the equality, $-a_0 q^n$, must also be divisible by p. Since p is coprime to q, then $a_0 q^n$ will be divisible by p only if a_0 is divisible by p. In other words, we justified that number p is an integer factor of the coefficient a_0, as was to be proved. In a similar way you can prove the second statement of the theorem. □

This theorem can be the starting point for finding the solutions of polynomial equations. As you find one of those rational roots, the next step will be to factor the original polynomial. That task is much simplified by way of division of the given polynomial by a linear polynomial (the polynomial long division method). The algorithm is similar to the algorithm of numbers division; you eliminate the highest degree variables step by step until you get 0 in the very last difference or a linear polynomial as a remainder.

Problem 1. Solve the equation

$$2x^3 + 4x^2 - 3x - 5 = 0.$$

Solution. First, note that $a_0 = -5$ and $a_3 = 2$. Therefore, the possible rational solutions are to be searched for in the set of numbers $-\frac{1}{2}, -\frac{5}{2}, -1, -5$. As soon as you realize that $x = -1$ satisfies the equation, the cubic polynomial can be factored into linear and quadratic polynomial factors. In order to do that, let's divide $2x^3 + 4x^2 - 3x - 5$ by $(x + 1)$:

$$
\begin{array}{r}
x + 1 \\
2x^2 + 2x - 5 \,\overline{\big)\, 2x^3 + 4x^2 - 3x - 5} \\
\underline{2x^3 + 2x^2} \\
2x^2 - 3x \\
\underline{2x^2 + 2x} \\
-5x - 5 \\
\underline{-5x - 5} \\
0
\end{array}
$$

Thus,

$$(x + 1)(2x^2 + 2x - 5) = 0.$$
$$x = -1 \text{ or } 2x^2 + 2x - 5 = 0.$$

Now solve the equation $2x^2 + 2x - 5 = 0$.

$$D = 4 + 40 = 44.$$
$$x = \frac{-2 \pm \sqrt{44}}{4},$$
$$x = \frac{-1 + \sqrt{11}}{2}, \text{ or } x = \frac{-1 - \sqrt{11}}{2}.$$

Answer: $-1, \frac{-1 \pm \sqrt{11}}{2}$.

Problem 2. Determine if $(x^2 + 1)$ is a factor of $P(x) = x^4 + x^3 + x^2 + 2x + 1$.

Solution. To answer the problem's question, we will apply again the long division algorithm:

$$
\begin{array}{r}
x^2 + 1 \\
x^2 + x \enclose{longdiv}{x^4 + x^3 + x^2 + 2x + 1} \\
\underline{x^4 + x^2} \\
x^3 + 2x + 1 \\
\underline{x^3 + x} \\
x + 1
\end{array}
$$

As we can see, there is a remainder $(x + 1)$. Therefore, $x^2 + 1$ is not a factor of $P(x)$, or, in other words, it is impossible to factor completely $P(x)$ with one of the factors being $(x^2 + 1)$. Instead we conclude that

$$x^4 + x^3 + x^2 + 2x + 1 = (x^2 + 1)(x^2 + x) + x + 1.$$

The great French mathematician René Descartes (1596–1650) introduced an important technique for establishing the number of positive and negative roots of a polynomial, known as *Descartes's Rule of Signs*:

For any polynomial $P(x)$ with real coefficients the number of its positive real roots is equal to the number of changes in sign of the coefficients of $P(x)$ or is less than this by an even number. The number of negative real roots of $P(x)$ is equal to the number of changes in sign of the coefficients of $P(-x)$ or is less than this by an even number. This rule does not give you a tool to find the roots, however it helps to determine how many roots to expect. So, one may save time in finding the roots of a polynomial when applying the Rational Root Theorem along with Descartes's Rule of Signs. You don't need to test all possible solutions, but only those which would be identified as potential solutions after applying Descartes's Rule of Signs. For example, consider the polynomial $P(x) = x^3 + 2x^2 - 5x - 6$. The coefficients are 1, 2, −5, −6 with one sign change. Hence, there has to be one positive root. Consider now $P(-x) = -x^3 + 2x^2 + 5x - 6$. The coefficients are −1, 2, 5, −6 with two sign changes. Therefore, there are 0 or 2 negative roots. In fact, $x^3 + 2x^2 - 5x - 6 = (x + 1)(x - 2)(x + 3)$, so there is one positive root 2 and two negative roots −1 and −3.

When you divide a polynomial $P(x)$ by a nonzero polynomial divisor $D(x)$, you get a quotient polynomial $Q(x)$ and a remainder polynomial $R(x)$:

$$\frac{P(x)}{D(x)} = Q(x) + \frac{R(x)}{D(x)}.$$

Instead of applying the polynomial long division method, a short cut is often utilized when you need to divide the polynomial by the binomial of the form $(x - a)$. It's called *Horner's scheme* after the British mathematician William George Horner (1786–1837) or the *synthetic division* method.

Let's go back to problem 1 and demonstrate the Horner's scheme algorithm:

1. List the coefficients of the dividend polynomial in order of descending exponents. If any x term is missing, then its coefficient is 0.

2. Write the a value to the left of the vertical bar. In our case that number is -1.

3. The first coefficient is brought down. You need to multiply the first coefficient by the a value and write the result under the second coefficient: we multiply 2 by (-1) and need to write -2 under 4. The numbers on line 1 and on line 2 are to be added.

4. Multiply the previous sum by the a value and write the result under the third coefficient: $4 - 2 = 2$, therefore, 2 is multiplied by -1 and the result -2 is to be written under -3.

5. Repeat the above step for the remaining coefficient, 5 to be written under -5.

$$\begin{array}{r|rrrr} -1 & 2 & 4 & -3 & -5 \\ & & -2 & -2 & 5 \\ \hline & 2 & 2 & -5 \end{array}$$

The numbers in the last row are the coefficients of the quotient. At this point you may get back to your polynomial and write down the outcome:

$$2x^3 + 4x^2 - 3x - 5 = (x+1)(2x^2 + 2x - 5).$$

Problem 3. Solve the equation

$$x^3 - 2x + 1 = 0.$$

Solution. It is easy to see that 1 is a root of the equation. The next step is to apply Horner's scheme algorithm to factor our cubic polynomial:

$$\begin{array}{r|rrrr} 1 & 1 & 0 & -2 & 1 \\ & & 1 & 1 & -1 \\ \hline & 1 & 1 & -1 \end{array}$$

$$x^3 - 2x + 1 = (x-1)(x^2 + x - 1) = 0.$$

We already found one root $x = 1$. By solving the quadratic equation $x^2 + x - 1 = 0$, we get another two solutions:

$$D = 1 + 4 = 5,$$

$$x = \frac{-1 \pm \sqrt{5}}{2}.$$

Answer: 1, $\frac{-1\pm\sqrt{5}}{2}$.

When you divide a polynomial $P(x)$ by $(x - a)$ the reminder will not always be 0. The remainder in Horner's scheme plays a very important role, which is stated in *The Remainder Theorem*, known also as *Little Bézout's Theorem* after the French mathematician Étienne Bézout (1730–1783):

The remainder of the division of a polynomial $P(x)$ by linear polynomial $(x - a)$ is equal to $P(a)$.

Proof. Let

$$P(x) = a_n x^n + a_{n-1} x^{n-1} + \cdots + a_1 x + a_0.$$

Assume that r is the remainder of the division of $P(x)$ by $(x - a)$ with $Q(x)$ being a quotient: $P(x) = Q(x)(x - a) + r$. Then

$$P(a) = Q(a)(a - a) + r,$$
$$P(a) = 0 + r,$$
$$r = P(a),$$

which was to be proved.   □

Étienne Bézout, the member of the French Academy of Science, is well-known for his work related to polynomial equations. His name is associated not just with the above theorem, but with a few innovations in the methods of the solution of simultaneous polynomial equations with many variables. He is credited for the introduction of the notion of the "polynomial multiplier" and the invention of the determinant as the key element in connecting the coefficients in a square matrix with their utilization for the system's solutions.

We will call number a to be the root or the zero of a given polynomial $P(x)$ if $P(a) = 0$. Then, the statement "Find all zeros or roots of a polynomial $P(x)$" is equivalent to the statement "Solve the equation $P(x) = 0$."

Lemma 1 (The Factor Theorem). *A polynomial $P(x)$ has a factor $(x - a)$ if and only if $P(a) = 0$.*

Proof. Let

$$P(x) = a_n x^n + a_{n-1} x^{n-1} + \cdots + a_1 x + a_0.$$

Assume $(x - a)$ is a factor of $P(x)$. Then

$$P(x) = (x - a)(b_n x^{n-1} + b_{n-1} x^{n-2} + \cdots + b_1 x + b_0),$$

where the b_i $(i = 0, \ldots, n)$ are the coefficients of the second factor, polynomial of a degree $n - 1$. It follows that

$$P(a) = (a - a)(b_n a^{n-1} + b_{n-1} a^{n-2} + \cdots + b_1 a + b_0) = 0,$$

and the direct statement of the lemma is proved.   □

The converse statement is the consequence of the Little Bézout's Theorem.

Lemma 2. *If in any nth degree polynomial the sum of the odd coefficients equals the sum of the even coefficients, then one of its roots must be* -1.

Proof. Let

$$P(x) = a_n x^n + a_{n-1} x^{n-1} + \cdots + a_1 x + a_0 \quad \text{and} \quad \sum_{k=0} a_{2k+1} = \sum_{k=1} a_{2k}.$$

Assume that the degree of $P(x)$ is an even number (in case it is an odd number, the proof will be identical). Then

$$P(-1) = a_n - a_{n-1} + a_{n-2} - \cdots + a_2 - a_1 + a_0 = 0,$$

which means -1 is a zero or a root of the polynomial $P(x)$. $\qquad\square$

Problem 4. Solve the equation

$$x^5 - 4x^4 + 5x^3 - 5x^2 + 4x - 1 = 0.$$

Solution. It's easy to determine that $x = 1$ is a root of the equation. Then by the Little Bézout's Theorem the remainder of the division of the polynomial $x^5 - 4x^4 + 5x^3 - 5x^2 + 4x - 1$ by $x - 1$ must be 0 and we would be able to find the complete factorization of the left side of the equation. For that purpose, we will apply Horner's scheme:

$$
\begin{array}{c|cccccc}
1 & 1 & -4 & 5 & -5 & 4 & -1 \\
  &   & 1 & -3 & 2 & -3 & 1 \\
\hline
  & 1 & -3 & 2 & -3 & 1 & 0
\end{array}
$$

$$(x - 1)(x^4 - 3x^3 + 2x^2 - 3x + 1) = 0.$$

The next step is to solve the quartic equation $x^4 - 3x^3 + 2x^2 - 3x + 1 = 0$. As we learned in the previous chapter, this is a reciprocal equation. Because $x \neq 0$ (this is easy to verify by direct substitution), we can divide both sides by x^2:

$$x^2 - 3x + 2 - \frac{3}{x} + \frac{1}{x^2} = 0, \text{ or } \left(x^2 + \frac{1}{x^2}\right) - \left(3x + \frac{3}{x}\right) + 2 = 0.$$

Substituting $y = x + \frac{1}{x}$ gives $x^2 + \frac{1}{x^2} = y^2 - 2$. It follows

$$y^2 - 2 - 3y + 2 = 0,$$
$$y^2 - 3y = 0,$$
$$y(y - 3) = 0,$$
$$y = 0 \text{ or } y = 3.$$

Therefore,

$$x + \frac{1}{x} = 0 \text{ or } x + \frac{1}{x} = 3.$$

Modifying the first equation gives $x^2 + 1 = 0$. Since $x^2 + 1 > 0$ for any real value of x, then the first equation has no solutions.

Let's solve the second equation, $x + \frac{1}{x} = 3$.

$$x^2 - 3x + 1 = 0,$$
$$D = 9 - 4 = 5,$$
$$x = \frac{3 \pm \sqrt{5}}{2}.$$

Answer: 1, $\frac{3 \pm \sqrt{5}}{2}$.

It is worth noting that the result obtained in this problem can be extended to the solution of any reciprocal equations of an odd degree. Indeed, for example, the general reciprocal equation of degree 5

$$ax^5 + bx^4 + cx^3 + cx^2 + bx + a = 0 \quad (a \neq 0)$$

will be modified to the following equation (not necessarily by Horner's scheme, but by the grouping of its terms as well):

$$a(x^5 + 1) + bx(x^3 + 1) + cx^2(x + 1) = 0.$$

Obviously, $x + 1$ is a dividend of $x^5 + 1$ and it is a dividend of $x^3 + 1$:

$$x^3 + 1 = (x + 1)(x^2 - x + 1),$$
$$x^5 + 1 = (x + 1)(x^4 - x^3 + x^2 - x + 1).$$

Generally, for any odd power, $x^{2k+1} = (x + 1)(x^{2k} - x^{2k-1} + \cdots + x^2 - x + 1)$. So, you got the root $x = -1$, reduced the power of the equation, and obtained the reciprocal equation of an even power, the solution of which was discussed in the previous chapter.

Problem 5. Solve the equation

$$x^5 + 2x^4 - 4x^3 - 4x^2 - 5x - 6 = 0.$$

Solution. Notice that the sum of the odd coefficients equals the sum of the even coefficients: $1 - 4 - 5 = -8$ and $2 - 4 - 6 = -8$. Therefore, $x = -1$ must be a root of this equation. The next step is to use Horner's scheme for the factorization.

$$
\begin{array}{r|rrrrrr}
-1 & 1 & 2 & -4 & -4 & -5 & -6 \\
   &   & -1 & -1 & 5 & -1 & -6 \\
\hline
   & 1 & 1 & -5 & 1 & -6 & 0
\end{array}
$$

We get $(x+1)(x^4+x^3-5x^2+x-6) = 0$. Let's solve the equation $x^4+x^3-5x^2+x-6 = 0$. The possible rational roots have to be among the divisors of -6: $\pm 3, \pm 2, \pm 6, \pm 1$. The lucky number is 2. Indeed, $16+8-20+2-6 = 0$. We use Horner's scheme one more time:

$$
\begin{array}{r|rrrrr}
2 & 1 & 1 & -5 & 1 & -6 \\
  &   & 2 & 6 & 2 & 6 \\
\hline
  & 1 & 3 & 1 & 3 & 0
\end{array}
$$

$$(x-2)(x^3+3x^2+x+3) = 0.$$

The equation $x^3+3x^2+x+3 = 0$ is easily solvable by grouping its terms and further factoring:

$$x^2(x+3) + (x+3) = 0,$$
$$(x+3)(x^2+1) = 0.$$

Therefore, $x+3 = 0$ or $x^2+1 = 0$. Solving the first equation gives $x = -3$. The second equation $x^2+1 = 0$ has no real solutions.

Answer: $-1, 2, -3$.

The Little Bézout's Theorem is instrumental in the solution of many non-standard problems offered in various math competitions and Olympiads.

Problem 6. Prove that $(x-1)$ is the factor of

$$P(x) = (1+x+x^2+\cdots+x^{n-1})^2 - n^2 x^2.$$

Solution. $P(1) = (1+1+1^2+\cdots+1^{n-1})^2 - n^2 1^2 = n^2 - n^2 = 0$, which means that the remainder of division of $P(x)$ by $(x-1)$ is 0. Therefore, due to the Little Bézout's Theorem, $(x-1)$ is the factor of $P(x)$.

Problem 7. Prove that the quadratic polynomial (x^2-3x+2) is a factor of $P(x) = (x-2)^{2m} + (x-1)^m - 1$.

Solution. Applying Viète's formulas for factoring x^2-3x+2 (its roots are 1 and 2) gives

$$x^2-3x+2 = (x-2)(x-1).$$

To prove the problem's statement, it has to be verified that $P(x)$ has each $x-2$ and $x-1$ as its factors. Indeed, $P(1) = (-1)^{2m} + 0 - 1 = 1 - 1 = 0$,

$P(2) = 0^{2m} + 1^m - 1 = 1 - 1 = 0$. Since $P(x)$ is divisible by $(x - 2)$ and $(x - 1)$, then it must be divisible by their product $(x - 2)(x - 1) = x^2 - 3x + 2$, and the proof is completed.

Problem 8. Find the value of a parameter t such that the polynomials $x^4 + tx^2 + 1$ and $x^3 + tx + 1$ have a common root.

Solution. Let's denote $f(x) = x^4 + tx^2 + 1$ and $g(x) = x^3 + tx + 1$. It is given that $f(x)$ and $g(x)$ have a common root. It follows that the auxiliary polynomial $P(x)$ obtained as the difference between $f(x)$ and product of $g(x)$ by x, $P(x) = f(x) - g(x) \cdot x$, has to have the same root. Since

$$P(x) = f(x) - g(x) \cdot x = (x^4 + tx^2 + 1) - (x^3 + tx + 1) \cdot x$$
$$= x^4 + tx^2 + 1 - x^4 - tx^2 - x = 1 - x,$$

then, clearly, $x = 1$ is the one and only root of $P(x)$. It follows that $x = 1$ is a common root of $f(x)$ and $g(x)$, which implies that $f(1) = 0$ and $g(1) = 0$. Therefore, $f(1) = 1 + t + 1 = t + 2 = 0$ and $g(1) = 1 + t + 1 = t + 2 = 0$. From the last equalities we conclude that $t = -2$.

Problem 9. Find the coefficients p and q of the polynomial

$$P(x) = x^5 + 3x^3 + px^2 + qx + 3$$

if it is known that $(x + 1)$ is the remainder after the division of $P(x)$ by $(x^2 + 2)$.

Solution. Since the problem is about the division of $P(x)$ by $(x^2 + 2)$, we need to find the remainder of that division and then compare it to $(x + 1)$. It will lead to setting up simple equations for coefficients p and q.

$$
\begin{array}{r}
x^2 + 2 \\
\hline
x^3 + x + p\, \big)\, x^5 + 3x^3 + px^2 + qx + 3 \\
\underline{x^5 + 2x^3} \\
x^3 + px^2 \\
\underline{x^3 + 2x} \\
px^2 + qx - 2x + 3 \\
\underline{px^2 + 2p} \\
qx - 2x + 3 - 2p
\end{array}
$$

The remainder in the division is $qx - 2x + 3 - 2p$. It follows from the conditions of the problem that the following equality must hold true:

$$qx - 2x + 3 - 2p = x + 1,$$

or equivalently

$$(q - 2)x + (3 - 2p) = x + 1.$$

In the last linear equation, the respective coefficients on the left-hand and right-hand sides must be equal, which yields the system of two linear equations

$$q - 2 = 1,$$
$$3 - 2p = 1.$$

Therefore, $q = 3$, $p = 1$.

At the last stage of the solution we utilized the technique known as the *method of undefined coefficients*.

Method of undefined coefficients is based on the definition of identical polynomials. Two single-variable polynomials of the nth degree will be equal if and only if their respective coefficients by the same powers of variable are equal. If $P(x) = ax^4 + 3x^3 - 7x + 4$ and $Q(x) = 2x^4 + bx^3 - cx^2 + mx + n$, then $P(x) = Q(x)$ if and only if the respective coefficients are equal: $a = 2, b = 3$, $c = 0$, $m = -7$, $n = 4$. The technique proves to be efficient in simplifying solutions of many problems. For instance, the Factor Theorem along with the method of undefined coefficients help to connect together the roots and the coefficients of a cubic equation and get an elegant proof of Viète's formulas for a cubic polynomial. Assume we have a cubic polynomial $P(x) = x^3 + px^2 + qx + r$. Then $P(x) = (x - \alpha)(x - \beta)(x - \gamma)$ if and only if α, β, and γ are the roots of the equation $x^3 + px^2 + qx + r = 0$. It follows that

$$x^3 + px^2 + qx + r = (x - \alpha)(x - \beta)(x - \gamma) = x^3 - (\alpha + \beta + \gamma)x^2 + (\alpha\beta + \alpha\gamma + \beta\gamma)x - \alpha\beta\gamma.$$

Accordingly,

$$\alpha + \beta + \gamma = -p,$$
$$\alpha\beta + \alpha\gamma + \beta\gamma = q,$$
$$\alpha\beta\gamma = -r.$$

The above formulas represent Viète's formulas extended for the cubic equation (one of the separate cases of the general Viète's formulas mentioned in chapter 2). I would strongly recommend including them in your arsenal as a great weapon in overcoming many difficult problems, among which are systems of nonlinear equations with three variables, which will be discussed in one of the later chapters.

Problem 10. Prove that $(x + 1)(x + 2)(x + 3)(x + 4) + 1$ is a perfect square of a trinomial and find that trinomial.

Solution. If we manage to find numbers p and q such that

$$(x + 1)(x + 2)(x + 3)(x + 4) + 1 = (x^2 + px + q)^2,$$

then the problem will be solved. The given expression can be rewritten by regrouping the factors and then expanding to

$$(x + 1)(x + 4)(x + 2)(x + 3) + 1 = (x^2 + 5x + 4)(x^2 + 5x + 6) + 1$$
$$= x^4 + 10x^3 + 35x^2 + 50x + 25.$$

On the other hand,

$$(x^2 + px + q)^2 = x^4 + p^2x^2 + q^2 + 2x^3p + 2x^2q + 2pqx$$
$$= x^4 + 2x^3p + (p^2 + 2q)x^2 + 2pqx + q^2.$$

Therefore,

$$x^4 + 10x^3 + 35x^2 + 50x + 25 = x^4 + 2px^3 + (p^2 + 2q)x^2 + 2pqx + q^2.$$

Comparing the respective coefficients on the left-hand and right-hand sides, we get

$$2p = 10,$$
$$p^2 + 2q = 35,$$
$$2pq = 50,$$
$$q^2 = 25.$$

Solving the above system of equations gives $q = 5$ and $p = 5$. It follows that

$$x^2 + px + q = x^2 + 5x + 5, \quad \text{and,}$$
$$(x + 1)(x + 2)(x + 3)(x + 4) + 1 = (x^2 + 5x + 5)^2.$$

Problem 11. Determine whether the polynomials $x^5 - x - 1$ and $x^2 + ax + b$, where a and b are rational numbers, have common real roots.

Solution. Assume there is a common real root t of the given polynomials. It implies that $t^5 - t - 1 = 0$ and $t^2 + at + b = 0$ at the same time. The first equation can be rewritten as $t^5 = t + 1$. The second equation can be rewritten as $t^2 = -at - b$. Therefore,

$$t + 1 = t^5 = t \cdot (t^2)^2 = t(-at - b)^2$$
$$= t(a^2t^2 + 2abt + b^2)$$
$$= t(a^2(-at - b) + 2abt + b^2)$$
$$= -a^3t^2 - a^2bt + 2abt^2 + b^2t$$
$$= -a^3(-at - b) + 2ab(-at - b) + t(b^2 - a^2b)$$
$$= a^4t + a^3b - 2a^2bt - 2ab^2 + b^2t - a^2bt$$
$$= (a^4 - 3a^2b + b^2)t + a^3b - 2ab^2.$$

So, we obtain that

$$t + 1 = (a^4 - 3a^2b + b^2)t + a^3b - 2ab^2.$$

Using the method of undefined coefficients, we arrive at the equalities

$$a^4 - 3a^2b + b^2 = 1,$$
$$a^3b - 2ab^2 = 1.$$

Let's express b in terms of a from the above system of equations for a and b. Multiply the first equation by $2a$ and add to the second equation. It follows that

$$2a^5 - 6a^3b + 2ab^2 + a^3b - 2ab^2 = 2a + 1,$$
$$2a^5 - 5a^3b - 2a - 1 = 0,$$
$$b = \frac{2a^5 - 2a - 1}{5a^3}.$$

Substituting b into either of the equations in the system and making the simplifications (we leave this to the reader to verify) yields

$$a^{10} + 3a^6 - 11a^5 - 4a^2 - 4a - 1 = 0.$$

Applying the Rational Root Theorem, it is easy to verify that there is no rational a that satisfies the last equation; the only available options are 1 or -1, and neither satisfies the equation. This contradicts the given conditions that a and b are rational numbers and we arrive at the conclusion that the given polynomials do not have any common real roots.

As you study the Rational Root Theorem, the Little Bézout's Theorem, and their consequences, you come across unexpectedly with very interesting properties related to the division of real numbers; specifically, some generalizations for a division of a sum or difference of two powers with the natural exponent by the sum or difference of their bases.

1. The difference $x^n - c^n$ will always be divisible by $(x - c)$ for any natural number n.

Proof. Indeed, if we consider $P(x) = x^n - c^n$, then $P(c) = c^n - c^n = 0$ and c has to be a root of the polynomial. So $(x - c)$ has to be a factor.   □

2. The difference $x^n - c^n$ will always be divisible by $(x + c)$ for any even natural number n and $c \neq 0$.

Proof. Let $P(x) = x^n - c^n$. Then

$$P(-c) = (-c)^n - c^n = \begin{cases} 0, & \text{if } n \text{ is even} \\ -2c^n, & \text{if } n \text{ is odd.} \end{cases}$$ □

3. The sum $x^n + c^n$ will always be divisible by $(x + c)$ for any odd natural number n.

Proof. Let $P(x) = x^n + c^n$. Then $P(-c) = (-c)^n + c^n = 0$, if n is a natural odd number.   □

4. The sum $x^n + c^n$ will never be divisible by $(x - c)$ for any natural number n.

Proof. Let $P(x) = x^n + c^n$. Then $P(c) = c^n + c^n = 2c^n \neq 0$ for any natural n and $c \neq 0$.   □

Problem 12. Prove that each of the numbers 7, 13, and 181 is a factor of the number $3^{105} + 4^{105}$.

Proof. The easiest proof is for 7. Since the sum of the bases is 7, $3 + 4 = 7$, then due to the above property 3, 7 is a factor of $3^{105} + 4^{105}$.

To prove that 13 is a factor of the given sum of powers, we need to do some preliminary work: $3^{105} + 4^{105} = (3^3)^{35} + (4^3)^{35} = 27^{35} + 64^{35}$. Notice now that $27 + 64 = 91 = 13 \cdot 7$. According to property 3 the original sum is divisible by 91, and since 91 has 13 as a factor, 13 has to be a factor of the given sum as well. Similar logic will be applied in the last case for 181:

$$3^{105} + 4^{105} = (3^5)^{21} + (4^5)^{21} = 243^{21} + 1024^{21}.$$

Notice that $243 + 1024 = 1267 = 7 \cdot 181$. Therefore, 181 is a factor of $3^{105} + 4^{105}$.   □

Problem 13. Prove that for any natural number n, $3^{4n+4} - 4^{3n+3}$ is divisible by 17.

Proof. $3^{4n+4} - 4^{3n+3} = (3^4)^{n+1} - (4^3)^{n+1} = 81^{n+1} - 64^{n+1}$. Since $81 - 64 = 17$, then the statement of the problem holds true due to property 1.   □

Problem 14. Prove that for any natural number n, $7^{n+2} + 8^{2n+1}$ is divisible by 57.

Proof. This problem is more complicated than the previous two problems. Properties 1 through 4 cannot be automatically applied to solve it. It is advisable in this case to modify the sum of the given powers to get to the point when one of the properties might be utilized. The trick is to add and subtract the same number $8 \cdot 7^n$:

$$7^{n+2} + 8^{2n+1} = 49 \cdot 7^n + 8 \cdot 64^n = 49 \cdot 7^n + 8 \cdot 7^n - 8 \cdot 7^n + 8 \cdot 64^n$$
$$= 7^n(49 + 8) + 8(64^n - 7^n)$$
$$= 7^n \cdot 57 + 8(64^n - 7^n).$$

As we can see, each addend in the last expression is divisible by 57. Indeed, the first addend has 57 as the factor, and the second addend is divisible by 57 due to property 1, since $64 - 7 = 57$. Therefore, the sum of these two numbers will be divisible by 57 as well.   □

Problem 15. Prove that for any numbers a, b, c, and d, such that $a + c = b + d$, the sum $ab^{2n} + cd^{2n}$ is divisible by $(a + c)$ (and therefore, by $(b + d)$ as well) for any natural number n.

Proof. As in the solution of problem 14, we will add and subtract the same number derived as the product of one of the factors present in each addend, in this case such a number is cb^{2n}:

$$ab^{2n} + cd^{2n} = ab^{2n} + cb^{2n} - cb^{2n} + cd^{2n} = b^{2n}(a + c) + c(d^{2n} - b^{2n}). \quad (*)$$

The first addend $b^{2n}(a + c)$ is divisible by $(a + c)$ because it includes it as a factor. The second addend $c(d^{2n} - b^{2n})$ has $(d^{2n} - b^{2n})$ as a factor, which is also divisible by $(a+c)$. Indeed, since it is given that $a+c = b+d$, if we prove that $(d^{2n} - b^{2n})$ is divisible by $(b+d)$, that would imply it is divisible by $(a+c)$ as well. Since the degree $2n$ is an even number, we can apply property 2 and assert that for any natural number n, $(d^{2n} - b^{2n})$ will be divisible by $(b+d)$. So, each addend in $(*)$ is divisible by $(a+c)$. Thus, the sum is also divisible by $(a+c)$, which is what was to be proved. □

Problem 16. There are given two prime numbers p and q such that $q = p+2$. Prove that $p^q + q^p$ is divisible by $p+q$.

Proof. First, observe that being prime numbers not equal to 2, each number p and q has to be odd. Let's now consider the sum $p^q + q^p$. We will substitute for the exponent q its value as $q = p+2$ and add and subtract $p^2 \cdot q^p$:

$$\begin{aligned} p^q + q^p &= p^{p+2} + q^p + p^2 \cdot q^p - p^2 \cdot q^p \\ &= (p^p \cdot p^2 + p^2 \cdot q^p) + (q^p - p^2 \cdot q^p) \\ &= p^2(p^p + q^p) + q^p(1 - p^2) \\ &= p^2(p^p + q^p) + q^p(1 - p)(1 + p). \end{aligned} \qquad (1)$$

Since p is an odd number, the sum $(p^p + q^p)$ is divisible by $(p+q)$, therefore the first addend $p^2(p^p + q^p)$ is divisible by $(p+q)$. It is given that $q = p+2$, therefore $p+q = p+p+2 = 2p+2 = 2(p+1)$. One of the factors in the second addend is $1+p$. That implies that the whole product $q^p(1-p)(1+p)$ is divisible by $(1+p)$. Since p is an odd number, each of the numbers $(1-p)$ and $(1+p)$ is even, and their product must be divisible by 2. We see that the second addend, $q^p(1-p)(1+p)$, is divisible by $2(p+1)$ or, equivalently, is divisible by $(p+q)$. As each addend in (1) is divisible by $(p+q)$, we finally conclude that the sum $p^q + q^p$ is divisible by $(p+q)$ as well. □

It is worth noting that the Little Bézout's Theorem, which is most commonly referred to as the Polynomial Remainder Theorem, has nothing to do with and should not be confused with Bézout's Theorem. Bézout's Theorem's coverage is beyond the scope of our book; however, it is interesting that there exists some controversy regarding the priority status of the statement of the theorem. Etienne Bézout published the theorem in 1779 in his work *Theorie generale des equations algebraiques*. Some authors argue that the same ideas were first expressed by Isaac Newton in the proof of Lemma 28 of volume 1 in his work *Philosophiae Naturalis Principia Mathematica* published in 1687.

One of the greatest minds in human history, who made invaluable contributions in the fields of mathematics and physics, Sir Isaac Newton (1642–1726), is considered by many as one of the most outstanding and influential scientists of all times. The man changed the way people understand the universe. An exquisite description of his achievements is given by the famous 18th-century English poet Alexander Pope:

> Nature and Nature's laws lay hid in night.
> God said, Let Newton be! And all was light.

His brilliant ideas laid out the foundation of modern physics and mathematics. He had a huge impact in the study of gravitational forces, laws of motion, optic theories, and astronomy. He is credited with the invention and development of calculus. He was the first (simultaneously with Gottfried Wilhelm Leibniz (1646–1716)) to set up and fulfill the goal of describing the physical world with the help of equations. Newton's achievements in the fields of mathematics are next to none. Ironically, he was surrounded by various controversies throughout almost all of his life: the prolonged feud with Leibniz regarding the priority of calculus development; the animosity with astronomer John Flamsteed in regard to star catalogue records data use and various astronomical calculations accuracy; he was on poor terms with Robert Hooke due to their arguments about prioritizing some ideas in optics. Probably the Bézout's Theorem's priority status controversy is among others on that list. Considering Newton's accomplishments, it wasn't a critical issue. What's important is that nowadays we still use both theorems in problem solving to significantly simplify solutions.

To have some practice, we suggest readers try a few of the challenges below.

Exercises

Problem 17. Given $P(x) = x^3 + a_1 x^2 + a_2 x + a_3$, find the coefficients a_1, a_2, and a_3 such that $P(x)$ divides evenly into $(x - 1)$ and $(x + 2)$ and has the remainder 10 after its division by $(x + 1)$.

Problem 18. Find the coefficients p and q such that

$$P(x) = x^5 - 3x^4 + px^3 + qx^2 - 5x - 5$$

divides evenly into $x^2 - 1$.

Problem 19. Find the coefficients a, b, and c such that $x^4 - x^3 + ax^2 + bx + c$ divides evenly into $x^3 - 2x^2 - 5x + 6$.

Problem 20. Prove that for any natural number n, $9^{n+1} + 2^{6n+1}$ is divisible by 11.

Problem 21. Prove that for any natural number n, $5^{2n+1} + 2^{n+4} + 2^{n+1}$ is divisible by 23.

Problem 22. Given α, β, and γ are the roots of the equation $x^3 + px^2 + qx + r = 0$, express in terms of p, q, and r:

   a) $\alpha^2 + \beta^2 + \gamma^2$,

   b) $\alpha^3 + \beta^3 + \gamma^3$,

   c) $\alpha^4 + \beta^4 + \gamma^4$.

Chapter 5

Irrational Equations

Irrational or radical equations are equations containing variables under the radical sign. The general idea behind the solution of irrational equations is to eliminate the radical sign and convert an equation to a rational equation, which is equivalent to the original equation or is its consequence. Usually this goal is achieved by isolating a radical on one side of an equation and then raising both sides of an equation to the power needed to eliminate the radical: $(\sqrt[n]{f(x)})^n = f(x)$.

While solving irrational equations it is very important to clearly understand the meaning of the symbols used. Very often the errors are made because of incorrect treatment of the sign $\sqrt{}$. The expression $\sqrt{x}$ means square root of a number x and by definition it is always a nonnegative number, $\sqrt{x} \geq 0$. Equation $\sqrt{x-8} = -3$ does not have any solutions, because you have a nonnegative number on the left-hand side and a negative number on the right-hand side. There are no real numbers x satisfying that equation. The same is true for the nth root, $\sqrt[n]{x}$, for any even n. Speaking about the domain of an irrational equation, the real number x under the square root sign or nth root sign for an even n cannot be negative, $x \geq 0$. When you solve, for example, an equation $\sqrt{x} = 5$, its solution will be only $x = 25$.

If you raise both sides of an irrational equation to an odd power, then the new rational equation will be equivalent to the original equation. On the other hand, when you raise both sides of an irrational equation to an even power, then, generally speaking, the new rational equation will be a consequence of the original equation, not necessarily its equivalent. It is important to understand that raising both sides of an equation to a power and modifications of radicals will never lead to

losing roots; however, sometimes an extraneous solution (that is, a value that is not
a solution to the original equation) will result from raising both sides of an equation
to even power, even though the proper method has been followed. It is advisable
to determine the domain of the equation before you start solving it. As you finish
the solution, you must eliminate those roots that do not belong to the domain. To
make sure that there are no extraneous solutions, always check by substituting the
remaining solutions in the original equation. It is important to understand that from
$x = y$ it follows that $x^2 = y^2$, while the converse is not always correct. The true
statement is: if $x^2 = y^2$, then $|x| = |y|$.

The following algebraic joke is a good demonstration of this concept:
Prove that $2 \cdot 2 = 5$.

We start with the obvious true equality $16 - 36 = 25 - 45$. Adding $20\frac{1}{4}$ to
each side of the equality gives

$$16 - 36 + 20\frac{1}{4} = 25 - 45 + 20\frac{1}{4},$$

which can be rewritten as

$$4^2 - 2 \cdot 4 \cdot \frac{9}{2} + \left(\frac{9}{2}\right)^2 = 5^2 - 2 \cdot 5 \cdot \frac{9}{2} + \left(\frac{9}{2}\right)^2,$$

or equivalently

$$\left(4 - \frac{9}{2}\right)^2 = \left(5 - \frac{9}{2}\right)^2.$$

Now, if from the last equality we conclude that $4 - \frac{9}{2} = 5 - \frac{9}{2}$, it will result in the
equality $4 = 5$, or $2 \cdot 2 = 5$, which was requested to be proved.

From my experience, at this point many students were not able to explain this
outcome and looked puzzled and amazed. They did not realize that the confusing
result was achieved because of the incorrect assumption of the equivalence of two
equalities: $x^2 = y^2$ and $x = y$. In this example, $(-\frac{1}{2})^2 = (\frac{1}{2})^2$; however, it does
not mean that $-\frac{1}{2} = \frac{1}{2}$. This cute joke should serve as a warning for being extra
cautious working with irrational equations, solutions of which are achieved by
raising both sides to an even power.

There are some cases when analyzing the domain of the equation might
significantly help you to save time in getting the result. For example, suppose
that you need to solve the equation $\sqrt{x-3} + \sqrt{2-x} = 3$. An expression under a
square root cannot be negative. Therefore, the domain of this equation consists of
all real numbers such that

$$x - 3 \geq 0,$$
$$2 - x \geq 0.$$

From the first inequality it follows that $x \geq 3$. From the second inequality it follows
that $x \leq 2$. There are no real numbers satisfying each condition at the same time.

Therefore, this equation has no solutions.

Sometimes you should be able to simplify your task of solving an equation by analyzing its range. If you are asked to solve an equation

$$\sqrt{x-6} + \sqrt{2x+3} = -\sqrt{2-x} - 4,$$

you would be better off determining the range of expressions on the left and right sides before doing any modifications to it. It has no solutions. Indeed, on the left-hand side you have a nonnegative number as the sum of two nonnegative numbers (arithmetic or principal square root is always a nonnegative number) and on the right-hand side you have a negative number. Therefore, it is impossible to find any value of the variable x satisfying the equation.

The point is that you have to be very careful and always keep your eyes open when you approach an irrational equation. As we warned you, let's now get acquainted with some helpful hints and techniques for solving various irrational equations.

1   Irrational equations with one or few perfect squares

In the simplest case you would not even need to raise both sides to any power to get rid of a radical; just use its definition.

Problem 1. Solve the equation

$$\sqrt{(2x-1)^2} = x+3.$$

Solution. Since $\sqrt{a^2} = |a|$, then $\sqrt{(2x-1)^2} = |2x-1|$ and our equation is simplified to the equation

$$|2x-1| = x+3.$$

In order to solve this equation we recall the definition of the absolute value of x as

$$|x| = \begin{cases} -x, & x < 0, \\ x, & x \geq 0 \end{cases}$$

and consider two choices:

1) $x < \frac{1}{2}$, then $-2x+1 = x+3$, or equivalently, $-3x = 2$, from which $x = -\frac{2}{3}$.

2) $x \geq \frac{1}{2}$, then $2x - 1 = x + 3$, from which $x = 4$.

Answer: $-\frac{2}{3}$ and 4.

Problem 2. Solve the equation

$$\sqrt{x^2 + 4x + 4} + \sqrt{x^2 - 10x + 25} = 10.$$

Solution. Observing that each trinomial under the radical sign is a perfect square, we rewrite the equation as

$$\sqrt{(x+2)^2} + \sqrt{(x-5)^2} = 10,$$

which yields

$$|x + 2| + |x - 5| = 10.$$

Consider three scenarios:

1) $x < -2$, then $-x - 2 - x + 5 = 10$, or $-2x = 7$, from which $x = -3.5$.

2) $-2 \leq x < 5$, then $x + 2 - x + 5 = 10$, from which we get $0x = 3$, a wrong statement. Therefore, there are no solutions.

3) $x \geq 5$, then $x + 2 + x - 5 = 10$, or $2x = 13$, from which $x = 6.5$.

Answer: -3.5 and 6.5.

2 Irrational equations with one radical sign

There is no special trick in this case. Isolate the radical on one side of the equation. Raise both sides to the same power to get rid of the radical. Don't forget to check the solutions.

Problem 3. Solve the equation

$$\sqrt{x - 2} + 3 = 4.$$

Solution. First, determining the domain, we see that $x \geq 2$. Squaring both sides of the modified equation $\sqrt{x - 2} = 1$ gives

$$x - 2 = 1,$$
$$x = 3.$$
$$\sqrt{3 - 2} + 3 = 4$$

is correct. Therefore, $x = 3$ is the solution of the equation.

Answer: 3.

Problem 4. Solve the equation

$$\sqrt[5]{24x - 2x^3} = x.$$

Solution. To get rid of the radical we raise both sides of the equation to the power 5. It follows that

$$24x - 2x^3 = x^5,$$
$$x^5 + 2x^3 - 24x = 0,$$
$$x(x^4 + 2x^2 - 24) = 0,$$
$$x = 0 \text{ or } x^4 + 2x^2 - 24 = 0.$$

The second equation is biquadratic, so we use the substitution $y = x^2, y \geq 0$. It follows that $y^2 + 2y - 24 = 0$,

$$D = 4 + 96 = 100, \quad y = \frac{-2 \pm \sqrt{100}}{2},$$
$$y_1 = 4, \quad y_2 = -6.$$

Recalling that $y = x^2 \geq 0$, we reject the second root -6. The only satisfactory solution is $y = 4$. Solving the equation $x^2 = 4$ gives $x = 2$ or $x = -2$.

Answer: 0, 2, −2.

Problem 5. Solve the equation

$$\frac{x^3}{\sqrt{4 - x^2}} + x^2 - 4 = 0.$$

Solution. The domain of the equation is determined from the solutions of the inequality $4 - x^2 > 0$. It follows $x \in \,]-2, 2[$. Rewrite the equation as

$$x^3 = (4 - x^2) \cdot \sqrt{4 - x^2}.$$

Observing that $\sqrt{4 - x^2} = (4 - x^2)^{\frac{1}{2}}$, the right-hand side of the equation can be rewritten as

$$(4 - x^2) \cdot (4 - x^2)^{\frac{1}{2}} = (4 - x^2)^{\frac{3}{2}}$$

and the original equation transforms to $x^3 = (4 - x^2)^{\frac{3}{2}}$. Squaring both sides yields $x^6 = (4 - x^2)^3$. It follows that $x^2 = 4 - x^2$, from which $x^2 = 2$, and finally, $x = \pm\sqrt{2}$. Since during our solution we squared both sides of the equation, we need to verify each root, even though both roots belong to the equation's domain.

$$\frac{(\sqrt{2})^3}{\sqrt{4-(\sqrt{2})^2}} + (\sqrt{2})^2 - 4 = \frac{(\sqrt{2})^3}{\sqrt{2}} + 2 - 4 = 2 - 2 = 0.$$

Hence, $\sqrt{2}$ satisfies the equation.

$$\frac{(-\sqrt{2})^3}{\sqrt{4-(-\sqrt{2})^2}} + (-\sqrt{2})^2 - 4 = \frac{-(\sqrt{2})^3}{\sqrt{2}} + 2 - 4 = -2 - 2 = -4 \neq 0.$$

We see that $-\sqrt{2}$ has to be rejected.

Answer: $\sqrt{2}$.

3   Irrational equations with more than one radical

These are more complicated equations. Sometimes you need to go through the process of eliminating radicals not once, but a few times. It is important to verify the suitability of any interim roots in getting the final result and checking the solutions.

Problem 6. Solve the equation

$$\sqrt{x+2} - \sqrt[3]{3x+2} = 0.$$

Solution. The domain of this equation is $x \geq -2$. Rewrite the equation as

$$\sqrt{x+2} = \sqrt[3]{3x+2}$$

and raise both sides to the power 6. We get the equation

$$(x+2)^3 = (3x+2)^2,$$
$$x^3 + 6x^2 + 12x + 8 = 9x^2 + 12x + 4,$$
$$x^3 - 3x^2 + 4 = 0,$$
$$x^3 + 1 - 1 - 3x^2 + 4 = 0,$$
$$x^3 + 1 - 3x^2 + 3 = 0,$$
$$(x^3 + 1) - 3(x^2 - 1) = 0,$$
$$(x+1)(x^2 - x + 1) - 3(x+1)(x-1) = 0,$$
$$(x+1)(x^2 - x + 1 - 3x + 3) = 0,$$
$$(x+1)(x^2 - 4x + 4) = 0,$$
$$(x+1)(x-2)^2 = 0,$$
$$x = -1 \text{ or } x = 2.$$

Both roots belong to the domain of the original equation, but since we raised both sides to an even power, we still need to verify each root.

For $x = -1$ the original equation becomes $\sqrt{-1+2} - \sqrt[3]{-3+2} = 0$, which yields $2 = 0$, a wrong statement. Hence, -1 is an extraneous solution that arose from raising each side of the equation to the power 6 and has to be rejected.

If $x = 2$, then $\sqrt{2+2} - \sqrt[3]{6+2} = 2 - 2 = 0$, $0 = 0$ – true. Therefore, 2 is the only solution of the original equation.

Answer: 2.

Problem 7. Solve the equation

$$\sqrt{4-x} + \sqrt{x+5} = 3.$$

Solution. First, let's determine the domain of the equation. In order to do it we need to solve the system of two inequalities:

$$4 - x \geq 0,$$
$$x + 5 \geq 0,$$

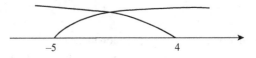

or equivalently,

$$x \leq 4,$$
$$x \geq -5.$$

The domain of the equation is $-5 \leq x \leq 4$. Squaring both sides of the equation gives

$$4 - x + 2\sqrt{4-x} \cdot \sqrt{x+5} + x + 5 = 9,$$
$$2\sqrt{20 - x - x^2} = 0,$$
$$20 - x - x^2 = 0,$$
$$x^2 + x - 20 = 0;$$

using Viète's formulas, we get the solutions $x_1 = 4$, $x_2 = -5$.

Checking the solutions, we see that each satisfies the equation. Indeed,

$$\sqrt{4-4} + \sqrt{4+5} = 0 + 3 = 3, \quad 3 = 3 - \text{true.}$$
$$\sqrt{4+5} + \sqrt{-5+5} = 3 + 0 = 3, \quad 3 = 3 - \text{true.}$$

Answer: −5, 4.

4   Equations with cubic radicals

The following well-known identities might be useful in the solutions of the types of equations considered in this section:

$$(a+b)^3 = a^3 + b^3 + 3ab(a+b), \qquad (1)$$
$$(a-b)^3 = a^3 - b^3 - 3ab(a-b). \qquad (2)$$

Problem 8. Solve the equation

$$\sqrt[3]{x+45} - \sqrt[3]{x-16} = 1.$$

Solution. Cubing both sides of the equation and using identity (2) yields

$$x + 45 - (x-16) - 3\sqrt[3]{x+45}\sqrt[3]{x-16}\left(\underbrace{\sqrt[3]{x+45} - \sqrt[3]{x-16}}_{=1}\right) = 1.$$

Noticing that the expression in the parentheses equals 1 (that's the idea behind using the suggested formulas for the cube of the sum and difference of two numbers) and simplifying, we obtain

$$3\sqrt[3]{x+45}\sqrt[3]{x-16} = 60,$$
$$\sqrt[3]{x+45}\sqrt[3]{x-16} = 20,$$

raising both sides to the power 3 yields

$$(x+45)(x-16) = 8000,$$
$$x^2 + 29x - 720 = 8000,$$
$$x^2 + 29x - 8720 = 0,$$
$$D = 841 + 34{,}880 = 35{,}721 = 189^2.$$
$$x = \frac{-29 \pm \sqrt{35{,}721}}{2},$$

then $x = 80$ or $x = -109$.

Answer: $-109, 80$.

Problem 9. Solve the equation

$$\sqrt[3]{x+5} + \sqrt[3]{x+6} = \sqrt[3]{2x+11}.$$

Solution. We will use the same trick as in the previous problem and cube both sides of the equation:

$$x+5+x+6+3\sqrt[3]{x+5}\sqrt[3]{x+6}\underbrace{\left(\sqrt[3]{x+5}+\sqrt[3]{x+6}\right)}_{=\sqrt[3]{2x+11}}=2x+11,$$

$$2x+11+3\sqrt[3]{x+5}\sqrt[3]{x+6}\sqrt[3]{2x+11}=2x+11,$$
$$3\sqrt[3]{x+5}\sqrt[3]{x+6}\sqrt[3]{2x+11}=0,$$
$$(x+5)(x+6)(2x+11)=0,$$
$$x+5=0, \text{ or } x+6=0, \text{ or } 2x+11=0.$$

Solving each of the equations leads to $x=-5$, or $x=-6$, or $x=-5.5$.

Answer: $-5, -6, -5.5$.

In the last equation the sum of the expressions under the radicals on the left-hand side equals the expression under the radical on the right-hand side. The solutions of the original equation then should be found among the solutions of the three equations when you equate each expression under the radical to 0. It will be the case in solving any similar equations. Indeed, as you raise both sides to power 3, you will always have like terms on both sides canceling out each other. As you just saw, the only remaining term on the left-hand side is the product of all three radicals. And 0 remains on the right-hand side, so by the zero-product property the equation is reduced to three linear equations.

5 Solution of irrational equations by substitution. Introduction of a new variable

Here you get to be creative and look for the appropriate substitution depending on the specifics of the given equation. Always look for some similar or the same terms and try to introduce the variable in such a way that the original equation becomes significantly simplified and transformed to something recognizable and easily solvable. We will show below some of the most commonly used techniques.

Problem 10. Solve the equation

$$\sqrt{x^2-3x+11}-4x^2+12x=11.$$

Solution. First, note that the domain of this equation consists of all real numbers. Indeed, since the discriminant of the quadratic trinomial under the square root is a negative number, $D=9-44=-35<0$, then $x^2-3x+11>0$, for any x.

Next, it's not hard to see that $-4x^2+12x=-4(x^2-3x)$. Therefore, there is a clear choice for the new variable substitution: $y=\sqrt{x^2-3x+11}$. Thus, $y^2=x^2-3x+11$, from which $x^2-3x=y^2-11$. Note that since y equals to the principal square root, it has to be a nonnegative number, $y \geq 0$. The original equation can be rewritten as $y - 4(y^2 - 11) = 11$, or equivalently,

$$4y^2 - y - 33 = 0,$$
$$D = 1 + 528 = 529,$$
$$y = \frac{1 \pm \sqrt{529}}{8},$$

so $y = 3$ or $y = -\frac{22}{8}$, which should be rejected due to the restriction that $y \geq 0$. So, the only suitable solution for y is $y = 3$. Hence,

$$\sqrt{x^2 - 3x + 11} = 3.$$

Squaring both sides gives

$$x^2 - 3x + 11 = 9,$$
$$x^2 - 3x + 2 = 0,$$

using Viète's formulas, we get the solutions $x = 1$ or $x = 2$.

Let's check the solutions:

$$\sqrt{1^2 - 3 + 11} - 4 + 12 = 3 + 8 = 11, \quad 11 = 11 - \text{true.}$$
$$\sqrt{2^2 - 6 + 11} - 16 + 24 = 3 + 8 = 11, \quad 11 = 11 - \text{true.}$$

So, we see that each number, 1 and 2, satisfies the original equation.

Answer: 1, 2.

Problem 11. Solve the equation

$$\sqrt[5]{\frac{16x}{x-1}} + \sqrt[5]{\frac{x-1}{16x}} = \frac{5}{2}.$$

Solution. The domain of the equation consists of all real numbers, except 0 and 1 ($x \neq 0$, $x \neq 1$). The substitution to use is obvious:

$$y = \sqrt[5]{\frac{16x}{x-1}}.$$

Then $y + \frac{1}{y} = \frac{5}{2}$, or equivalently,

$$2y^2 - 5y + 2 = 0,$$
$$D = 25 - 16 = 9,$$
$$y = \frac{5 \pm \sqrt{9}}{4},$$

then $y = 2$ or $y = \frac{1}{2}$.

Substituting the values of y for $\sqrt[5]{\frac{16x}{x-1}}$ gives

$$\sqrt[5]{\frac{16x}{x-1}} = 2 \text{ or } \sqrt[5]{\frac{16x}{x-1}} = \frac{1}{2}.$$

After raising both sides to the power 5, the first equation becomes $16x = 32x - 32$, from which $x = 2$.

Similarly, solving the second equation we get $512x = x - 1$, from which $x = -\frac{1}{511}$.

Answer: $-\frac{1}{511}$, 2.

Problem 12. Solve the equation

$$\sqrt[4]{(x-1)^2} - \sqrt[4]{(x+1)^2} = \frac{3}{2}\sqrt[4]{x^2-1}.$$

Solution. The domain of this equation is determined from the solutions of the inequality $x^2 - 1 \geq 0$ or equivalently, $(x-1)(x+1) \geq 0$, $x \in \,]-\infty, -1] \cup [1, +\infty[$. The trick in this case is to divide both sides of the equation by the radical on the right-hand side. Observing that $x \neq \pm 1$ (otherwise the left side does not equal to 0 and the right side will become 0), we may do it and will not lose any solutions. It follows that

$$\sqrt[4]{\frac{(x-1)^2}{x^2-1}} - \sqrt[4]{\frac{(x+1)^2}{x^2-1}} = \frac{3}{2},$$

$$\sqrt[4]{\frac{(x-1)^2}{(x-1)(x+1)}} - \sqrt[4]{\frac{(x+1)^2}{(x-1)(x+1)}} = \frac{3}{2},$$

$$\sqrt[4]{\frac{x-1}{x+1}} - \sqrt[4]{\frac{x+1}{x-1}} = \frac{3}{2}.$$

Substituting $y = \sqrt[4]{\frac{x-1}{x+1}} > 0$ yields $y - \frac{1}{y} = \frac{3}{2}$, or equivalently,

$$2y^2 - 3y - 2 = 0,$$
$$D = 9 + 16 = 25.$$

$y = \frac{3 \pm \sqrt{25}}{4}$, so $y = 2$ or $y = -\frac{1}{2}$. The value $y = -\frac{1}{2}$ is not satisfactory, since $y > 0$. Going back to the equation for x, we get $\sqrt[4]{\frac{x-1}{x+1}} = 2$, raising both sides to the power 4 gives $\frac{x-1}{x+1} = 16$, or $16x + 16 = x - 1$, from which $x = -\frac{17}{15}$. It belongs to the equation's domain. Now we need to check the solution. We have on the left side

$$\sqrt[4]{\left(-\frac{17}{15}-1\right)^2}-\sqrt[4]{\left(-\frac{17}{15}+1\right)^2}=\sqrt{\frac{32}{15}}-\sqrt{\frac{2}{15}}$$

$$=4\cdot\sqrt{\frac{2}{15}}-\sqrt{\frac{2}{15}}=3\cdot\sqrt{\frac{2}{15}}.$$

Calculating the right side, we get

$$\frac{3}{2}\cdot\sqrt[4]{\left(-\frac{17}{15}\right)^2}-1=\frac{3}{2}\cdot\sqrt{\frac{8}{15}}=3\cdot\sqrt{\frac{2}{15}}.$$

Since $3\cdot\sqrt{\frac{2}{15}}=3\cdot\sqrt{\frac{2}{15}}$, we conclude that $-\frac{17}{15}$ is a valid solution of the equation.

Answer: $-\frac{17}{15}$.

Problem 13. Solve the equation

$$\sqrt[4]{x+41}+\sqrt[4]{41-x}=4.$$

Solution. First, find the domain of the equation:

$$x+41\geq 0,$$
$$41-x\geq 0,$$

or equivalently,

$$x\geq -41,$$
$$x\leq 41.$$

The domain of the equation is $-41\leq x\leq 41$.

What distinguishes this equation from other equations is that the sum of the expressions under the radical signs is a constant. It is $x+41+41-x=82$. A very efficient technique will be to introduce two new variables (non-trivial approach, isn't it?): $\sqrt[4]{x+41}=u\geq 0$ and $\sqrt[4]{41-x}=v\geq 0$. It follows that $x+41=u^4$ and $41-x=v^4$, so we see that $u^4+v^4=82$.

Let's solve the following system of two equations:

$$u+v=4,$$
$$u^4+v^4=82.$$

Express v in terms of u from the first equation and substitute into the second equation.

$$v=4-u,$$
$$u^4+(4-u)^4=82.$$

To solve the second equation, we need one more substitution $u = y - \frac{0-4}{2} = y + 2$. It follows that $(y+2)^4 + (y-2)^4 = 82$. Expanding parentheses and simplifying gives

$$2y^4 + 48y^2 + 32 = 82,$$
$$y^4 + 24y^2 - 25 = 0.$$

Introduce another substitution, $z = y^2 \geq 0$. We obtain the equation $z^2 + 24z - 25 = 0$. Using Viète's formulas, the solutions are $z = 1$ or $z = -25$. The value $z = -25$ is not satisfactory and has to be rejected since z has to be a nonnegative number. Therefore, $y^2 = 1$ and so $y = \pm 1$.

If $y = 1$, then $u = y + 2 = 3$ and $v = 4 - u = 1$, which leads to $\sqrt[4]{x+41} = 3$. Raising to the fourth power gives

$$x + 41 = 81,$$
$$x = 40.$$

If $y = -1$, then $u = y + 2 = 1$ and $v = 4 - u = 3$, which leads to $\sqrt[4]{x+41} = 1$,

$$x + 41 = 1,$$
$$x = -40.$$

Let's now check each solution:

$$\sqrt[4]{40+41} + \sqrt[4]{41-40} = \sqrt[4]{81} + \sqrt[4]{1} = 3 + 1 = 4, \quad 4 = 4 - \text{true.}$$
$$\sqrt[4]{-40+41} + \sqrt[4]{41+40} = \sqrt[4]{1} + \sqrt[4]{81} = 1 + 3 = 4, \quad 4 = 4 - \text{true.}$$

Answer: $-40, 40$.

Problem 14. Solve the equation

$$\sqrt{x\sqrt[5]{x}} - \sqrt[5]{x\sqrt{x}} = 56.$$

Solution. The domain of the equation is the set of all nonnegative numbers, $x \geq 0$. Let's modify the equation and instead of radicals use fractional exponents:

$$x^{\frac{3}{5}} - x^{\frac{3}{10}} = 56.$$

The substitution is $x^{\frac{3}{10}} = y$, where $y \geq 0$. It follows that

$$y^2 - y - 56 = 0,$$
$$D = 1 + 224 = 225,$$
$$y = \frac{1 \pm \sqrt{225}}{2}.$$

Thus, $y = 8$ or $y = -7$. The value $y = -7$ is not satisfactory, since $y \geq 0$. Therefore, $x^{\frac{3}{10}} = 8$, from which $x = \sqrt[3]{8^{10}} = 1024$.

Answer: 1024.

6 Auxiliary function introduction

Sometimes, when solving an irrational equation with addition or subtraction of radicals on the left-hand side and a constant or linear binomial on the right-hand side, the method of "conjugate expression" might be helpful in simplifying the solution. The idea behind it is to use the identity $(a-b)(a+b) = a^2 - b^2$. This method is effective when after dividing the difference of the expressions under the radicals on the left-hand side by the expression staying on the right-hand side the resulting quotient is a constant number or a linear polynomial.

Problem 15. Solve the equation

$$\sqrt{2x^2 + 3x + 5} + \sqrt{2x^2 - 3x + 5} = 3x.$$

Solution. First, observe that since we have the sum of two nonnegative numbers on the left-hand side of the equation, then the right-hand side has to be a nonnegative number as well, which means that x has to be a nonnegative number, $x \geq 0$. The domain of the equation will be all real numbers (since discriminant of each quadratic polynomial under the radical is negative; we leave it to the reader to verify).

Let's find the difference of the expressions under the radicals:

$$2x^2 + 3x + 5 - (2x^2 - 3x + 5) = 2x^2 + 3x + 5 - 2x^2 + 3x - 5 = 6x.$$

Note that $6x : 3x = 2$. Therefore, we introduce the auxiliary function as the conjugate expression to the left-hand side $f(x) = \sqrt{2x^2 + 3x + 5} - \sqrt{2x^2 - 3x + 5}$ and multiply both sides of the equation by $f(x)$:

$$(\sqrt{2x^2 + 3x + 5} + \sqrt{2x^2 - 3x + 5})(\sqrt{2x^2 + 3x + 5} - \sqrt{2x^2 - 3x + 5})$$
$$= 3x(\sqrt{2x^2 + 3x + 5} - \sqrt{2x^2 - 3x + 5}).$$

We get the difference of squares on the left-hand side, therefore,

$$2x^2 + 3x + 5 - (2x^2 - 3x + 5) = 3x(\sqrt{2x^2 + 3x + 5} - \sqrt{2x^2 - 3x + 5}),$$
$$6x = 3x(\sqrt{2x^2 + 3x + 5} - \sqrt{2x^2 - 3x + 5}).$$

Obviously, $x \neq 0$ (it is not the solution of the original equation, $\sqrt{0^2 + 0 + 5} + \sqrt{0^2 - 0 + 5} = 0$ is wrong, since $\sqrt{5} + \sqrt{5} \neq 0$), then we can divide both sides by $3x$ and get

$$\sqrt{2x^2 + 3x + 5} - \sqrt{2x^2 - 3x + 5} = 2. \qquad (1)$$

Recalling the original equation $\sqrt{2x^2 + 3x + 5} + \sqrt{2x^2 - 3x + 5} = 3x$ and adding it to (1) gives

$$2\sqrt{2x^2 + 3x + 5} = 2 + 3x.$$

Squaring both sides leads to

$$4(2x^2 + 3x + 5) = 4 + 12x + 9x^2,$$
$$8x^2 + 12x + 20 = 4 + 12x + 9x^2,$$
$$x^2 = 16,$$

from which $x = \pm 4$. The negative solution $x = -4$ does not satisfy the range of the permissible values for x, $x \geq 0$, and has to be rejected.

We need to check the other solution: If $x = 4$, then

$$\sqrt{32 + 12 + 5} + \sqrt{32 - 12 + 5} = 7 + 5 = 12 = 3 \cdot 4 - \text{a correct statement,}$$

so 4 is the solution of the equation.

Answer: 4.

Problem 16. Solve the equation

$$\sqrt{9x^2 - 12x + 11} + \sqrt{5x^2 - 8x + 10} = 2x - 1.$$

Solution. Notice that

$$9x^2 - 12x + 11 - (5x^2 - 8x + 10) = 9x^2 - 12x + 11 - 5x^2 + 8x - 10$$
$$= 4x^2 - 4x + 1 = (2x - 1)^2.$$

Dividing $(2x - 1)^2$ by $(2x - 1)$, the expression on the right-hand side of the equation, gives $2x - 1$. So, we can apply the suggested technique and introduce the auxiliary function as the conjugate expression to the left-hand side,

$$f(x) = \sqrt{9x^2 - 12x + 11} - \sqrt{5x^2 - 8x + 10},$$

and multiply both sides of the equation by $f(x)$:

$$(\sqrt{9x^2 - 12x + 11} + \sqrt{5x^2 - 8x + 10})(\sqrt{9x^2 - 12x + 11} - \sqrt{5x^2 - 8x + 10})$$
$$= (\sqrt{9x^2 - 12x + 11} - \sqrt{5x^2 - 8x + 10})(2x - 1).$$

We have the difference of squares on the left side. Therefore, the equation transforms into

$$9x^2 - 12x + 11 - 5x^2 + 8x - 10$$
$$= (\sqrt{9x^2 - 12x + 11} - \sqrt{5x^2 - 8x + 10})(2x - 1),$$
$$(2x - 1)^2 = (\sqrt{9x^2 - 12x + 11} - \sqrt{5x^2 - 8x + 10})(2x - 1).$$

Since it is easy to verify that $x \neq \frac{1}{2}$, then dividing both sides by $2x - 1$ gives

$$\sqrt{9x^2 - 12x + 11} - \sqrt{5x^2 - 8x + 10} = 2x - 1. \qquad (1)$$

As in the previous problem, we recall the original equation

$$\sqrt{9x^2 - 12x + 11} + \sqrt{5x^2 - 8x + 10} = 2x - 1.$$

Subtracting (1) from the original equation gives

$$2\sqrt{5x^2 - 8x + 10} = 0,$$

or equivalently,

$$5x^2 - 8x + 10 = 0.$$
$$D = 64 - 200 = -136 < 0.$$

The last quadratic equation has no real solutions. Therefore, we conclude that the original equation has no real solutions.

In some problems, when you are dealing with the inverse operations, the application of the properties of inverse functions may be worthwhile and helpful.

Problem 17. Solve the equation

$$\sqrt[3]{3x + 9} = 27(x + 1)^3 - 6.$$

Solution. Let's rewrite the equation as

$$\sqrt[3]{(3x + 3) + 6} = (3x + 3)^3 - 6.$$

Letting $t = 3x + 3$ and introducing the functions $f(t) = \sqrt[3]{t + 6}$ and $g(t) = t^3 - 6$, the original equation can be rewritten as $f(t) = g(t)$. It's not hard to see that $f(t)$ and $g(t)$ are inverse functions. Indeed, $f(g(t)) = \sqrt[3]{g(t) + 6} = \sqrt[3]{t^3 - 6 + 6} = \sqrt[3]{t^3} = t$, for all real values of t. Hence, by definition, $g(t)$ is the inverse of $f(t)$ for all real values of t. By the properties of inverse functions, the graph of g can be obtained from the graph of f by switching the positions of the X and Y axes. Since both functions are increasing functions, this is equivalent to reflecting the graph of g across the line $y = x$ to obtain the graph of f. That implies that solutions of the equation $f(t) = g(t)$ must be located among the abscissas of points on the line $y = x$. Hence, to find those solutions, we have to consider the equation $\sqrt[3]{t + 6} = t$. Cubing both sides yields $t + 6 = t^3$, which can be rewritten as $t^3 - t - 6 = 0$ and is easily solved by factoring.

$$t^3 - t - 8 + 2 = 0,$$
$$(t^3 - 8) - (t - 2) = 0,$$
$$(t - 2)(t^2 + 2t + 4) - (t - 2) = 0,$$
$$(t - 2)(t^2 + 2t + 3) = 0.$$

Therefore, either $t - 2 = 0$ or $t^2 + 2t + 3 = 0$. From the first equation, $t = 2$. The discriminant of the second quadratic equation is negative, $D = 4 - 12 = -8$.

So, it has no real solutions. Recalling that $t = 3x + 3$, we get $3x + 3 = 2$, from which $x = -\frac{1}{3}$.

Answer: $-\frac{1}{3}$.

Exercises

Problem 18. Solve the equation $\sqrt{x-5} + \sqrt{x+3} = \sqrt{2x+4}$.

Problem 19. Solve the equation $x - \sqrt[3]{x^2 - x - 1} = 1$.

Problem 20. Solve the equation $\sqrt{x + \sqrt{x+11}} + \sqrt{x - \sqrt{x+11}} = 4$.

Problem 21. Solve the equation $\sqrt{\frac{x+5}{x}} + 4\sqrt{\frac{x}{x+5}} = 4$.

Problem 22. Solve the equation $\sqrt[4]{x+8} - \sqrt[4]{x-8} = 2$.

Problem 23. Solve the equation $\sqrt{x^3 + x^2 - 1} + \sqrt{x^3 + x^2 + 2} = 3$.

Problem 24. Solve the equation $\frac{(34-x)\sqrt[3]{x+1} - (x+1)\sqrt[3]{34-x}}{\sqrt[3]{34-x} - \sqrt[3]{x+1}} = 30$.

Chapter 6

Trigonometric Identities

Trigonometric identities in problem solving

Greek words *trigonon*—"triangle" and *metron*—"measure" form the name "trigonometry," one of the branches of mathematics, which historically originated from applications of geometry to astronomical calculations and studies. The ancient Greek mathematician and astronomer Hipparchus of Nicaea (c. 190–c. 120 BC) is regarded as the father of trigonometry. He was credited with the calculation and construction of the first trigonometric tables used in various astronomical computations. Starting from the studies of relationships involving lengths of the sides and angle measures in right triangles, mostly utilized in astronomy and navigation during ancient times, trigonometry eventually grew into a separate branch of mathematics. It is involved in various calculations and applications, not just in pure mathematics, but in such fields as physics, land surveying and geodesy, engineering, astronomy, music theory, acoustics, optics, chemistry, cryptology, architecture, computer graphics, meteorology, biology, and even ecology. The list can be endless, as it is hard to imagine any scientific research not utilizing trigonometric applications in one way or another.

There is a common perception though among high school students about trigonometric functions as being used in geometry only to relate the angles of a triangle to its sides. For many their application of those functions is restricted to the calculation of the elements of a right triangle. There is not much time devoted to the study of trigonometric functions and trigonometric equations in the high school curriculum. In this and the next chapter we will try to fill that hole to some degree and offer an insight into the application of trigonometric functions in algebra, demonstrate various techniques for solving trigonometric equations, and give their connections to other topics covered throughout the book.

To begin, let's recall the basic trigonometric functions as they relate to a right triangle. These definitions are applied to angles between $0°$ and $90°$ (or 0 and $\frac{\pi}{2}$ radians):

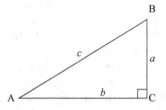

$\sin \angle A = \frac{a}{c}$, the ratio of the opposite leg to the hypotenuse

$\cos \angle A = \frac{b}{c}$, the ratio of the adjacent leg to the hypotenuse

$\tan \angle A = \frac{a}{b}$, the ratio of the opposite to the adjacent leg

$\cot \angle A = \frac{b}{a}$, the ratio of the adjacent to the opposite leg.

The Cartesian coordinate system allows taking another view of trigonometric functions and extending them to all real numbers.

Consider a point $A(x_A, y_A)$ in the Cartesian coordinate system located on the unit circle (circle with radius 1) with center O in the origin. Let α be the angle between the X-axis and the straight line connecting A to the origin, then by definition,

The sine of α is the ordinate of point A, $\sin \alpha = y_A$;

The cosine of α is the abscissa of point A, $\cos \alpha = x_A$;

The tangent α is the ratio of $\sin \alpha$ to $\cos \alpha$, $\tan \alpha = \frac{\sin \alpha}{\cos \alpha}$ (by definition, then $\tan \alpha$ is the slope of the straight line OA); and

The cotangent α is the function reciprocal to the tangent, $\cot \alpha = \frac{1}{\tan \alpha} = \frac{\cos \alpha}{\sin \alpha}$.

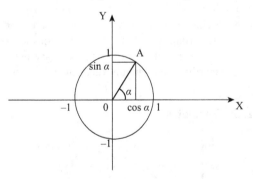

Considering the right triangle formed by point A located on the unit circle, base of the perpendicular dropped from A to X-axis, and the origin, one can easily see that the above definitions are equivalent to the respective definitions using the ratios of the lengths of two sides of a right triangle. It also immediately implies one of the most important trigonometric identities, the so-called *Pythagorean Identity*: $\sin^2\alpha + \cos^2\alpha = 1$.

The most important properties of the trigonometric functions $f(x) = \sin x$, $f(x) = \cos x$, $f(x) = \tan x$, $f(x) = \cot x$ are derived from their definitions.

Trigonometric functions are periodic functions, which repeat their values at a specific point in regular intervals or periods. The period for the trigonometric functions sine and cosine is 2π radians or $360°$; the period for the trigonometric functions tangent and cotangent is π radians or $180°$. It is very important to understand when you solve trigonometric equations that as you get one solution, you must include all the other solutions repeated in the interval period. The domain of the functions $f(x) = \sin x$ and $f(x) = \cos x$ is all real numbers. Since $|\sin x| \le 1$ and $|\cos x| \le 1$, the range of the functions $f(x) = \sin x$ and $f(x) = \cos x$ consists of real numbers from -1 through 1, including the end points. The ranges of the functions $f(x) = \tan x$ and $f(x) = \cot x$ are all real numbers. The domain of $f(x) = \tan x$ is all real numbers except $x = \frac{\pi}{2} + \pi n$ and the domain of $f(x) = \cot x$ is all real numbers except $x = \pi n$, where n is any integer. In order not to repeat it every time in the solutions of all the following equations, please remember that when we write the answer and add the period multiplied by some number n, that number is assumed to be any integer, which will be written as $n \in Z$. The function $f(x) = \cos x$ is an even function, i.e., $\cos(-x) = \cos x$, for all real x; $f(x) = \sin x$ is an odd function, i.e., $\sin(-x) = -\sin x$, for all real x; $f(x) = \tan x$ is an odd function, i.e., $\tan(-x) = -\tan x$, for all real $x \ne \frac{\pi}{2}(2n+1)$; $f(x) = \cot x$ is an odd function, i.e., $\cot(-x) = -\cot x$, for all real $x \ne \pi n, n \in Z$.

Solutions of trigonometric equations rely heavily on manipulations with various trigonometric identities and properties of trigonometric functions.

Before proceeding to the main topic of the book, solving the equations, we need to recall a few very important identities, which will be broadly utilized in our studies. We invite readers to prove the identities below by themselves; it would be a good practice before we dig into the trigonometric equations world. All the exercises offered in this chapter in one way or another will rely on the outlined major identities listed. It is important to note that the identities are only true provided that all the arguments for the trigonometric functions have permissible values. For example, any identity involving the tangent function will be unusable if one of the angles has value $\frac{\pi}{2}$.

Sum and Difference Formulas

$$\sin(x+y) = \sin x \cos y + \sin y \cos x,$$
$$\sin(x-y) = \sin x \cos y - \sin y \cos x,$$
$$\cos(x+y) = \cos x \cos y - \sin x \sin y,$$
$$\cos(x-y) = \cos x \cos y + \sin x \sin y,$$
$$\tan(x-y) = \frac{\tan x - \tan y}{1 + \tan x \tan y},$$

$$\tan(x+y) = \frac{\tan x + \tan y}{1 - \tan x \tan y},$$

$$\cot(x+y) = \frac{\cot x \cot y - 1}{\cot y + \cot x},$$

$$\cot(x-y) = \frac{\cot x \cot y + 1}{\cot y - \cot x}.$$

Double-Angle Formulas

$$\sin 2x = 2 \sin x \cos x,$$

$$\cos 2x = \cos^2 x - \sin^2 x,$$

$$\tan 2x = \frac{2 \tan x}{1 - \tan^2 x},$$

$$\cot 2x = \frac{\cot^2 x - 1}{2 \cot x}.$$

Sum-to-Product Formulas

$$\sin x + \sin y = 2 \sin \frac{x+y}{2} \cos \frac{x-y}{2},$$

$$\sin x - \sin y = 2 \sin \frac{x-y}{2} \cos \frac{x+y}{2},$$

$$\cos x + \cos y = 2 \cos \frac{x+y}{2} \cos \frac{x-y}{2},$$

$$\cos x - \cos y = 2 \sin \frac{x+y}{2} \sin \frac{y-x}{2} = -2 \sin \frac{x+y}{2} \sin \frac{x-y}{2},$$

$$\tan x \pm \tan y = \frac{\sin(x \pm y)}{\cos x \cos y},$$

$$\cot x \pm \cot y = \frac{\sin(y \pm x)}{\sin x \sin y}.$$

Product-to-Sum Formulas

$$\sin x \sin y = \frac{1}{2}(\cos(x-y) - \cos(x+y)),$$

$$\sin x \cos y = \frac{1}{2}(\sin(x+y) + \sin(x-y)),$$

$$\cos x \cos y = \frac{1}{2}(\cos(x-y) + \cos(x+y)).$$

Power-Reducing Formulas

$$\sin^2 x = \frac{1}{2} - \frac{1}{2} \cos 2x,$$

$$\cos^2 x = \frac{1}{2} + \frac{1}{2} \cos 2x.$$

Half-Angle Formulas

$$\left|\sin\frac{x}{2}\right| = \sqrt{\frac{1-\cos x}{2}}, \quad \left|\cos\frac{x}{2}\right| = \sqrt{\frac{1+\cos x}{2}},$$

$$\left|\tan\frac{x}{2}\right| = \sqrt{\frac{1-\cos x}{1+\cos x}}, \quad \left|\cot\frac{x}{2}\right| = \sqrt{\frac{1+\cos x}{1-\cos x}}.$$

$$\tan\frac{x}{2} = \frac{\sin x}{1+\cos x} = \frac{1-\cos x}{\sin x}, \quad \cot\frac{x}{2} = \frac{1+\cos x}{\sin x} = \frac{\sin x}{1-\cos x}.$$

Universal Trigonometric Substitution

$$\sin x = \frac{2\tan\frac{x}{2}}{1+\tan^2\frac{x}{2}}, \quad \cos x = \frac{1-\tan^2\frac{x}{2}}{1+\tan^2\frac{x}{2}}, \quad \tan x = \frac{2\tan\frac{x}{2}}{1-\tan^2\frac{x}{2}}.$$

The following *Cofunction identities* follow directly from the definition of the trigonometric functions: $\sin\left(\frac{\pi}{2}-x\right) = \cos x$, $\cos\left(\frac{\pi}{2}-x\right) = \sin x$, $\tan\left(\frac{\pi}{2}-x\right) = \cot x$, and $\cot\left(\frac{\pi}{2}-x\right) = \tan x$. It follows that sine and cosine, tangent and cotangent are complementary functions, or sometimes they are called cofunctions.

As a general rule you may always apply the following for the trigonometric functions of angles calculated as addition or subtraction from $\frac{\pi}{2}$ and π (along with the multiples of $\frac{\pi}{2}$ and π):

1. If you add or subtract an angle α from any multiple of $\frac{\pi}{2}$, then $\sin\alpha$ will convert to $\cos\alpha$ and vice versa, $\tan\alpha$ will convert to $\cot\alpha$ and vice versa (in other words, the function will change to its cofunction); the sign of the new function is determined by the quadrant in which the angle ends.

2. If you add or subtract an angle from any multiple of π, then the function will not change; the sign of the new function is determined by the quadrant in which the angle ends.

Examples:

a) $\sin\left(\frac{\pi}{2}+x\right) = \cos x$. The angle $\left(\frac{\pi}{2}+x\right)$ is located in the second quadrant, where sine is positive.

b) $\cos\left(\frac{3\pi}{2}-x\right) = -\sin x$. The angle $\left(\frac{3\pi}{2}-x\right)$ is located in the third quadrant, where cosine is negative.

c) $\tan(\pi+x) = \tan x$. The angle $(\pi+x)$ is located in the third quadrant, where tangent is positive.

d) $\cot(2\pi+x) = \cot x$. The angle $(2\pi+x)$ is located in the first quadrant, where cotangent is positive.

Finally, a few commonly used values of trigonometric functions for specific angles:

$$\sin \frac{\pi}{4} = \cos \frac{\pi}{4} = \frac{\sqrt{2}}{2}, \quad \sin \frac{\pi}{3} = \cos \frac{\pi}{6} = \frac{\sqrt{3}}{2}, \quad \sin \frac{\pi}{6} = \cos \frac{\pi}{3} = \frac{1}{2},$$

$$\sin \frac{\pi}{2} = \cos 2\pi = 1, \quad \cos \frac{\pi}{2} = \sin 2\pi = 0, \quad \tan \frac{\pi}{4} = \cot \frac{\pi}{4} = 1,$$

$$\tan \frac{\pi}{3} = \cot \frac{\pi}{6} = \sqrt{3}, \quad \tan \frac{\pi}{6} = \cot \frac{\pi}{3} = \frac{\sqrt{3}}{3}.$$

Problem 1. Find the expression of $\sin 3x$ and $\cos 3x$ in terms of $\sin x$ and $\cos x$.

Solution.

$$\begin{aligned}
\sin 3x &= \sin(2x + x) = \sin 2x \cos x + \sin x \cos 2x \\
&= 2 \sin x \cos x \cos x + \sin x (\cos^2 x - \sin^2 x) \\
&= 2 \sin x \cos^2 x + \sin x (1 - 2 \sin^2 x) \\
&= 2 \sin x (1 - \sin^2 x) + \sin x (1 - 2 \sin^2 x) \\
&= 2 \sin x - 2 \sin^3 x + \sin x - 2 \sin^3 x \\
&= 3 \sin x - 4 \sin^3 x.
\end{aligned}$$

$$\begin{aligned}
\cos 3x &= \cos(2x + x) = \cos 2x \cos x - \sin 2x \sin x \\
&= (\cos^2 x - \sin^2 x) \cos x - 2 \sin x \cos x \sin x \\
&= \cos^3 x - \cos x (1 - \cos^2 x) - 2 \cos x \sin^2 x \\
&= \cos^3 x - \cos x + \cos^3 x - 2 \cos x (1 - \cos^2 x) \\
&= \cos^3 x - \cos x + \cos^3 x - 2 \cos x + 2 \cos^3 x \\
&= 4 \cos^3 x - 3 \cos x.
\end{aligned}$$

Let's emphasize the results, because we will be making use of them in the future:

$$\sin 3x = 3 \sin x - 4 \sin^3 x.$$
$$\cos 3x = 4 \cos^3 x - 3 \cos x.$$

Problem 2. Evaluate without using a calculator the value of $\sin \frac{\pi}{10} = \sin 18°$.

Solution. First, notice that applying the expression $\sin x = \cos \left(\frac{\pi}{2} - x \right)$ and the formulas for the cosine of a triple angle from the previous problem gives

$$\sin \frac{\pi}{5} = \cos \left(\frac{\pi}{2} - \frac{\pi}{5} \right) = \cos \frac{3\pi}{10} = 4 \cos^3 \frac{\pi}{10} - 3 \cos \frac{\pi}{10}. \qquad (1)$$

Secondly, applying the trigonometric formulas for a double angle gives

$$\sin \frac{\pi}{5} = \sin \frac{2\pi}{10} = 2 \sin \frac{\pi}{10} \cos \frac{\pi}{10}. \qquad (2)$$

Comparing (1) and (2) yields

$$2 \sin \frac{\pi}{10} \cos \frac{\pi}{10} = 4 \cos^3 \frac{\pi}{10} - 3 \cos \frac{\pi}{10},$$

and, after factoring,

$$2 \sin \frac{\pi}{10} \cos \frac{\pi}{10} = \cos \frac{\pi}{10} \left(4 \cos^2 \frac{\pi}{10} - 3 \right).$$

Dividing both sides by $\cos \frac{\pi}{10}$ gives

$$2 \sin \frac{\pi}{10} = 4 \cos^2 \frac{\pi}{10} - 3.$$

Substituting the expression for $\cos \frac{\pi}{10}$ in terms of $\sin \frac{\pi}{10}$ using the Pythagorean Identity as $\cos^2 \frac{\pi}{10} = 1 - \sin^2 \frac{\pi}{10}$ gives

$$2 \sin \frac{\pi}{10} = 4 \left(1 - \sin^2 \frac{\pi}{10} \right) - 3,$$
$$4 \sin^2 \frac{\pi}{10} + 2 \sin \frac{\pi}{10} - 1 = 0.$$

Substituting $y = \sin \frac{\pi}{10}$ results in the equation

$$4y^2 + 2y - 1 = 0.$$

Solving this quadratic equation gives

$$D = 4 + 16 = 20,$$
$$y = \frac{-2 \pm \sqrt{20}}{8} = \frac{-1 \pm \sqrt{5}}{4}.$$

Since $\frac{\pi}{10}$ is located in the first quadrant, its sine must be a positive number. Therefore, we consider only the positive solution of the above quadratic equation for

$$y = \sin \frac{\pi}{10} = \frac{-1 + \sqrt{5}}{4}.$$

Answer: $\sin \frac{\pi}{10} = \sin 18° = \frac{-1+\sqrt{5}}{4}$.

Using this result, we invite readers to find without a calculator the values of $\sin \frac{\pi}{5}$, $\cos \frac{\pi}{5}$, $\tan \frac{\pi}{5}$, and $\cot \frac{\pi}{5}$.

Problem 3. Find the sum of angles α and β, if $\tan\alpha$ and $\tan\beta$ are the roots of the equation $6x^2 - 5x + 1 = 0$.

Solution. If $\tan\alpha$ and $\tan\beta$ are the roots of the equation $6x^2 - 5x + 1 = 0$, then according to Viète's formulas

$$\tan\alpha + \tan\beta = \frac{5}{6},$$

$$\tan\alpha\tan\beta = \frac{1}{6}.$$

Applying the formula for the tangent of the sum of two angles gives

$$\tan(\alpha + \beta) = \frac{\tan\alpha + \tan\beta}{1 - \tan\alpha\tan\beta} = \frac{\frac{5}{6}}{1 - \frac{1}{6}} = 1,$$

therefore,

$$\alpha + \beta = \frac{\pi}{4} + \pi n, \quad n \in Z.$$

Problem 4. Prove that the roots of the equation $x + x^{-1} = 2\cos 40°$ satisfy also the equation $x^4 + x^{-4} = 2\cos 160°$.

Solution. Instead of solving the first equation and verifying that its roots satisfy the second equation, we will take a different approach, which is much more elegant and efficient.

Squaring twice both sides of the equation $x + x^{-1} = 2\cos 40°$ gives the equation $x^4 + x^{-4} = 2\cos 160°$, the roots of which should satisfy the original equation as well. Indeed,

$$(x + x^{-1})^2 = 4\cos^2 40°,$$
$$x^2 + 2x \cdot x^{-1} + x^{-2} = 4\cos^2 40°,$$
$$x^2 + 2 + x^{-2} = 4\cos^2 40°,$$
$$x^2 + x^{-2} = 4\cos^2 40° - 2,$$
$$x^2 + x^{-2} = 2(2\cos^2 40° - 1).$$

The expression in the parentheses on the right-hand side can be replaced using the formula for the cosine of a double angle, $2\cos^2 40° - 1 = \cos 80°$. It follows that

$$x^2 + x^{-2} = 2\cos 80°.$$

Squaring both sides of the last equation gives

$$x^4 + 2 + x^{-4} = 4\cos^2 80°,$$
$$x^4 + x^{-4} = 4\cos^2 80° - 2,$$
$$x^4 + x^{-4} = 2(2\cos^2 80° - 1).$$

Using again the formula for the cosine of a double angle, $2\cos^2 80° - 1 = \cos 160°$, we get $x^4 + x^{-4} = 2\cos 160°$. So, the last equation was obtained by squaring the first equation twice. Therefore, the roots of the first equation must satisfy the second equation as well. The statement of the problem was justified by relatively simple trigonometric manipulations, without actually solving either of the equations.

The following problem will present us with two more important formulas, which are worth remembering in addition to the identities listed above.

Problem 5. Find the expression through trigonometric functions of a double angle x for the following sums:

$$\sin^4 x + \cos^4 x,$$
$$\sin^6 x + \cos^6 x.$$

Solution. Squaring both sides of the Pythagorean Identity $\sin^2 x + \cos^2 x = 1$ yields

$$\sin^4 x + 2\sin^2 x \cos^2 x + \cos^4 x = 1,$$
$$\sin^4 x + \cos^4 x = 1 - 2\sin^2 x \cos^2 x,$$
$$\sin^4 x + \cos^4 x = 1 - \frac{1}{2}\sin^2 2x.$$

Cubing both sides of the Pythagorean Identity and using the well-known formula for the cube of the sum of two numbers $(a+b)^3 = a^3 + 3ab(a+b) + b^3$ gives

$$(\sin^2 x + \cos^2 x)^3 = 1^3,$$
$$\sin^6 x + 3\sin^2 x \cos^2 x \left(\underbrace{\sin^2 x + \cos^2 x}_{=1} \right) + \cos^6 x = 1,$$
$$\sin^6 x + 3\sin^2 x \cos^2 x + \cos^6 x = 1,$$
$$\sin^6 x + \cos^6 x = 1 - 3\sin^2 x \cos^2 x,$$
$$\sin^6 x + \cos^6 x = 1 - \frac{3}{4}\sin^2 2x.$$

Answer:

$$\sin^4 x + \cos^4 x = 1 - \frac{1}{2}\sin^2 2x;$$
$$\sin^6 x + \cos^6 x = 1 - \frac{3}{4}\sin^2 2x.$$

Problem 6. Prove the identity $\sin x \cdot \sin \left(\frac{\pi}{3} - x \right) \cdot \sin \left(\frac{\pi}{3} + x \right) = \frac{1}{4}\sin 3x$.

Proof. Applying the formula for the product of the sines of two angles to the second and the third factors on the left-hand side and recalling that $\cos \frac{2\pi}{3} = -\frac{1}{2}$, and the fact that the cosine is an even function and the sine is an odd function gives

$$\sin x \cdot \sin\left(\frac{\pi}{3} - x\right) \cdot \sin\left(\frac{\pi}{3} - x\right)$$

$$= \sin x \cdot \frac{1}{2}\left(\cos\left(\frac{\pi}{3} - x - \frac{\pi}{3} - x\right) - \cos\left(\frac{\pi}{3} - x + \frac{\pi}{3} + x\right)\right)$$

$$= \frac{1}{2}\sin x\left(\cos 2x - \cos\frac{2\pi}{3}\right) = \frac{1}{2}\sin x\left(\cos 2x + \frac{1}{2}\right)$$

$$= \frac{1}{2}\sin x \cos 2x + \frac{1}{4}\sin x.$$

Applying the identity for the product

$$\sin\alpha\cos\beta = \frac{1}{2}(\sin(\alpha+\beta) + \sin(\alpha-\beta))$$

allows further modifying the last expression to

$$\frac{1}{2}\cdot\frac{1}{2}(\sin(x+2x) + \sin(x-2x)) + \frac{1}{4}\sin x = \frac{1}{4}\sin 3x - \frac{1}{4}\sin x + \frac{1}{4}\sin x$$

$$= \frac{1}{4}\sin 3x,$$

as was to be proved. $\square$

Problem 7. Prove that $\sin 20° \cdot \sin 40° \cdot \sin 60° \cdot \sin 80° = \frac{3}{16}$.

Solution. Converting radians into degrees yields $60° = \frac{\pi}{3}$ and $20° = \frac{\pi}{9}$. Rearranging the factors and applying the formula proved in the previous problem gives

$$\sin 20° \cdot \sin 40° \cdot \sin 60° \cdot \sin 80° = \sin 60° \cdot \sin 20° \cdot \sin 40° \cdot \sin 80°$$

$$= \frac{\sqrt{3}}{2}\cdot\sin\frac{\pi}{9}\cdot\sin\left(\frac{\pi}{3} - \frac{\pi}{9}\right)\cdot\sin\left(\frac{\pi}{3} + \frac{\pi}{9}\right)$$

$$= \frac{\sqrt{3}}{2}\cdot\frac{1}{4}\sin\frac{3\pi}{9}$$

$$= \frac{\sqrt{3}}{2}\cdot\frac{1}{4}\sin\frac{\pi}{3}$$

$$= \frac{\sqrt{3}}{2}\cdot\frac{1}{4}\cdot\frac{\sqrt{3}}{2}$$

$$= \frac{3}{16},$$

as was to be proved.

We invite readers to prove on their own identities similar to the one proved in problem 6 for the other trigonometric functions:

$$\cos x \cdot \cos\left(\frac{\pi}{3} - x\right) \cdot \cos\left(\frac{\pi}{3} + x\right) = \frac{1}{4}\cos 3x$$

$$\tan x \cdot \tan\left(\frac{\pi}{3} - x\right) \cdot \tan\left(\frac{\pi}{3} + x\right) = \tan 3x$$

$$\cot x \cdot \cot\left(\frac{\pi}{3} - x\right) \cdot \cot\left(\frac{\pi}{3} + x\right) = \cot 3x.$$

Problem 8. Given the function $f(x) = \sin^4 x + \cos^4 x$. Find $f(\alpha)$ knowing that $\sin 2\alpha = \frac{2}{3}$.

Solution. As was proved in problem 5,

$$\sin^4 x + \cos^4 x = 1 - \frac{1}{2}\sin^2 2x,$$

so $f(x) = 1 - \frac{1}{2}\sin^2 2x$ and respectively

$$f(\alpha) = 1 - \frac{1}{2} \cdot \left(\frac{2}{3}\right)^2 = 1 - \frac{1}{2} \cdot \frac{4}{9} = \frac{7}{9}.$$

Problem 9. Given $\cos 2x = m$. Evaluate $\cos^8 x - \sin^8 x$.

Solution. Applying the formulas for the difference of squares of two numbers, the Pythagorean Identity $\sin^2 x + \cos^2 x = 1$, and the formula from problem 5 we get

$$\cos^8 x - \sin^8 x = (\cos^4 x)^2 - (\sin^4 x)^2 = (\cos^4 x - \sin^4 x)(\cos^4 x + \sin^4 x)$$

$$= \underbrace{(\cos^2 x - \sin^2 x)}_{\cos 2x}\underbrace{(\cos^2 x + \sin^2 x)}_{1}(\cos^4 x + \sin^4 x)$$

$$= \cos 2x \underbrace{(\cos^4 x + \sin^4 x)}_{1 - \frac{1}{2}\sin^2 2x}$$

$$= \cos 2x\left(1 - \frac{1}{2}\sin^2 2x\right) = m\left(1 - \frac{1}{2}(1 - \cos^2 2x)\right)$$

$$= m\left(1 - \frac{1}{2}(1 - m^2)\right) = m\left(1 - \frac{1}{2} + \frac{1}{2}m^2\right)$$

$$= m\left(\frac{1}{2} + \frac{1}{2}m^2\right) = \frac{m(1 + m^2)}{2}.$$

Problem 10. Evaluate

$$\frac{\sin 24° \cos 6° - \sin 6° \sin 66°}{\sin 21° \cos 39° - \sin 39° \cos 21°}.$$

Solution.

$$\frac{\sin 24° \cos 6° - \sin 6° \sin 66°}{\sin 21° \cos 39° - \sin 39° \cos 21°} = \frac{\sin 24° \cos 6° - \sin 6° \sin(90° - 24°)}{\sin 21° \cos 39° - \sin 39° \cos 21°}$$

$$= \frac{\sin 24° \cos 6° - \sin 6° \cos 24°}{\sin 21° \cos 39° - \sin 39° \cos 21°}$$

$$= \frac{\sin(24° - 6°)}{\sin(21° - 39°)} = \frac{\sin 18°}{\sin(-18°)}$$

$$= \frac{\sin 18°}{-\sin 18°} = -1.$$

Problem 11. Evaluate

$$\frac{1}{2 \sin 10°} - 2 \sin 70°.$$

Solution. If instead of an angle of $10°$, we would have an angle of $20°$, the problem will be easily solved. Since $20°$ and $70°$ are complementary angles, we would be able to apply the equalities $\sin(90° - 70°) = \cos 70°$ and $\cos(90° - 20°) = \sin 20°$. So, the plan is to get to these angles and utilize them in further manipulations. Observing that $2 \sin 10° \cos 10° = \sin 20°$ and multiplying numerator and denominator of the minuend in the given expression by the same number of $\cos 10°$ we get

$$\frac{\cos 10°}{2 \sin 10° \cos 10°} - 2 \sin 70° = \frac{\cos 10°}{\sin 20°} - 2 \sin 70° = \frac{\cos 10°}{\sin(90° - 70°)} - 2 \sin 70°$$

$$= \frac{\cos 10°}{\cos 70°} - 2 \sin 70° = \frac{\cos 10° - 2 \sin 70° \cos 70°}{\cos 70°} =$$

$$= \frac{\cos 10° - \sin 140°}{\cos 70°} = \frac{\cos 10° - \sin(90° + 50°)}{\cos 70°}$$

$$= \frac{\cos 10° - \cos 50°}{\cos(90° - 20°)} = \frac{2 \sin \frac{(50° - 10°)}{2} \cdot \sin \frac{(10° + 50°)}{2}}{\sin 20°}$$

$$= \frac{2 \sin 20° \cdot \sin 30°}{\sin 20°} = 2 \sin 30° = 2 \cdot \frac{1}{2} = 1.$$

Problem 12. Given $\tan x + \cot x = a$, find $\tan^4 x + \cot^4 x$.

Solution. Squaring the sum $(\tan x + \cot x)$ gives

$$(\tan x + \cot x)^2 = \tan^2 x + 2 \tan x \cot x + \cot^2 x = \tan^2 x + \cot^2 x + 2. \quad (1)$$

On the other hand, using the conditions of the problem,

$$(\tan x + \cot x)^2 = a^2. \quad (2)$$

Comparing (1) and (2) gives

$$\tan^2 x + \cot^2 x = a^2 - 2.$$

Squaring both sides of the last equality yields

$$(\tan^2 x + \cot^2 x)^2 = (a^2 - 2)^2,$$
$$\tan^4 x + \cot^4 x + 2\underbrace{\tan^2 x \cot^2 x}_{=1} = (a^2 - 2)^2,$$
$$\tan^4 x + \cot^4 x + 2 = a^4 - 4a^2 + 4,$$
$$\tan^4 x + \cot^4 x = a^4 - 4a^2 + 4 - 2,$$
$$\tan^4 x + \cot^4 x = a^4 - 4a^2 + 2.$$

Problem 13. Given $x + \frac{1}{x} = 2\cos y$ $(x \neq 0)$. Prove that for any natural number n the following equality holds true: $x^n + \frac{1}{x^n} = 2\cos ny$.

Proof. Let's verify the equality for $n = 2$. Squaring both sides of the given equality and applying the formula $\cos 2y = 2\cos^2 y - 1$ gives

$$\left(x + \frac{1}{x}\right)^2 = 4\cos^2 y, \text{ or } x^2 + \underbrace{2 \cdot x \cdot \frac{1}{x}}_{2} + \frac{1}{x^2} = 4\cos^2 y,$$

which yields

$$x^2 + \frac{1}{x^2} = 4\cos^2 y - 2 = 2(2\cos^2 y - 1) = 2\cos 2y.$$

So, we see that for $n = 2$ the equality holds true.

Assume now that it holds for any $n \le k$. If we prove that on this assumption the equality holds for $n = k + 1$, then by mathematical induction it will hold for any natural number n.

Our assumption is that for $n = k - 1$ and for $n = k$:

$$x^{k-1} + \frac{1}{x^{k-1}} = 2\cos(k-1)y \text{ and } x^k + \frac{1}{x^k} = 2\cos ky.$$

We have to prove that

$$x^{k+1} + \frac{1}{x^{k+1}} = 2\cos(k+1)y.$$

Consider

$$\left(x^k + \frac{1}{x^k}\right)\left(x + \frac{1}{x}\right) = x^{k+1} + \frac{1}{x^{k-1}} + x^{k-1} + \frac{1}{x^{k+1}}.$$

Expressing $x^{k+1} + \frac{1}{x^{k+1}}$ from the last equality and using the formula for the cosine of the sum of two angles leads to

$$x^{k+1} + \frac{1}{x^{k+1}} = \left(x^k + \frac{1}{x^k}\right)\left(x + \frac{1}{x}\right) - \left(\frac{1}{x^{k-1}} + x^{k-1}\right)$$
$$= 2\cos ky \cdot 2\cos y - 2\cos(k-1)y$$

$$= 2(2\cos ky \cdot \cos y - \cos(ky - y))$$
$$= 2(2\cos ky \cdot \cos y - \cos ky \cdot \cos y - \sin ky \sin y)$$
$$= 2\big(\underbrace{\cos ky \cdot \cos y - \sin ky \sin y}_{\cos(ky+y)}\big) = 2\cos(k+1)y.$$

As we see, the statement holds for $n = k+1$. Hence, by mathematical induction we infer that the given statement is correct for all natural numbers and the equality is proved. □

The application of trigonometric functions in geometry goes way beyond just the calculation of the lengths of sides and angles measures in a triangle. As a matter of fact, many geometrical problems have either pure trigonometric solutions or solutions that are based on trigonometric identities. Let's look at some examples.

Problem 14. The angles of triangle ABC satisfy the equality

$$\cos^2 A + \cos^2 B + \cos^2 C = 1.$$

Prove that ABC is a right triangle.

Proof. For this geometrical problem there is no need even to make a picture. It's irrelevant. You just have to apply a few trigonometric identities to modify the given expression for the angles of the triangle ABC.

$$\cos^2 A + \cos^2 B + \cos^2 C = 1,$$

then

$$\frac{1+\cos 2A}{2} + \frac{1+\cos 2B}{2} + \cos^2 C = 1,$$

from which

$$1 + \cos 2A + 1 + \cos 2B + 2\cos^2 C = 2.$$

Grouping the terms and applying the formula for the sum of cosines gives

$$(\cos 2A + \cos 2B) + 2\cos^2 C = 0,$$
$$2\cos\frac{2A+2B}{2} \cdot \cos\frac{2A-2B}{2} + 2\cos^2 C = 0,$$
$$2\cos(A+B)\cdot\cos(A-B) + 2\cos^2 C = 0. \tag{1}$$

The sum of the angles in a triangle equals $180°$, therefore $A + B = 180° - C$, $A + C = 180° - B$, and $C + B = 180° - A$. Substituting the expression for $(A+B)$ into (1) and recalling that for any angle α, $\cos(180° - a) = -\cos\alpha$ yields

$$2\cos(180° - C)\cdot\cos(A - B) + 2\cos^2 C = 0,$$
$$-2\cos C \cdot \cos(A - B) + 2\cos^2 C = 0;$$

dividing by 2 and factoring for $\cos C$ gives

$$\cos C(-\cos(A - B) + \cos C) = 0,$$
$$\cos C(\cos(A - B) - \cos C) = 0;$$

applying the formula for the difference of cosines gives

$$\cos C\left(2\sin\frac{A - B + C}{2} \cdot \sin\frac{C - A + B}{2}\right) = 0;$$

substituting $A + C = 180° - B$ and $C + B = 180° - A$ gives

$$\cos C \cdot \sin\frac{180° - 2B}{2} \cdot \sin\frac{180° - 2A}{2} = 0,$$
$$\cos C \cdot \sin(90° - B) \cdot \sin(90° - A) = 0,$$
$$\cos C \cdot \cos B \cdot \cos A = 0.$$

The last equality implies that at least one factor must equal to 0, which means that at least one angle of the given triangle must equal 90°. Thus ABC is a right triangle and the proof is completed.   □

Problem 15. The angles of triangle ABC satisfy the equality $\frac{\sin^2 A}{\sin^2 B} = \frac{\tan A}{\tan B}$. Prove that this triangle is either isosceles ($\angle A = \angle B$) or is a right triangle with the angle $\angle C = 90°$.

Proof. Let's modify the given equality $\frac{\sin^2 A}{\sin^2 B} = \frac{\tan A}{\tan B}$ using the definition of a tangent:

$$\frac{\sin^2 A}{\sin^2 B} - \frac{\sin A \cos B}{\sin B \cos A} = 0,$$
$$\frac{\sin A}{\sin B} \cdot \left(\frac{\sin A}{\sin B} - \frac{\cos B}{\cos A}\right) = 0,$$
$$\frac{\sin A}{\sin B} \cdot \frac{\sin A \cos A - \sin B \cos B}{\sin B \cos A} = 0.$$

Since $\sin A \neq 0$, then for the product in the last equality to equal 0, the second factor must equal 0:

$$\frac{\sin A \cos A - \sin B \cos B}{\sin B \cos A} = 0.$$

It follows that

$$\sin A \cos A - \sin B \cos B = 0.$$

Multiplying both sides of the last equality by 2 and recalling the formula for the sine of a double angle gives

$$\sin 2A - \sin 2B = 0.$$

Using the formula for the difference of sines of two angles yields

$$2\sin(A - B) \cdot \cos(A + B) = 0,$$

which implies that $\sin(A - B) = 0$, or $\cos(A + B) = 0$. From the first equality, $A - B = 0$, or $A = B$, which implies triangle ABC is isosceles. From the second equality, $A + B = 90°$, which means that the third angle C in the triangle ABC must be a right angle, $C = 180° - (A + B) = 180° - 90° = 90°$. So, we see that the statement of the problem is correct. Under the given conditions, the triangle is either isosceles ($\angle A = \angle B$) or is a right triangle with $\angle C = 90°$. □

Trigonometry in construction problems; cyclotomic equations

At this point let's change the focus of our discussion and turn to geometrical problems of compass and straightedge constructions of segments with given lengths and angles with given measures. We assume readers are familiar with performing the basic constructions with these classic tools.

Since Euclid's time ancient Greeks knew how to perform the basic compass-straightedge constructions (the compass is used to draw circles, and the straightedge to draw straight lines only) of segments with the sum, difference, product, ratio, and even square roots of given lengths. For example, the construction problem of the segment with the length of $\sqrt{5}$ is reduced to the construction of a right triangle with the legs 2 and 1. According to the Pythagorean Theorem, the hypotenuse of that triangle has a length of $\sqrt{5}$. The same logic applies for the constructions of segments with lengths expressed by many other irrational numbers. You repeat the process of right triangle constructions until you get to the needed one with the hypotenuse of the requested length. Let's say we need to construct a segment with length $\sqrt{14}$.

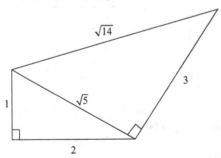

First, we construct a segment with length $\sqrt{5}$ and then build a right triangle with the legs $\sqrt{5}$ and 3. The hypotenuse of that triangle has length $\sqrt{(\sqrt{5})^2 + 3^2} = \sqrt{14}$.

Generally, the constructible numbers include all rational numbers, all numbers that can be formed from them using the basic arithmetic operations and the extraction of square roots, and irrational numbers that are the solutions to some quadratic equations (real, not imaginary, solutions) with rational coefficients that are irreducible over the set of rational numbers. Speaking in algebraic terms, every irrational number that is constructible can be expressed as a root of a polynomial of degree 2 with some rational coefficients. The rigorous proof of this assertion is given by Galois theory and is not going to be presented in this book.

For the constructability of angles, the problem is solved for those angles whose trigonometric function (any function—sine, cosine, tangent or cotangent) is constructible as a number. It follows directly from the definition of the trigonometric functions as ratios of the lengths of the sides in a right triangle.

For example, is it possible to make compass-straightedge construction of an angle of 15°? In order to answer that question let's try to find $\cos 15°$ and see if it is a constructible number. We know that $\cos 30° = \frac{\sqrt{3}}{2}$. Using this value and the formula for the cosine of a double angle gives

$$\cos^2 15° - \sin^2 15° = \cos 30° = \frac{\sqrt{3}}{2}.$$

On the other hand,

$$\cos^2 15° - \sin^2 15° = \cos^2 15° - (1 - \cos^2 15°) = 2\cos^2 15° - 1.$$

Thus, $2\cos^2 15° - 1 = \frac{\sqrt{3}}{2}$, from which $\cos 15° = \sqrt{\frac{\sqrt{3}+2}{4}}$. Going one step further, multiplying the numerator and denominator of the fraction under the square root by 4, we complete the square in the numerator and simplify the expression for $\cos 15°$. Indeed,

$$\frac{\sqrt{3}+2}{4} = \frac{4\sqrt{3}+8}{16} = \frac{6+2\cdot 2\sqrt{3}+2}{16} = \frac{(\sqrt{6}+\sqrt{2})^2}{16},$$

which yields

$$\cos 15° = \sqrt{\frac{(\sqrt{6}+\sqrt{2})^2}{16}} = \frac{\sqrt{6}+\sqrt{2}}{4}.$$

This number clearly is constructible, and the answer to the question regarding the constructability of angle 15° is positive.

One of the most famous construction problems not solved by the ancient Greeks with the compass and straightedge was *trisecting the angle*: dividing a given angle into three equal angles. The impossibility of such a construction can be clarified by trigonometry through cubic equation analyses. Assume that the angle that we need to trisect has the value $3x$. Now that we are armed with the formula for the cosine of a triple angle from problem 1 above, we will make use of it here: $\cos 3x = 4\cos^3 x - 3\cos x$. Multiplying both sides by 2 gives $2\cos 3x = 8\cos^3 x - 6\cos x$. If we let $a = 2\cos 3x$ and $y = 2\cos x$, then $a = y^3 - 3y$, or equivalently,

$$y^3 - 3y - a = 0. \tag{$*$}$$

To prove that the trisection of a given angle with compass and straightedge is generally impossible, it suffices to indicate at least one angle for which it will not be possible. The angle of 60° will serve that purpose. First, note that 60° itself is constructible (an equilateral triangle with the given side is easily constructible by the compass and straightedge). However, we will show that it is impossible to construct an angle of 20°, which would imply the impossibility of trisecting a 60° angle. Indeed, if $3x = 60°$, then $\cos 3x = \frac{1}{2}$ and $(*)$ will be transformed into $y^3 - 3y - 1 = 0$.

According to the Rational Root Theorem, the rational solutions of this equation can be only 1 or -1. However, neither of these numbers satisfies the equation. Therefore, it has no rational solutions. In other words, the polynomial $y^3 - 3y - 1$ is irreducible to a lower degree over the field of rational numbers and the minimal polynomial for $\cos 20°$ is of degree 3. If a 60° angle is trisectible, then the degree of a minimal polynomial of $\cos 20°$ over the field of rational numbers would be two, which contradicts with the obtained result. Hence, an angle of 60° is not trisectible with compass and straightedge. It's not hard to find many angles for which the equation $(*)$ is not solvable in radicals, so the trisection of an angle problem generally is not solvable by means of compass and straightedge. Usually, if the rational root test reveals a rational root of the cubic equation the cubic polynomial can be factored as a product of linear and quadratic polynomials that can be solved for the remaining two roots in terms of a square root; then all of these roots are constructible with compass and straightedge since they are expressible by no higher than square roots. On the other hand, if the rational root test shows that no rational root exists, the angle $\frac{x}{3}$ is not constructible, and the angle x is not trisectible with the classic construction tools.

From the lengths of the sides and angle constructions the next step would be to pose the question regarding the constructability of regular polygons, that is, polygons with equal angles and equal sides. Would it be possible to construct any regular polygon using a straightedge and compass? That question is equivalent to the question of how to divide the circle into n equal parts. We may conclude from our analysis of the trisectibility of an angle of 60°, that an angle of 20°, and an angle of 40°, are not trisectible with compass and straightedge either, which leads to the conclusion that the regular 18-gon and regular 9-gon cannot be constructed with those tools. By the same logic, an angle of 15° is constructible. Thus, it is possible to construct a regular 24-gon.

Let's consider the problem of constructing a regular 10-gon. Why would we want to bring a pure geometrical problem in here? How would it be relevant for our study of equations?

Well, we just saw the original geometrical problem of trisecting an angle was reduced to a purely algebraic equation solution problem. We are about to see another vivid and interesting example of close links between geometrical issues and

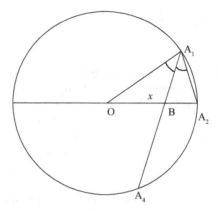

their algebraic interpretation through equations. The goal is to perform a compass-straightedge construction of a regular 10-gon. We will arrive at the desired result if we figure out how to divide the circumference of a circle into 10 equal parts. Assume the problem is solved and we have a regular 10-gon inscribed in the unit circle with center O (the radius of the circle equals 1). Points A_1 and A_2 are adjacent vertices of our regular 10-gon. $\angle A_1 O A_2 = 360° : 10 = 36°$, then, obviously, in the isosceles triangle $O A_1 A_2$ each angle by the base $A_1 A_2$ equals $\frac{180° - 36°}{2} = 72°$, $\angle O A_1 A_2 = \angle O A_2 A_1 = 72°$. Draw the bisector $A_1 B$ of the angle $O A_1 A_2$. $\angle O A_1 B = \angle B A_1 A_2 = 72° : 2 = 36°$. Therefore, $O B A_1$ and $B A_1 A_2$ are isosceles triangles ($O B = B A_1 = A_1 A_2$). Since triangles $A_1 O A_2$ and $B A_1 A_2$ have respective equal angles, they are similar. If we let $OB = x$, then $B A_2 = 1 - x$ and from the similarity of the above triangles we get

$$\frac{x}{1-x} = \frac{1}{x}, \text{ or equivalently, } x^2 + x - 1 = 0.$$

Solving this quadratic equation gives

$$x = \frac{-1 + \sqrt{5}}{2} \quad \text{or} \quad x = \frac{-1 - \sqrt{5}}{2}.$$

Since $x = OB = B A_1 = A_1 A_2$, the positive root $\frac{-1+\sqrt{5}}{2}$ represents the length of the side of the regular 10-gon. Therefore, the problem of dividing the circumference into 10 equal parts is reduced to constructing a segment of length $\frac{-1+\sqrt{5}}{2}$, which is compass-straightedge solvable, as we saw before.

Extending $A_1 B$ till its intersection with the circle at point A_4, we get the fourth vertex of the regular 10-gon. The third vertex has to be located at the midpoint of the arc $A_4 A_2$. We basically solved not just the problem of the compass-straightedge construction of a regular 10-gon, but the problem of inscribing a regular pentagon in the circle as well. Indeed, $A_4 A_2$, obtained as described, has to be the side of a regular pentagon inscribed in the circle. We suggest the reader prove this fact and find the length of $A_4 A_2$.

In the following chapters we will come across and talk in detail about the mysterious number $\varphi = \frac{1+\sqrt{5}}{2}$ known as the so-called *golden ratio*, conjugate to which is $\frac{-1+\sqrt{5}}{2}$. The ancient Greeks knew this number and used it in measurements and constructions related to regular pentagrams and pentagons. The first mention of the golden ratio as "extreme and mean ratio" was found more than 2000 years ago in Euclid's *Elements*. If you notice that the product of the golden ratio and its conjugate equals 1, $\frac{1+\sqrt{5}}{2} \cdot \frac{-1+\sqrt{5}}{2} = \frac{5-1}{4} = 1$, then the length of the side of a regular decagon is the reciprocal of the golden ratio $\frac{1}{\varphi}$. The triangle $A_1 O A_2$ is called a *golden triangle*. When we draw the bisector of $\angle O A_1 A_2$, we have created two more golden triangles $O B A_1$ and $B A_1 A_2$.

As you saw, this construction problem was solved with the help of a quadratic equation and we are back to our equations world. We managed to set it up without referring to trigonometric functions.

Generally, it turned out that the answer to the question of how to divide a circle into n equal parts, which is equivalent to answering the question regarding the constructability of a regular n-gon, is reduced to constructing a length expressed as a trigonometric function of an angle $\frac{2\pi}{n}$, more specifically, $\cos \frac{2\pi}{n}$. As n increases, the problem becomes more complicated. As it was mentioned in chapter 1, Carl Gauss solved the problem of the compass-straightedge construction of a regular 17-gon and discovered that it was possible due to the constructability of the angle of $\frac{2\pi}{17}$ radians. Solving the equation $x^{17} = 1$, he found the cosine of $\frac{2\pi}{17}$ in terms of arithmetic operations and square roots of integers as

$$\cos \frac{2\pi}{17} = -\frac{1}{16} + \frac{1}{16}\sqrt{17} + \frac{1}{16}\sqrt{34 - 2\sqrt{17}} + \frac{1}{8}\sqrt{17 + 3\sqrt{17} - \sqrt{34 - 2\sqrt{17}} - 2\sqrt{34 + 2\sqrt{17}}}.$$

Amazingly, the geometric classical problem of a regular n-gon construction has evolved into the algebraic problem of finding the roots of the so-called cyclotomic equation $x^n - 1 = 0$, where n is a natural number. The term "cyclotomy", interpreted as a division of the circle into equal parts, ties these two seemingly unrelated issues together. The constructions of regular triangles, squares, pentagons, and regular polygons with double the number of sides of those were known to ancient Greeks. However, the question regarding the constructability of any regular n-gon stayed open for more than 2000 years. It was answered only at the end of the 18th century upon Carl Gauss's solution of the regular 17-gon construction, which was one of the first steps in developing the theory of so-called Gaussian periods, and Gauss's arrival at sufficient conditions for the constructability of regular polygons. He believed that the conditions were also necessary, but did not provide a proof. Such a proof was given by the French mathematician Pierre Laurent Wantzel (1814–1848). It is particularly important to emphasize that one can't underestimate the usefulness and importance of trigonometry applications in these discoveries. It was well reflected not just in Gauss's and Wantzel's work, but in accomplishments of their predecessors as well.

Even though we agreed to work with real numbers and try to avoid discussion of complex numbers, speaking about cyclotomic equations is impossible without mentioning a remarkable relationship significant in the algebraic theory of equations, known as de Moivre's formula (after French mathematician Abraham de Moivre (1667–1754))

$$(\cos x + i \sin x)^n = \cos(nx) + i \sin(nx), \quad \text{where } i = \sqrt{-1}.$$

In the field of complex numbers, expressed as $a + bi$ (where a and b are real numbers), the cyclotomic equation $x^n - 1 = 0$ has 1 as a root with other $n - 1$ roots,

$$x_k = \cos\left(\frac{2\pi k}{n}\right) + i \sin\left(\frac{2\pi k}{n}\right), \quad k = 0, 1, \ldots, n - 1.$$

These solutions are called the roots of unity. Geometrically, in a two-dimensional coordinate system, they represent the points $\left(\cos\left(\frac{2\pi k}{n}\right), \sin\left(\frac{2\pi k}{n}\right)\right)$ dividing the unit circle into n equal parts and, therefore, being the vertices of a regular n-gon inscribed in that circle.

The equation $x^n - 1 = 0$ can be rewritten as

$$(x - 1)(x^{n-1} + x^{n-2} + \cdots + x + 1) = 0.$$

de Moivre obtained solutions to the equation $x^{n-1} + x^{n-2} + \cdots + x + 1 = 0$ for prime powers 2, 3, 5, and 7.

Let's demonstrate here the application of de Moivre's formula, for example, for the regular pentagon constructability problem, so you can compare it to the geometrical solution examined above.

Factoring $x^5 - 1$ as $x^5 - 1 = (x - 1)(x^4 + x^3 + x^2 + x + 1)$, consider the equation $x^4 + x^3 + x^2 + x + 1 = 0$. Dividing both sides by x^2 ($x \neq 0$) and regrouping the terms, we obtain the reciprocal equation studied in chapter 3:

$$x^2 + \frac{1}{x^2} + x + \frac{1}{x} + 1 = 0.$$

Making the substitution $y = x + \frac{1}{x}$ transforms the equation into $y^2 + y - 1 = 0$. The roots of the last equation are $y = \frac{-1 \pm \sqrt{5}}{2}$. So, to find x we have to solve two equations

$$x + \frac{1}{x} = \frac{-1 + \sqrt{5}}{2} \tag{1}$$

and

$$x + \frac{1}{x} = \frac{-1 - \sqrt{5}}{2}. \tag{2}$$

Instead of doing this, let's use the roots of unity for the equation $x^5 - 1 = 0$ expressed as

$$x_k = \cos\left(\frac{2\pi k}{5}\right) + i \sin\left(\frac{2\pi k}{5}\right), \quad \text{for } k = 0, 1, 2, 3, 4.$$

Even though doing some manipulations with complex numbers below, we don't have to have a discussion of complex number properties. We merely have to observe that $i^2 = -1$.

For simplicity let's denote $\alpha = \frac{2\pi k}{5}$. Then $x = \cos\alpha + i \sin\alpha$, so $\frac{1}{x} = \frac{1}{\cos\alpha + i \sin\alpha}$. Multiplying numerator and denominator by $\cos\alpha - i \sin\alpha$ gives

$$
\begin{aligned}
\frac{1}{x} &= \frac{1}{\cos\alpha + i \sin\alpha} = \frac{\cos\alpha - i \sin\alpha}{(\cos\alpha + i \sin\alpha)(\cos\alpha - i \sin\alpha)} \\
&= \frac{\cos\alpha - i \sin\alpha}{\cos^2\alpha - i^2 \cdot \sin^2\alpha} = \frac{\cos\alpha - i \sin\alpha}{\cos^2\alpha - (-1) \cdot \sin^2\alpha} \\
&= \frac{\cos\alpha - i \sin\alpha}{\cos^2\alpha + \sin^2\alpha} = \frac{\cos\alpha - i \sin\alpha}{1} = \cos\alpha - i \sin\alpha.
\end{aligned}
$$

Therefore, $x = \cos\alpha + i \sin\alpha$ and $\frac{1}{x} = \cos\alpha - i \sin\alpha$, which yields

$$x + \frac{1}{x} = \cos\alpha + i \sin\alpha + \cos\alpha - i \sin\alpha = 2\cos\alpha = 2\cos\frac{2\pi k}{5}.$$

For $k = 1$, making a substitution from (1), we obtain $2\cos\frac{2\pi}{5} = \frac{-1+\sqrt{5}}{2}$, (the negative value $\frac{-1-\sqrt{5}}{2}$ from (2) has to be rejected since $\cos\frac{2\pi}{5} > 0$), from which $\cos\frac{2\pi}{5} = \frac{-1+\sqrt{5}}{4}$. The last value is clearly constructible, and bearing this in mind, the proof of regular pentagon constructability appears to need no further explanation.

de Moivre's work on determining radical expressions for the roots of unity of prime values of n for the cyclotomic equations was extended and further developed by another French mathematician, Alexandre-Théophile Vandermonde (1735–1796). He gave the radical solutions for the cyclotomic equation of the eleventh power. He used the trigonometric identity

$$\cos x \cdot \cos y = \frac{1}{2}(\cos(x + y) + \cos(x - y))$$

to obtain the relations between the roots and to express the eleventh roots of unity in terms of radicals.

Credit for solving the cyclotomic equations in terms of radicals for any prime n higher than 11 belongs to Carl Gauss. He concluded also that a regular n-gon is constructible if any root of the nth cyclotomic polynomial

$$P_n(x) = x^{n-1} + x^{n-2} + \cdots + x + 1$$

is constructible and demonstrated that the issue for the constructability of a regular n-gon amounts to expressing the roots of the equation

$$x^{n-1} + x^{n-2} + \cdots + x + 1 = 0$$

in integers or through arithmetic operations and extracting square roots from integers.

Gauss established that the solution of the equation $P_n(x) = 0$ can be reduced to solving a series of quadratic equations whenever n is a Fermat prime (a number of the form $2^{2^n} + 1$). Speaking about seventeenth roots of unity, since $17 = 2^{2^2} + 1 = 2^4 + 1$, the solutions of the equation $x^{17} - 1 = 0$ can be calculated by solving four successive quadratic equations. Because the solution of a quadratic equation is expressed through arithmetic operations and extraction of square roots, Gauss concluded that the regular 17-gon is constructible by straightedge and compass.

In modern times the sufficient and necessary conditions for the constructability of regular polygons are stated in the Gauss-Wantzel Theorem:

A regular n-gon can be constructed with compass and straightedge if and only if n, the number of its sides, is the product of a power of 2 and any number of distinct Fermat primes: $n = 2^k \cdot p_1 \cdot p_2 \cdots p_m$, ($k$ and m are natural numbers; the p_i are distinct Fermat primes).

Considering a pure geometrical problem and raising a question regarding the conditions for the constructability of a regular n-gon, we come across cyclotomic equations and the issues related to their solvability in radicals. Through applying trigonometric identities and the properties of trigonometric functions, cyclotomic equations present a great tool in connecting geometrical construction problems with number theory and algebra.

As we discussed in the previous chapter, there is no formula or algorithm for solving a general equation of degree greater than 4. It's interesting to observe that the cyclotomic equations present the simplest instance of a closed solution of the quintic equation by radicals, and Galois Theory explains why the cyclotomic equation is solvable by radicals for any natural n. Galois Theory also clarifies that constructible lengths must come from base lengths by solutions of some sequence of quadratic equations.

The detailed investigation of the regular polygons constructability and its connections to algebraic equations and to number theory is beyond the scope of our book. We've merely grazed the surface of the fascinating issues related to classical geometrical constructions and their algebraic interpretations. We hope that the reader will explore further.

As we have some practice working with trigonometric identities and are acquainted with the properties of trigonometric functions, it's time to proceed to the next chapter for the study of trigonometric equations.

Exercises

Problem 16. Evaluate without the use of a calculator the value of cos 18°.

Problem 17. Evaluate without the use of a calculator the value of

$$\sin 10° \cdot \sin 20° \cdot \sin 30° \cdot \sin 40° \cdot \sin 50° \cdot \sin 60° \cdot \sin 70° \cdot \sin 80°.$$

Problem 18. Evaluate $\sin^3 x - \cos^3 x$, if $\sin x - \cos x = n$.

Problem 19. Does there exist an angle γ such that $\cos \gamma = a + \frac{1}{a}$ $(a \neq 0)$?

Problem 20. Evaluate $\sin(x + y) \sin(x - y)$, if $\sin x = -\frac{1}{3}$, and $\sin y = -\frac{1}{2}$ where $\frac{3\pi}{2} < x < 2\pi$ and $\frac{3\pi}{2} < y < 2\pi$.

Problem 21. Does there exist an angle α such that $\tan \alpha = 2 + \sqrt{3}$ and $\cot \alpha = 2 - \sqrt{3}$?

Problem 22. Prove that for all x such that $0 < x < \frac{\pi}{4}$ the following equality holds true:

$$1 - \tan x + \tan^2 x - \tan^3 x + \cdots = \frac{\sqrt{2}\cos x}{2\sin\left(\frac{\pi}{4} + x\right)}.$$

Problem 23. Prove that for any n the following identities are true:

$$\cos nx = 2 \cdot \cos x \cdot \cos(n - 1)x - \cos(n - 2)x;$$
$$\sin nx = 2 \cdot \cos x \cdot \sin(n - 1)x - \sin(n - 2)x.$$

(As you prove these identities, you should be able to get the identities for the sine and cosine of double and triple angles as their direct corollaries.)

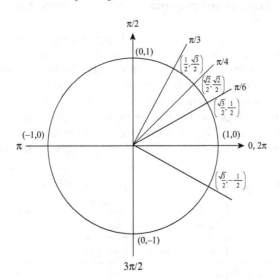

Chapter 7

Trigonometric Equations

Tackling difficult problems in our journey through the book, the most efficient strategy suggests using multiple techniques, integrating algebra, geometry, and trigonometry. When solving algebraic or geometric problems, one often comes across setting up trigonometric equations. Trigonometry helps us establish links, build connections between algebraic and geometrical issues, and enlighten and simplify solutions. In solving trigonometric equations, we broadly utilize various trigonometric identities and properties of the trigonometric functions covered in the previous chapter.

Generally speaking, since each of the trigonometric functions is periodic and therefore is not one-to-one, none of the trigonometric functions have an inverse function. However, considering restricted domains for basic trigonometric functions, we can define the inverse function for each. Those functions are used when you solve trigonometric equations. According to the definition of an inverse function, its domain coincides with the range of the invertible function and the range of the inverse function is the domain of the invertible function. The table below summarizes the definitions and major properties of inverse trigonometric functions:

$\alpha = \arcsin m$	$\alpha = \arccos m$	$\alpha = \arctan m$	$\alpha = \text{arccot}\, m$
domain: $-1 \leq m \leq 1$	$-1 \leq m \leq 1$	$-\infty < m < \infty$	$-\infty < m < \infty$
$\sin \alpha = m$	$\cos \alpha = m$	$\tan \alpha = m$	$\cot \alpha = m$
$\sin(\arcsin m) = m$	$\cos(\arccos m) = m$	$\tan(\arctan m) = m$	$\cot(\text{arccot}\, m) = m$
range: $-\frac{\pi}{2} \leq \alpha \leq \frac{\pi}{2}$	$0 \leq \alpha \leq \pi$	$-\frac{\pi}{2} < \alpha < \frac{\pi}{2}$	$0 < \alpha < \pi$

Before proceeding to solutions of trigonometric equations, let's go over some simple exercises to get acquainted with inverse trigonometric functions and their properties. It is important to remember that the properties of inverse functions in the table work both ways. They can be expressed as $f(f^{-1}(x)) = x$ and $f^{-1}(f(x)) = x$ and hold only in restricted domains. For x-values outside these domains they do not hold. For example, $\arcsin(\sin \pi)$ is equal to 0, not π.

Recalling the values of trigonometric functions of some angles, it's easy to find the values of the inverse functions:

$$\sin \frac{\pi}{4} = \cos \frac{\pi}{4} = \frac{\sqrt{2}}{2}, \text{ therefore } \arcsin \frac{\sqrt{2}}{2} = \arccos \frac{\sqrt{2}}{2} = \frac{\pi}{4},$$

$$\sin \frac{\pi}{3} = \cos \frac{\pi}{6} = \frac{\sqrt{3}}{2}, \text{ therefore } \arcsin \frac{\sqrt{3}}{2} = \frac{\pi}{3}, \arccos \frac{\sqrt{3}}{2} = \frac{\pi}{6},$$

$$\sin \frac{\pi}{6} = \cos \frac{\pi}{3} = \frac{1}{2}, \text{ therefore } \arcsin \frac{1}{2} = \frac{\pi}{6}, \arccos \frac{1}{2} = \frac{\pi}{3},$$

$$\sin \frac{\pi}{2} = \cos 0 = 1, \text{ therefore } \arcsin 1 = \frac{\pi}{2}, \arccos 1 = 0,$$

$$\cos \frac{\pi}{2} = \sin 0 = 0, \text{ therefore } \arccos 0 = \frac{\pi}{2}, \arcsin 0 = 0,$$

$$\tan \frac{\pi}{4} = \cot \frac{\pi}{4} = 1, \text{ therefore } \arctan 1 = \operatorname{arccot} 1 = \frac{\pi}{4},$$

$$\tan \frac{\pi}{3} = \cot \frac{\pi}{6} = \sqrt{3}, \text{ therefore } \arctan \sqrt{3} = \frac{\pi}{3}, \operatorname{arccot} \sqrt{3} = \frac{\pi}{6},$$

$$\tan \frac{\pi}{6} = \cot \frac{\pi}{3} = \frac{\sqrt{3}}{3}, \text{ therefore } \arctan \frac{\sqrt{3}}{3} = \frac{\pi}{6}, \operatorname{arccot} \frac{\sqrt{3}}{3} = \frac{\pi}{3}.$$

A few more important relationships (we invite readers to prove them on their own):

$$\arcsin(-x) = -\arcsin x, \ \arccos(-x) = \pi - \arccos x, \ \arctan(-x) = -\arctan x,$$

$$\operatorname{arccot}(-x) = \pi - \operatorname{arccot} x, \ \arcsin x + \arccos x = \frac{\pi}{2}, \ \arctan x + \operatorname{arccot} x = \frac{\pi}{2}.$$

Problem 1. Evaluate $\cot \left(\arccos\left(-\frac{1}{3}\right)\right)$.

Solution. By definition, $\arccos\left(-\frac{1}{3}\right) = \alpha$ implies that $\cos \alpha = -\frac{1}{3}$. In the interval $\left[\frac{\pi}{2}, \pi\right]$ we have

$$\cot \alpha = \frac{\cos \alpha}{\sin \alpha} = \frac{\cos \alpha}{\sqrt{1 - \cos^2 \alpha}} = \frac{-\frac{1}{3}}{\sqrt{1 - \frac{1}{9}}} = \frac{-\frac{1}{3}}{\frac{\sqrt{8}}{3}} = -\frac{1}{\sqrt{8}} = -\frac{\sqrt{8}}{8}.$$

Problem 2. Evaluate $\arcsin \frac{1}{3} + \arcsin \frac{3}{4}$.

Solution. Let's introduce angles x and y and rephrase the problem. If $\arcsin \frac{1}{3} = x$, then $\sin x = \frac{1}{3}$ and $0 \leq x \leq \frac{\pi}{2}$. If $\arcsin \frac{3}{4} = y$, then $\sin y = \frac{3}{4}$ and $0 \leq y \leq \frac{\pi}{2}$. Our goal is to find the sum $x + y$. We would be able to solve the problem if we manage to evaluate the sine or cosine of the angle $(x + y)$. In this case it does not matter which function to pick. Let's find $\cos(x + y)$. Recalling the identity $\cos(x + y) = \cos x \cos y - \sin x \sin y$, we see that to complete the calculations we need to find the values of $\cos x$ and $\cos y$.

$$\cos x = \sqrt{1 - \sin^2 x} = \sqrt{1 - \frac{1}{9}} = \frac{\sqrt{8}}{3}, \quad 0 \le x \le \frac{\pi}{2}.$$

$$\cos y = \sqrt{1 - \sin^2 y} = \sqrt{1 - \frac{9}{16}} = \frac{\sqrt{7}}{4}, \quad 0 \le y \le \frac{\pi}{2}.$$

Substituting the values into the expression for $\cos(x + y)$ yields

$$\cos(x + y) = \frac{\sqrt{8}}{3} \cdot \frac{\sqrt{7}}{4} - \frac{1}{3} \cdot \frac{3}{4} = \frac{\sqrt{56} - 3}{12} = \frac{2\sqrt{14} - 3}{12}.$$

Therefore,

$$x + y = \arccos \frac{2\sqrt{14} - 3}{12}.$$

Problem 3. Verify the equality $\sin\left(2 \arctan \frac{1}{2}\right) + \tan\left(\frac{1}{2} \arcsin \frac{15}{17}\right) = \frac{7}{5}$.

Solution. Once again, as in the previous two problems, by introducing two angles x and y and rephrasing the problem's question we make it easily understandable. By definition, $\arctan \frac{1}{2} = x$ implies that $\tan x = \frac{1}{2}$, where $0 \le x \le \frac{\pi}{4}$ and so $0 \le 2x \le \frac{\pi}{2}$. By definition, $\arcsin \frac{15}{17} = y$ implies that $\sin y = \frac{15}{17}$, where $0 \le y \le \frac{\pi}{2}$. Our goal is to find the sum $\sin 2x + \tan \frac{y}{2}$. Using the identity $\sin 2x = 2 \sin x \cos x$, $\sin 2x$ will be evaluated if we manage to find the values of $\sin x$ and $\cos x$ knowing the value of $\tan x = \frac{1}{2}$. By definition, $\tan x = \frac{\sin x}{\cos x} = \frac{\sin x}{\sqrt{1 - \sin^2 x}}$. Solving the equation $\frac{\sin x}{\sqrt{1 - \sin^2 x}} = \frac{1}{2}$ gives $2 \sin x = \sqrt{1 - \sin^2 x}$. Squaring both sides gives $4 \sin^2 x = 1 - \sin^2 x$, $5 \sin^2 x = 1$, from which $\sin x = \pm \frac{1}{\sqrt{5}}$. Since only the positive root of the equation satisfies the conditions of the problem, we get $\sin x = \frac{1}{\sqrt{5}}$. Then $\cos x = \sqrt{1 - \sin^2 x} = \frac{2}{\sqrt{5}}$. It follows that

$$\sin 2x = 2 \sin x \cos x = 2 \cdot \frac{1}{\sqrt{5}} \cdot \frac{2}{\sqrt{5}} = \frac{4}{5}.$$

The second task is to get the value of $\tan \frac{y}{2}$ knowing that $\sin y = \frac{15}{17}$. The identity $\tan \frac{y}{2} = \frac{\sin y}{1 + \cos y}$ leads to

$$\tan \frac{y}{2} = \frac{\frac{15}{17}}{1 + \sqrt{1 - \left(\frac{15}{17}\right)^2}} = \frac{\frac{15}{17}}{1 + \frac{8}{17}} = \frac{3}{5}.$$

The last step is to find the sum $\sin 2x + \tan \frac{y}{2}$: $\sin 2x + \tan \frac{y}{2} = \frac{4}{5} + \frac{3}{5} = \frac{7}{5}$. $\frac{7}{5}$ is the number given on the right-hand side of the original equality, $\frac{7}{5} = \frac{7}{5}$. Thus, the equality holds true.

When solving trigonometric equations, it is important to always pay attention to the domain and range of an equation (and functions). Since each trigonometric function is periodic, trigonometric equations usually have an infinite number of solutions unless there are some restrictions on the set of solutions. It is worth noting that the way you write an answer to a trigonometric equation often depends on the method of its solution. Sometimes the proof of the equivalency of two differently written answers might become a complicated problem on its own. It is also critical to carefully follow all the manipulations and simplifications leading to the final outcome to make sure you do not lose any solutions or exclude the extraneous solutions. Usually this happens when you narrow the domain of a new equation in comparison to the original equation or when you get to an equation with a broader domain. The basic strategy for solving a trigonometric equation is to apply trigonometric identities and algebraic techniques to reduce the given equation to an equivalent but simpler and more manageable equation. The three major methods for trigonometric equation solutions are:

1. Modifying the trigonometric equation to some type of algebraic equation. In other words, the original equation has to be transformed into some polynomial of only one trigonometric function as a variable.

2. Modifying the trigonometric equation to a simple trigonometric equation (or equations) of one variable.

3. Applying the properties of trigonometric functions for assessment of left-hand and right-hand sides of an equation to analyze the transition from the equation to a system of simple manageable equations. Usually this is achieved by comparing the ranges of each side of an equation.

Before proceeding to the discussion of each method, let's list the general solutions for the simple trigonometric equations we will be making use of in all of the following problems:

$\sin x = a$:

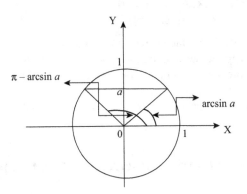

When $|a| > 1$, there are no solutions;

$a = -1$, then $x = -\frac{\pi}{2} + 2\pi k$, $k \in Z$ (k is an integer);

$a = 0$, then $x = \pi k$, $k \in Z$;

$a = 1$, then $x = \frac{\pi}{2} + 2\pi k$, $k \in Z$;

$|a| < 1$, then there are two solutions, $x = \arcsin a + 2\pi k$ and $x = \pi - \arcsin a + 2\pi k = -\arcsin a + \pi(2k + 1)$, $k \in Z$.

 These solutions are expressed by one formula as $x = (-1)^k \arcsin a + \pi k$, $k \in Z$.

$\cos x = a$:

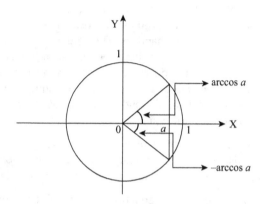

When $|a| > 1$, there are no solutions;

$a = -1$, then $x = \pi + 2\pi k$, $k \in Z$;

$a = 0$, then $x = \frac{\pi}{2} + \pi k$, $k \in Z$;

$a = 1$, then $x = 2\pi k$, $k \in Z$;

$|a| < 1$, then $x = \pm\arccos a + 2\pi k$, $k \in Z$.

$\tan x = a$:

$x = \arctan a + \pi k$, $k \in Z$, for any $a \neq 0$.

$a = 0$, then $x = \pi k$, $k \in Z$;

$\cot x = a$:

$x = \text{arccot}\, a + \pi k$, $k \in Z$, for any $a \neq 0$.

$a = 0$, then $x = \frac{\pi}{2} + \pi k$, $k \in Z$.

 We'll see that sometimes it becomes a real challenge to properly give an answer to some trigonometric equations and do it in the most efficient and simple way. But before we get to those complexities, it is critical to understand and memorize the standard basic expressions listed above.

 At this point we are ready to embark on the analysis of trigonometric equations.

Transformation of a trigonometric equation to a polynomial

Problem 4. Solve the equation

$$\tan x + \tan 2x - \tan 3x = 0.$$

Solution. As in many trigonometric equations, we need to do some preliminary work. The goal is to get to one trigonometric function of the same angle, in this case $\tan x$. First, recall the identity for the tangent of a double angle given in the previous chapter

$$\tan 2x = \frac{2\tan x}{1 - \tan^2 x}. \tag{1}$$

Secondly, in order to express $\tan 3x$ in terms of $\tan x$, we will use the trigonometric identities for a triple angle obtained in the previous chapter. Using the definition of tangent and dividing numerator and denominator by $\cos^3 x$ ($\cos x \neq 0$), we obtain

$$\tan 3x = \frac{\sin 3x}{\cos 3x} = \frac{3\sin x - 4\sin^3 x}{4\cos^3 x - 3\cos x}$$

$$= \frac{3 \cdot \frac{\tan x}{\cos^2 x} - 4\tan^3 x}{4 - 3 \cdot \frac{1}{\cos^2 x}} = \frac{3\tan x \cdot \frac{1}{\cos^2 x} - 4\tan^3 x}{4 - 3 \cdot \frac{1}{\cos^2 x}}.$$

Observing that

$$\frac{1}{\cos^2 x} = \frac{\cos^2 x + \sin^2 x}{\cos^2 x} = 1 + \frac{\sin^2 x}{\cos^2 x} = 1 + \tan^2 x$$

and substituting this value into the last equality leads to

$$\tan 3x = \frac{3\tan x(1 + \tan^2 x) - 4\tan^3 x}{4 - 3(1 + \tan^2 x)} = \frac{3\tan x + 3\tan^3 x - 4\tan^3 x}{4 - 3 - 3\tan^2 x}$$

$$= \frac{3\tan x - \tan^3 x}{1 - 3\tan^2 x}. \tag{2}$$

Substituting (1) and (2) into the original equation gives

$$\tan x + \frac{2\tan x}{1 - \tan^2 x} - \frac{3\tan x - \tan^3 x}{1 - 3\tan^2 x} = 0.$$

Let $\tan x = y$. It follows

$$y + \frac{2y}{1 - y^2} - \frac{3y - y^3}{1 - 3y^2} = 0,$$

$$\frac{(y - y^3)(1 - 3y^2) + 2y(1 - 3y^2) - (3y - y^3)(1 - y^2)}{(1 - y^2)(1 - 3y^2)} = 0,$$

$$\frac{y - 3y^3 - y^3 + 3y^5 + 2y - 6y^3 - 3y + y^3 + 3y^3 - y^5}{(1 - y^2)(1 - 3y^2)} = 0.$$

After combining the like terms in the numerator, we get

$$2y^5 - 6y^3 = 0,$$

$$y \neq \pm 1,$$

$$y \neq \pm\frac{1}{\sqrt{3}}.$$

Solving the equation $2y^5 - 6y^3 = 0$ gives

$$2y^3(y^2 - 3) = 0,$$
$$y^3 = 0 \text{ or } y^2 - 3 = 0.$$

It follows that $y = 0$, $y = \sqrt{3}$, or $y = -\sqrt{3}$. Going back to the variable x and substituting the found values into $\tan x = y$ gives three simple trigonometric equations:

$$\tan x = 0, \quad \tan x = \sqrt{3}, \quad \text{or} \quad \tan x = -\sqrt{3}.$$

Solutions of the first equation are $x = \pi k$, $k \in Z$.
Solutions of the second equation are $x = \frac{\pi}{3} + \pi n$, $n \in Z$.
Solutions of the third equation are $x = -\frac{\pi}{3} + \pi m$, $m \in Z$.
    When you get a few different sets of solutions, as above, you should see if it's possible to combine them in one formula. Analyzing the locations of the answers on the unit circle enables us to express all solutions as $x = \frac{\pi p}{3}$, $p \in Z$.

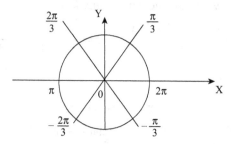

Answer: $\frac{\pi p}{3}$, $p \in Z$.

Problem 5. Solve the equation

$$\cos 2x - 5 \sin x - 3 = 0.$$

Solution. Using the identity $\cos 2x = \cos^2 x - \sin^2 x$ allows transforming this equation to a trigonometric equation of one function with the same argument:

$$\cos^2 x - \sin^2 x - 5 \sin x - 3 = 0,$$
$$1 - \sin^2 x - \sin^2 x - 5 \sin x - 3 = 0,$$
$$1 - 2 \sin^2 x - 5 \sin x - 3 = 0,$$
$$2 \sin^2 x + 5 \sin x + 2 = 0.$$

Substituting $y = \sin x$ yields the quadratic equation $2y^2 + 5y + 2 = 0$.

$$D = 25 - 16 = 9,$$

$$y = \frac{-5 \pm \sqrt{9}}{4},$$

$$y_1 = -\frac{1}{2}, \quad y_2 = -2.$$

Substituting $\sin x$ for y gives two equations: $\sin x = -\frac{1}{2}$ and $\sin x = -2$.

Solutions of the first equation:

$$x = (-1)^k \cdot \left(-\frac{\pi}{6}\right) + \pi k = (-1)^{k+1} \cdot \frac{\pi}{6} + \pi k, \quad k \in Z.$$

The second equation has no solutions because the absolute value of a sine cannot exceed 1.

Answer: $(-1)^{k+1} \cdot \frac{\pi}{6} + \pi k, k \in Z.$

Problem 6. Solve the equation

$$\sin^4 2x + \cos^4 2x = \sin 2x \cos 2x.$$

Solution. There are a few different approaches to this problem. One of them is to complete the square on the left-hand side. Adding and subtracting $2 \sin^2 2x \cos^2 2x$ yields

$$\sin^4 2x + 2 \sin^2 2x \cos^2 2x + \cos^4 2x - 2 \sin^2 2x \cos^2 2x = \sin 2x \cos 2x,$$

$$\underbrace{(\sin^2 2x + \cos^2 2x)^2}_{1} - 2 \sin^2 2x \cos^2 2x = \sin 2x \cos 2x,$$

$$1 - 2 \sin^2 2x \cos^2 2x = \sin 2x \cos 2x,$$

$$1 - \frac{1}{2} \cdot 4 \sin^2 2x \cos^2 2x = \frac{1}{2} \cdot 2 \sin 2x \cos 2x,$$

$$1 - \frac{1}{2} \sin^2 4x = \frac{1}{2} \sin 4x,$$

$$\sin^2 4x + \sin 4x - 2 = 0.$$

Substituting $y = \sin 4x$ gives the quadratic equation $y^2 + y - 2 = 0$, solutions of which by Viète's formulas are $y_1 = 1$, $y_2 = -2$. Therefore, $\sin 4x = 1$ or $\sin 4x = -2$.

Solutions of the first equation:

$$4x = \frac{\pi}{2} + 2\pi k, \quad k \in Z,$$

$$x = \frac{\pi}{8} + \frac{\pi k}{2}, \quad k \in Z.$$

The second equation has no solutions because the absolute value of a sine cannot exceed 1.

Answer: $\frac{\pi}{8} + \frac{\pi k}{2}, k \in Z.$

Transformation into trigonometric equation of one function of one angle

Problem 7. Solve the equation

$$2(\cos 4x - \sin x \cos 3x) = \sin 4x + \sin 2x.$$

Solution. In examining the equation, not only do we have different functions involved, namely sine and cosine, we also have different arguments to contend with, namely $4x$, $3x$, $2x$, and x. Our goal is to make a transition from the original equation to an equation of one trigonometric function of the same argument. Using the identity $\sin x \cos y = \frac{1}{2}(\sin(x + y) + \sin(x - y))$ to modify the product of $\sin x \cos 3x$ gives $\sin x \cos 3x = \frac{1}{2}(\sin 4x + \sin(-2x))$. Hence, the equation can be rewritten as

$$2\left(\cos 4x - \frac{1}{2}(\sin 4x + \sin(-2x))\right) - \sin 4x - \sin 2x = 0,$$

$$2\cos 4x - (\sin 4x - \sin 2x) - \sin 4x - \sin 2x = 0,$$

$$2\cos 4x - \sin 4x + \sin 2x - \sin 4x - \sin 2x = 0,$$

$$2\cos 4x - 2\sin 4x = 0,$$

$$\cos 4x - \sin 4x = 0.$$

Clearly, $\cos 4x \neq 0$ (hopefully readers can easily verify that fact). Then, dividing both sides by $\cos 4x$ gives

$$\tan 4x - 1 = 0,$$

$$\tan 4x = 1,$$

$$4x = \frac{\pi}{4} + \pi k, \quad k \in Z,$$

which yields

$$x = \frac{\pi}{16} + \frac{\pi k}{4}, \quad k \in Z.$$

Answer: $\frac{\pi}{16} + \frac{\pi k}{4}, k \in Z.$

Problem 8. Solve the equation

$$\sin^2 x + \cos^2 3x = 1.$$

Solution. We'll reduce the power of each function on the left side of the equation by making use of the identities $\sin^2 x = \frac{1}{2} - \frac{1}{2}\cos 2x$ and $\cos^2 3x = \frac{1}{2} + \frac{1}{2}\cos 6x$

and rewrite the equation as

$$\frac{1}{2} - \frac{1}{2}\cos 2x + \frac{1}{2} + \frac{1}{2}\cos 6x = 1,$$

$$\frac{1}{2}(\cos 6x - \cos 2x) = 0,$$

$$\cos 6x - \cos 2x = 0.$$

Using the identity for the difference of two cosines yields $-\sin 4x \sin 2x = 0$. Hence, $\sin 4x = 0$ or $\sin 2x = 0$.

Solutions of the first equation are $x = \frac{\pi k}{4}, k \in Z$.

Solutions of the second equation are $x = \frac{\pi n}{2}, n \in Z$.

Clearly, the second set of solutions is contained in the first set of solutions. Therefore, the answer to this problem is $x = \frac{\pi k}{4}, k \in Z$.

Problem 9. Solve the equation

$$\sin x \cos x \cos 2x \cos 8x = \frac{1}{4}\sin 12x.$$

Solution. Multiplying both sides of the equation by 4 and using the identity for the sine of a double angle twice gives

$$2 \cdot (2\sin x \cos x)\cos 2x \cos 8x = 4 \cdot \frac{1}{4}\sin 12x,$$

$$2\sin 2x \cos 2x \cos 8x = \sin 12x,$$

$$(2\sin 2x \cos 2x)\cos 8x = \sin 12x,$$

$$\sin 4x \cos 8x = \sin 12x,$$

$$\sin 4x \cos 8x = \sin(4x + 8x),$$

using the identity for the sine of the sum on the right-hand side yields

$$\sin 4x \cos 8x = \sin 4x \cos 8x + \sin 8x \cos 4x,$$

$$\sin 8x \cos 4x = 0.$$

It follows that $\sin 8x = 0$ or $\cos 4x = 0$.

From the first equation $8x = \pi k, k \in Z$, so, $x = \frac{\pi k}{8}, k \in Z$.

From the second equation $4x = \frac{\pi}{2} + \pi n, n \in Z$, which yields $x = \frac{\pi}{8} + \frac{\pi n}{4} = \frac{\pi(1+2n)}{8}, n \in Z$.

The solutions of the second equation satisfy the first equation as well. Therefore, the answer to this problem is $x = \frac{\pi k}{8}, k \in Z$.

Before proceeding any further with the analyses of trigonometric equations, I suggest we examine one very interesting problem. Its result provides a great non-standard technique called *the auxiliary angle introduction*.

Problem 10. For any real numbers a and b prove the inequality

$$-\sqrt{a^2+b^2} \le a\cos x + b\sin x \le \sqrt{a^2+b^2}.$$

Solution. First, notice that if $a = b = 0$, then the inequality holds true. Let's then consider numbers $a \ne 0$ and $b \ne 0$ and divide all the sides of inequality by the same positive number $\sqrt{a^2+b^2}$:

$$-1 \le \frac{a}{\sqrt{a^2+b^2}}\cos x + \frac{b}{\sqrt{a^2+b^2}}\sin x \le 1. \qquad (*)$$

Clearly, for all real numbers $a \ne 0$, $\frac{a}{\sqrt{a^2+b^2}} \le 1$. Indeed, $\frac{a}{\sqrt{a^2+b^2}} < \frac{a}{\sqrt{a^2}} < 1$. Similarly, $\frac{b}{\sqrt{a^2+b^2}} < 1$. Since the range of the sine function consists of all real numbers not exceeding 1 in absolute value, then there must exist a unique angle φ such that $0 \le \varphi < 2\pi$ and a $\sin\varphi = \frac{a}{\sqrt{a^2+b^2}}$. We know from the Pythagorean Identity that $\sin^2 x + \cos^2 x = 1$ for any real number x. The converse statement is also correct. Therefore, for the above mentioned angle φ its cosine equals $\frac{b}{\sqrt{a^2+b^2}}$, $\cos\varphi = \frac{b}{\sqrt{a^2+b^2}}$. To prove that statement, we merely have to observe that

$$\left(\frac{a}{\sqrt{a^2+b^2}}\right)^2 + \left(\frac{b}{\sqrt{a^2+b^2}}\right)^2 = \frac{a^2+b^2}{a^2+b^2} = 1.$$

Substituting the expressions for $\sin\varphi$ and $\cos\varphi$ into $(*)$ allows us, instead of proving the original inequality, to prove

$$-1 \le \sin\varphi\cos x + \cos\varphi\sin x \le 1.$$

From the trigonometric identity for the sine of the sum of two angles it follows that $\sin\varphi\cos x + \cos\varphi\sin x = \sin(\varphi+x)$. So, the original inequality is simplified to $-1 \le \sin(\varphi+x) \le 1$, which is a true statement because the range of the sine function is the set of all real numbers not exceeding 1 in absolute value, $|\sin(\varphi+x)| \le 1$, and so we conclude that the original inequality holds true.

Problem 11. Solve the equation

$$\sin x + 7\cos x + 7 = 0.$$

Solution. The goal is to transform this equation to an equation of one function with one argument. It can be achieved by two different methods.

Method 1. Use universal trigonometric substitution for the tangent of a half-angle:

$$\sin x = \frac{2\tan\frac{x}{2}}{1+\tan^2\frac{x}{2}}; \quad \cos x = \frac{1-\tan^2\frac{x}{2}}{1+\tan^2\frac{x}{2}} \quad \text{(see chapter 6)}.$$

The above substitutions yield the equation

$$\frac{2\tan\frac{x}{2}}{1+\tan^2\frac{x}{2}} + 7\cdot\frac{1-\tan^2\frac{x}{2}}{1+\tan^2\frac{x}{2}} + 7 = 0.$$

Letting $y = \tan\frac{x}{2}$ and substituting it in the last equation gives

$$\frac{2y}{1+y^2} + 7\cdot\frac{1-y^2}{1+y^2} + 7 = 0,$$
$$2y + 7(1-y^2) + 7(1+y^2) = 0,$$
$$2y + 7 - 7y^2 + 7 + 7y^2 = 0,$$
$$2y = -14,$$
$$y = -7.$$

Substituting this value for $\tan\frac{x}{2}$ gives $\tan\frac{x}{2} = -7$. Therefore, $\frac{x}{2} = -\arctan 7 + \pi k$, from which the solutions are $x = -2\arctan 7 + 2\pi k$, $k \in Z$.

Are we done? Not really. Universal trigonometric substitution is a great tool and justifies its name "universal" because it gives the ability to modify the original equation to an equation with one function $\tan\frac{x}{2}$. However, after the substitution, the new equation has a different domain than the original one. One utilizes universal trigonometric substitution assuming that $x \neq \pi + 2\pi k$, $k \in Z$ (angles for which $\tan\frac{x}{2}$ exists). So, we must now check if those excluded values of x would satisfy the original equation. In this case they will: $\sin\pi + 7\cos\pi + 7 = 0 + 7\cdot(-1) + 7 = 0.$ $0 = 0$ – a true statement. Therefore, the full answer to the original problem will be

$$x = -2\arctan 7 + 2\pi n, \ n \in Z; \quad x = \pi + 2\pi k, \ k \in Z.$$

Method 2. This time we will refer to the introduction of an auxiliary angle, as was explained in problem 10, and solve the equation without changing its domain.

$$\sin x + 7\cos x + 7 = 0.$$

Dividing both sides of the equation by $\sqrt{1+7^2} = \sqrt{50}$ gives

$$\frac{1}{\sqrt{50}}\sin x + \frac{7}{\sqrt{50}}\cos x = -\frac{7}{\sqrt{50}}.$$

There must exist an angle φ, $0 \leq \varphi < 2\pi$, such that $\sin\varphi = \frac{7}{\sqrt{50}}$ and $\cos\varphi = \frac{1}{\sqrt{50}}$. Then, the equation can be rewritten as $\cos\varphi\sin x + \sin\varphi\cos x = -\frac{7}{\sqrt{50}}$, or $\sin(x + \varphi) = -\frac{7}{\sqrt{50}}$, from which $x + \varphi = (-1)^k \arcsin\left(-\frac{7}{\sqrt{50}}\right) + \pi k, \; k \in \mathbb{Z}$.

Recalling that $\varphi = \arcsin\frac{7}{\sqrt{50}}$ and that $\sin x$ is an odd function gives

$$x = -\arcsin\frac{7}{\sqrt{50}} + (-1)^{k+1}\arcsin\frac{7}{\sqrt{50}} + \pi k, \quad k \in \mathbb{Z}.$$

Let's verify that this answer agrees with the result we got by using Method 1. Indeed, if k is an odd number, $k = 2n + 1$, then

$$x = -\arcsin\frac{7}{\sqrt{50}} + \arcsin\frac{7}{\sqrt{50}} + \pi(2n+1) = \pi + 2\pi n, \quad n \in \mathbb{Z}.$$

This is exactly the second solution obtained in Method 1.

If k is an even number, $k = 2n$, then

$$x = -\arcsin\frac{7}{\sqrt{50}} - \arcsin\frac{7}{\sqrt{50}} + 2\pi n = -2\arcsin\frac{7}{\sqrt{50}} + 2\pi n, \quad n \in \mathbb{Z}.$$

So, now we need to see if this solution agrees with the first solution obtained in Method 1. We have to prove that $-2\arcsin\frac{7}{\sqrt{50}} = -2\arctan 7$, or, which is the same, that $\arcsin\frac{7}{\sqrt{50}} = \arctan 7$. If $\arcsin\frac{7}{\sqrt{50}} = \varphi$, then $\sin\varphi = \frac{7}{\sqrt{50}}$ and $\cos\varphi = \frac{1}{\sqrt{50}}$ (since $\cos\varphi = \sqrt{1 - \sin^2\varphi}$). Thus, $\tan\varphi = \frac{\sin\varphi}{\cos\varphi} = \frac{\frac{7}{\sqrt{50}}}{\frac{1}{\sqrt{50}}} = 7$, which implies that $\arctan 7 = \varphi$. It follows that $\arcsin\frac{7}{\sqrt{50}} = \arctan 7$ and we see that these solutions in each case coincide as well.

I want to point out the important conclusions from problem 11:

1. Depending on the method of the solution (universal trigonometric substitution, for example), you have to analyze the domains of the original and newly transformed equation. If you narrowed the domain of the original equation, you must check the potentially missing solutions.

2. You may get a different presentation of the solutions. This does not necessarily mean that an error was made. You should verify the equivalence of outcomes (as we did, comparing $\arcsin\frac{7}{\sqrt{50}}$ to $\arctan 7$).

In addition to auxiliary angle substitution and universal trigonometric substitution in terms of the tangent of a half-angle, another interesting technique appears in applying two more helpful substitutions $y = \sin x + \cos x$ and $z = \cos 2x$. Squaring both sides of the equality $y = \sin x + \cos x$ and using the Pythagorean Identity along with the identity for the sine of a double angle gives

$$y^2 = \underbrace{\sin^2 x + \cos^2 x}_{1} + \underbrace{2\sin x\cos x}_{\sin 2x} = 1 + \sin 2x,$$

from which $\sin 2x = y^2 - 1$. So, we see that this substitution will be efficient in solving equations containing expressions with $(\sin x + \cos x)$ and $\sin 2x$. The substitution $z = \cos 2x$ proves useful in solving equations containing the expressions with $\cos 2x$, $\cos^2 x$, and $\sin^2 x$. Indeed, recalling that $\cos 2x = \cos^2 x - \sin^2 x = 2\cos^2 x - 1 = 1 - 2\sin^2 x$ gives two more expressions in terms or z, $\cos^2 x = \frac{1+z}{2}$, and $\sin^2 x = \frac{1-z}{2}$.

Problem 12. Solve the equation

$$2(\sin x + \cos x) + \sin 2x + 1 = 0.$$

Solution. Introducing the substitution $y = \sin x + \cos x$ and noticing that $\sin 2x = y^2 - 1$ (as shown above) gives $2y + y^2 - 1 + 1 = 0$, or equivalently, $y^2 + 2y = 0$, which leads to $y(y+2) = 0$. Therefore, $y = 0$ or $y = -2$. Substituting these values back into the expression for x yields two equations

$$\sin x + \cos x = 0 \text{ or } \sin x + \cos x = -2.$$

Clearly, $\cos x \neq 0$ (otherwise $\sin x = 1$ or $\sin x = -1$ and $\sin x + \cos x \neq 0$) and dividing both sides of the first equation by $\cos x$ yields $\tan x + 1 = 0$ or equivalently, $\tan x = -1$, from which $x = -\frac{\pi}{4} + \pi k$, $k \in Z$.

It's easy to prove that the second equation, $\sin x + \cos x = -2$, has no solutions. Indeed, dividing both sides by $\sqrt{2}$ gives $\frac{1}{\sqrt{2}}\sin x + \frac{1}{\sqrt{2}}\cos x = \frac{-2}{\sqrt{2}}$; this can be rewritten as $\cos\frac{\pi}{4}\sin x + \sin\frac{\pi}{4}\cos x = -\sqrt{2}$, or equivalently, $\sin\left(x + \frac{\pi}{4}\right) = -\sqrt{2}$. Since $|-\sqrt{2}| > 1$, the last equation has no solutions.

Answer: $-\frac{\pi}{4} + \pi k$, $k \in Z$,

Problem 13. Solve the equation

$$\cos 2x + 4\sin^4 x = 8\cos^6 x.$$

Solution. Using the substitution $z = \cos 2x$ along with the expressions derived above, $\cos^2 x = \frac{1+z}{2}$ and $\sin^2 x = \frac{1-z}{2}$, leads to the equation

$$z + 4 \cdot \left(\frac{1-z}{2}\right)^2 = 8 \cdot \left(\frac{1+z}{2}\right)^3.$$

Rewrite the equation as

$$z + (1-z)^2 = (1+z)^3,$$
$$z + 1 - 2z + z^2 - 1 - 3z - 3z^2 - z^3 = 0,$$
$$z^3 + 2z^2 + 4z = 0,$$
$$z(z^2 + 2z + 4) = 0.$$
$$z = 0 \text{ or } z^2 + 2z + 4 = 0.$$

The second quadratic equation has no solutions because its discriminant $D = -12 < 0$. Substituting $z = 0$ back into $z = \cos 2x$ gives $\cos 2x = 0$, from which $2x = \frac{\pi}{2} + \pi k$, $k \in Z$ and finally, $x = \frac{\pi}{4} + \frac{\pi k}{2}$, $k \in Z$.

Answer: $\frac{\pi}{4} + \frac{\pi k}{2}$, $k \in Z$.

Application of major properties of trigonometric functions for assessment of both sides of an equation

Sometimes, when one solves an equation $f(x) = g(x)$, the ranges of the functions $f(x)$ and $g(x)$ might have just a few common elements or do not intersect at all. If there are no common elements, then clearly the equation has no solutions. If there is a common element, then it has to satisfy both conditions and one should be able to set up a system of equations. Solutions of such a system will be solutions of the original equation. This great technique is also applicable to many other, not necessarily trigonometric equations, as will be demonstrated in later chapters.

Problem 14. Solve the equation

$$2\cos\frac{x}{10} = 2^x + 2^{-x}.$$

Solution. First, note that $2\cos\frac{x}{10} \le 2$ (since $\left|\cos\frac{x}{10}\right| \le 1$ for any real x). On the other hand, $2^x + 2^{-x} \ge 2$, because $2^x > 0$ and $2^{-x} > 0$ and for any positive number a, $a + \frac{1}{a} \ge 2$:

$$a + \frac{1}{a} - 2 = \frac{a^2 - 2a + 1}{a} = \frac{(a-1)^2}{a} \ge 0.$$

Therefore, $a + \frac{1}{a} \ge 2$ for any $a > 0$.

The original equality is possible only when the two equalities hold true at the same time:

$$2\cos\frac{x}{10} = 2,$$
$$2^x + 2^{-x} = 2.$$

Solving the first equation gives

$$2\cos\frac{x}{10} = 2,$$
$$\cos\frac{x}{10} = 1,$$
$$\frac{x}{10} = 2\pi k,\ k \in Z,\ \text{or } x = 20\pi k,\ k \in Z.$$

Solving the second equation by making the substitution $2^x = y > 0$ yields

$$y + \frac{1}{y} = 2,\ \text{or}$$
$$y^2 - 2y + 1 = 0,$$
$$(y-1)^2 = 0,$$

from which $y = 1$, so, then $2^x = 1$, and finally, $x = 0$. Comparing the outcomes of the first and second equations, we conclude that the only common solution is $x = 0$.

Answer: 0.

Problem 15. Solve the equation

$$\sqrt[5]{2(1+\sqrt{x-2})-x} + \sqrt[5]{4(\sqrt{x+1}-1)-x} = \frac{2}{|\cos \pi x|}.$$

Solution. Clearly, the right-hand side is greater or equal than 2, $\frac{2}{|\cos \pi x|} \geq 2$, because for any real x, $|\cos \pi x| \leq 1$. Let's prove that the left-hand side of the equation does not exceed 2. Completing the square leads to

$$2(1+\sqrt{x-2})-x = 2+2\sqrt{x-2}-x = 1-(1-2\sqrt{x-2}+x-2)$$
$$= 1-(1-\sqrt{x-2})^2 \leq 1.$$

In a similar way, $4(\sqrt{x+1}-1)-x = 1-(\sqrt{x+1}-2)^2 \leq 1$. Therefore, the left-hand side of the equation cannot exceed 2. This implies that the original equality is possible only when the following three equalities hold at the same time:

$$\frac{2}{|\cos \pi x|} = 2,$$
$$1-(1-\sqrt{x-2})^2 = 1,$$
$$1-(\sqrt{x+1}-2)^2 = 1.$$

Solving the first equation $\frac{2}{\cos \pi x} = 2$ gives

$$|\cos \pi x| = 1,$$
$$\pi x = \pi k, \ k \in Z.$$

Then $x = k, k \in Z$.

Solving the second equation $1-(1-\sqrt{x-2})^2 = 1$ and noting that $x \geq 2$ (the domain of this equation) gives

$$1-\sqrt{x-2} = 0,$$
$$\sqrt{x-2} = 1,$$
$$x = 3.$$

Solving the third equation $1-(\sqrt{x+1}-2)^2 = 1$ and noting that $x \geq -1$ (the domain of this equation) gives

$$\sqrt{x+1}-2 = 0,$$
$$\sqrt{x+1} = 2,$$
$$x = 3.$$

Comparing all three solutions, we select the only common solution, $x = 3$.

Answer: 3.

Problem 16. Solve the equation

$$\sin^{1965} x + \cos^{1965} x = 1.$$

Solution. The given equation can be rewritten as

$$1 - (\sin^{1965} x + \cos^{1965} x) = 0.$$

Using the Pythagorean Identity and regrouping terms gives

$$\underbrace{\sin^2 x + \cos^2 x}_{1} - \sin^{1965} x - \cos^{1965} x = 0,$$

$$\sin^2 x - \sin^{1965} x + \cos^2 x - \cos^{1965} x = 0,$$

$$(\sin^2 x - \sin^{1965} x) + (\cos^2 x - \cos^{1965} x) = 0,$$

$$\sin^2 x (1 - \sin^{1963} x) + \cos^2 x (1 - \cos^{1963} x) = 0.$$

Analyzing the last equation and referring to the range of the functions sine and cosine, $|\sin x| \le 1$ and $|\cos x| \le 1$ for any real x, we get $1 - \sin^{1963} x \ge 0$ and $1 - \cos^{1963} x \ge 0$. Hence, $\sin^2 x (1 - \sin^{1963} x) \ge 0$ and $\cos^2 x (1 - \cos^{1963} x) \ge 0$. It follows that the left-hand side is a nonnegative number. Equaling to 0 would be possible if and only if every addend equals 0:

$$\sin^2 x (1 - \sin^{1963} x) = 0 \quad \text{and} \quad \cos^2 x (1 - \cos^{1963} x) = 0.$$

The last two equations will have solutions only when the following systems have solutions

$$\begin{array}{ccc} \sin x = 0, & & \sin x = 1, \\ & \text{or} & \\ \cos x = 1 & & \cos x = 0. \end{array}$$

Solutions of the first system:

$$x = \pi l, \quad l \in Z$$
$$x = 2\pi k, \quad k \in Z.$$

Combining the results gives $x = 2\pi k, k \in Z$.
    Solutions of the second system:

$$x = \frac{\pi}{2} + 2\pi n, \quad n \in Z,$$
$$x = \frac{\pi}{2} + \pi m, \quad m \in Z,$$

from which the combined solution is $x = \frac{\pi}{2} + 2\pi n, n \in Z$.

Answer: $x = 2\pi k, k \in Z, x = \frac{\pi}{2} + 2\pi n, n \in Z$.

    We will close this chapter with two historical examples from the 16th century demonstrating the relevance of trigonometry to solving algebraic equations.

Both are associated with the work of François Viète, whose outstanding mathematical accomplishments have been mentioned in previous chapters.

The first is a very interesting technique developed by François Viète and extended by René Descartes for the trigonometric solution of the cubic equation with three real roots. Viète was solving a geometric problem of the division of an angle into an odd number of equal parts. He described how the chord AB of a trisected arc associated with the chord BD (see Figure 1) of a circle of a given radius r, is a solution of a cubic equation having the form $y^3 + py + q = 0$. Viète trisected the arc AB into equal arcs AC = CD = BD. Denoting angle DOB as φ, $\angle DOB = \varphi$, we get $\angle AOB = 3\varphi$. In each of the isosceles triangles AOB and DOB we draw altitudes OH and OM respectively to the sides AB and DB (see Figure 2).

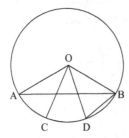   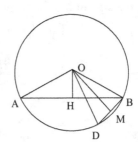

Figure 1   **Figure 2**

It follows that $\angle HOB = \frac{3}{2}\varphi$, and $\angle MOB = \frac{\varphi}{2}$. Let AB = b, DB = c, and OB = r (the radius of the circle).

In the right triangle OMB,

$$\sin \frac{\varphi}{2} = \frac{MB}{OB} = \frac{DB}{2OB} = \frac{c}{2r}.$$

In the right triangle OHB,

$$\sin \frac{3\varphi}{2} = \frac{HB}{OB} = \frac{AB}{2OB} = \frac{b}{2r}.$$

Recalling the identity $\sin 3\alpha = 3\sin\alpha - 4\sin^3\alpha$ (see chapter 6), we can state that $\sin \frac{3\varphi}{2} = 3\sin\frac{\varphi}{2} - 4\sin^3\frac{\varphi}{2}$. Substituting the expressions of $\sin\frac{3\varphi}{2}$ and $\sin\frac{\varphi}{2}$ determined from triangles HOB and MOB gives $\frac{b}{2r} = 3 \cdot \frac{c}{2r} - 4 \cdot \left(\frac{c}{2r}\right)^3$. The last equation can be rewritten as $c^3 - 3r^2c + r^2b = 0$, which establishes the relationship of chords of multiple arcs to the chord of a simple arc. In order to get three distinct roots of the last cubic equation Viète restricted the length of AB $< 2r$. Geometrically this means that the chord cannot coincide with the diameter of the circle. Viète modified the associated cubic equation to an equivalent equation presented in a trigonometric form and solved it.

Consider $y^3 + py + q = 0$, where p and q are nonzero real numbers. Introduce the substitution $y = u \cos x$, where the angle x, as selected by Viète, satisfies $0 < x < \pi$. Then the equation can be rewritten as

$$u^3 \cos^3 x + pu \cos x + q = 0. \tag{1}$$

Now let's select u in such a way that we can utilize the previously obtained identity (see chapter 6) for a cosine of a triple angle,

$$4 \cos^3 x - 3 \cos x = \cos 3x. \tag{2}$$

Note that making the substitution $y = u \sin x$ with the further application of the formula for the sine of a triple angle will lead to the similar results. Choosing $u = 2\sqrt{-\frac{p}{3}}$ (so, $u^3 = -\frac{8p}{3}\sqrt{-\frac{p}{3}}$) and dividing both sides of (1) by $\frac{u^3}{4}$ gives

$$4 \cos^3 x + \frac{4p}{u^2} \cos x + \frac{4q}{u^3} = 0$$

or equivalently,

$$4 \cos^3 x - \frac{4p}{\frac{4p}{3}} \cos x + \frac{4q}{-\frac{8p}{3}\sqrt{-\frac{p}{3}}} = 0,$$

which simplifies to

$$4 \cos^3 x - 3 \cos x - \frac{3q}{2p}\sqrt{-\frac{3}{p}} = 0.$$

Rewriting the last equation as

$$4 \cos^3 x - 3 \cos x = \frac{3q}{2p}\sqrt{-\frac{3}{p}}$$

and comparing it to (2) yields $\cos 3x = \frac{3q}{2p}\sqrt{-\frac{3}{p}}$, from which

$$3x = \arccos\left(\frac{3q}{2p}\sqrt{-\frac{3}{p}}\right) + 2\pi k, \quad k \in \mathbf{Z}.$$

It follows that, for $0 < x < \pi$ and for $\left|\frac{3q}{2p}\sqrt{-\frac{3}{p}}\right| < 1$ (a cosine cannot exceed 1 in absolute value), we get

$$x = \frac{1}{3}\arccos\left(\frac{3q}{2p}\sqrt{-\frac{3}{p}}\right) + \frac{2\pi k}{3}, \quad k \in \mathbf{Z}.$$

When $k = 0, 1, 2$, and $p < 0$ we obtain three distinct real roots of the original cubic equation as

$$y_1 = 2\sqrt{-\frac{p}{3}} \cdot \cos\left(\frac{1}{3}\arccos\left(\frac{3q}{2p}\sqrt{-\frac{3}{p}}\right)\right), \quad \text{when } k = 0,$$

$$y_2 = 2\sqrt{-\frac{p}{3}} \cdot \cos\left(\frac{1}{3}\arccos\left(\frac{3q}{2p}\sqrt{-\frac{3}{p}}\right) + \frac{2\pi}{3}\right), \quad \text{when } k = 1,$$

$$y_3 = 2\sqrt{-\frac{p}{3}} \cdot \cos\left(\frac{1}{3}\arccos\left(\frac{3q}{2p}\sqrt{-\frac{3}{p}}\right) - \frac{2\pi}{3}\right), \quad \text{when } k = 2$$

$\Bigg($ indeed, letting

$$\alpha = \frac{1}{3}\arccos\left(\frac{3q}{2p}\sqrt{-\frac{3}{p}}\right)$$

gives

$$\cos\left(\alpha + \frac{4\pi}{3}\right) = \cos\left(\alpha + \underbrace{2\pi - \frac{2\pi}{3}}_{=\frac{4\pi}{3}}\right) = \cos\left(\left(\alpha - \frac{2\pi}{3}\right) + 2\pi\right)$$

$$= \cos\left(\alpha - \frac{2\pi}{3}\right)\Bigg).$$

The second problem we are going to illustrate was posed by the Flemish mathematician Adriaan van Roomen (1561–1615) in 1593. According to some historians, the problem was mentioned by the Dutch ambassador during a meeting with King Henry IV of France. The Dutch ambassador referred to Adriaan van Roomen's published survey of the outstanding living European mathematicians, in which a complicated equation was proposed for their solution. No living French mathematicians were included in the survey because the author did not believe any of them were capable of solving this equation:

$$x^{45} - 45x^{43} + 945x^{41} - 12300x^{39} + 111150x^{37} - 740259x^{35} + 3764565x^{33}$$
$$- 14945040x^{31} + 46955700x^{29} - 117679100x^{27} + 236030652x^{25}$$
$$- 378658800x^{23} + 483841800x^{21} - 488494125x^{19} + 384942375x^{17}$$
$$- 232676280x^{15} + 105306075x^{13} - 34512075x^{11} + 7811375x^{9}$$
$$- 1138500x^{7} + 95634x^{5} - 3795x^{3} + 45x = \sqrt{\frac{7}{4} - \sqrt{\frac{5}{16}} - \sqrt{\frac{15}{8}} - \sqrt{\frac{45}{64}}}.$$

The problem was shown to François Viète, who managed to solve it by applying the substitution $x = 2\sin\alpha$.

We truly hope that will not be the last sentence you read in the book. Just by looking at van Roomen's frightful equation one might get confused and even feel intimidated, which is totally understandable. We are not going to go over the solution in detail; another whole chapter would need to be devoted to that exercise. However, to give you an idea about Viète's brilliant approach to the solution, we will cover the major steps. The justifications of the details we leave to the ambitious reader.

Viète linked the equation's solution to the geometrical problem of dividing an arc into 45 parts. He realized that the polynomial on the left-hand side represents some expression of $2\sin(45\alpha)$ in terms of $2\sin\alpha$. Hence, making the substitution $x = 2\sin\alpha$ will lead to the solution as soon as an angle α is found such that the expression on the right-hand side equals $2\sin\alpha$.

First, Viète proved that the expression on the right-hand side of the equation equals $2\sin\frac{\pi}{15}$,

$$\sqrt{\frac{7}{4} - \sqrt{\frac{5}{16}} - \sqrt{\frac{15}{8}} - \sqrt{\frac{45}{64}}} = 2\sin\frac{\pi}{15}.$$

Secondly, considering the formula $(-1)^i \cdot \frac{n}{n-i} \cdot \frac{(n-i)!}{i!(n-2i)!} x^{n-2i}$ (where the factorial of a natural number m by definition is the product of numbers from 1 through m, $m! = 1 \cdot 2 \cdot \ldots \cdot m$ and $0! = 1$) and calculating its expression for each $i = 0, 1, \ldots, 22$ and $n = 45$ gives

$$(-1)^0 \cdot \frac{45}{45} \cdot \frac{45!}{0!45!} x^{45} = x^{45}, \text{ for } i = 0,$$

$$(-1)^1 \cdot \frac{45}{44} \cdot \frac{44!}{1!43!} x^{43} = -45x^{43}, \text{ for } i = 1,$$

$$(-1)^2 \cdot \frac{45}{43} \cdot \frac{43!}{2!41!} x^{41} = 945x^{41}, \text{ for } i = 2,$$

and doing similar calculations for $i = 3, \ldots, 22$, one may verify that the formula expresses each term on the left-hand side of the equation. Let's denote

$$\sum_{i=0}^{\lfloor\frac{n}{2}\rfloor}(-1)^i \frac{n}{n-i} \cdot \frac{(n-i)!}{i!(n-2i)!} x^{n-2i} = f_n(x),$$

where the floor function $\lfloor\frac{n}{2}\rfloor$ is the largest integer less than or equal to the ratio $\frac{n}{2}$. The left-hand side of the equation can be expressed by the above formula for $n = 45$ (the floor function $\lfloor\frac{n}{2}\rfloor$ is used because there are 22 different terms in the equation). Expressing it in algebra terms, Viète proved that for any odd $n \geq 1$

$$2\sin(n\alpha) = (-1)^{\frac{n-1}{2}} \cdot f_n(2\sin\alpha).$$

Hence, when $n = 45$,

$$2\sin(45\alpha) = (-1)^{\frac{45-1}{2}} \cdot f_{45}(2\sin\alpha) = f_{45}(2\sin\alpha).$$

Recalling that $f_{45}(2\sin\alpha) = 2\sin\frac{\pi}{15}$, the original equation is reduced to $2\sin(45\alpha) = 2\sin\frac{\pi}{15}$, or equivalently, $\sin(45\alpha) = \sin\frac{\pi}{15}$, from which

$$45\alpha = (-1)^k \underbrace{\arcsin(\sin\frac{\pi}{15})}_{=\frac{\pi}{15}} + \pi k, \quad k \in Z.$$

Dividing both sides by 45 gives $\alpha = (-1)^k \cdot \frac{1}{45} \cdot \frac{\pi}{15} + \frac{1}{45} \cdot \pi k$, $k \in Z$. It follows that $\alpha = (-1)^k \frac{\pi}{675} + \frac{\pi k}{45}$, $k \in Z$. Going back to the substitution $x = 2\sin\alpha$, we see that $x = 2\sin\left((-1)^k \frac{\pi}{675} + \frac{\pi k}{45}\right)$. Using this formula, it's not hard to get all 45 distinct solutions of the equation for $k = 0, 1, \ldots, 44$:

$$k = 0, \text{ then } x_1 = 2\sin\frac{\pi}{675};$$

$$k = 1, \text{ then } x_2 = 2\sin\left(-\frac{\pi}{675} + \frac{\pi}{45}\right);$$

$$\ldots;$$

$$k = 44, \text{ then } x_{45} = 2\sin\left(\frac{\pi}{675} + \frac{44\pi}{45}\right).$$

When $k = 45$, we get again the first root of the equation:

$$x = 2\sin\left(-\frac{\pi}{675} + \frac{45\pi}{45}\right) = 2\sin\left(-\frac{\pi}{675} + \pi\right) = 2\sin\frac{\pi}{675} = x_1.$$

In a similar way we obtain the repetitive roots for every next integer values of k. So, for all $k \geq 45$ there are no new solutions. Viète himself presented only the positive solutions (indicating in passing though that it also has 22 negative solutions), since the idea of negative numbers was accepted by European mathematicians only by the 17th century.

How amazing is this solution?! No wonder van Roomen was so impressed after receiving it from Viète that he traveled to France to meet Viète and to seek his friendship. It is interesting to emphasize the fact that this 45th degree equation has 45 roots. We may guess it was probably one of the first equations showing that polynomial equations of degree n have a maximum of n solutions, which was proved later on in the fundamental theorem of algebra. The approach of introducing a trigonometric function as a substitution for a variable in an equation is not only very powerful and useful technique, but it also demonstrates how important trigonometry's role is in connecting algebra and geometry and emphasizes the significance of trigonometrical identities and equations in problem solving, making the solutions elegant and, more importantly, manageable.

Exercises

Problem 17. Solve the equation $\frac{1}{2}(\cos 5x + \cos 7x) - \cos^2 2x + \sin^2 3x = 0$.

Problem 18. Solve the equation $2(\cos 4x - \sin x \cos 3x) = \sin 4x + \sin 2x$.

Problem 19. Solve the equation $\tan(70° + x) + \tan(20° - x) = 2$.

Problem 20. Evaluate $\cos(2 \arctan 2) - \sin(4 \arctan 3)$.

Problem 21. Solve the equation $\sin^2 4x + \cos^2 x = 2 \sin 4x \cos^4 x$.

Problem 22. Solve the equation $\sin x + \sqrt{3} \cos x = 1$.

Problem 23. Solve the equation $\cos x + 2 \cos 2x + \cdots + n \cos nx = \frac{n^2 + n}{2}$, where n is a natural number.

Trigonometric functions graphs

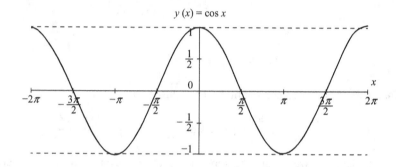

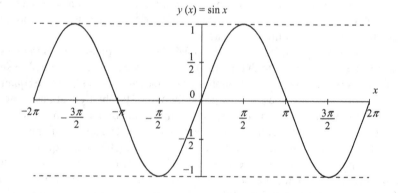

$y(x) = \tan x$

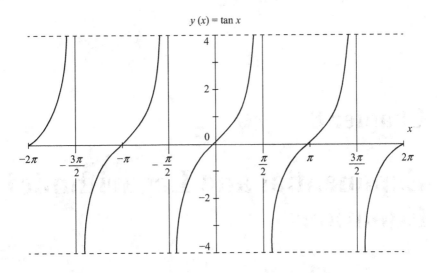

$y(x) = \cot an\ x$

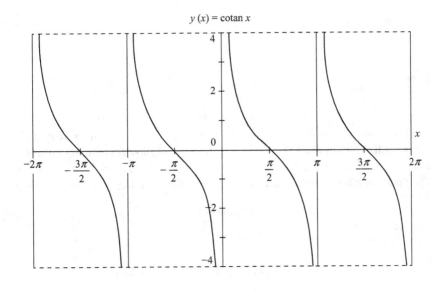

Chapter 8

Exponential and Logarithmic Equations

The function $f(x) = a^x$, where the base constant $a > 0$, is called an exponential function. Its domain is all real numbers; its range is all positive real numbers. If $a = 1$, then $a^x = 1$ for any x. If $x = 0$, then by definition, $a^x = 1$ for any $a > 0$. The major properties of exponential functions follow directly from the definition:

$$a^m a^n = a^{m+n},$$

$$a^m \div a^n = a^{m-n},$$

$$(a^m)^n = a^{mn},$$

$$a^m b^m = (ab)^m,$$

$$a^m \div b^m = \left(\frac{a}{b}\right)^m,$$

$$a^{-m} = \frac{1}{a^m}.$$

If the base of an exponential function $a > 1$, then $f(x) = a^x$ is a growth function. If $0 < a < 1$, then $f(x) = a^x$ is a decay function.

The *logarithm* of b with base a is the number x such that $a^x = b$ for any numbers a and b such that $a > 0$ and $b > 0$, $a \neq 1$. It is written as $\log_a b = x$. So, the statement $\log_a b = x$ is equivalent to the statement $a^x = b$. The function

$f(x) = \log_a x$ is the inverse of the exponential function $f(x) = a^x$. The domain of a logarithmic function is all positive real numbers; its range is all real numbers. Depending on the value of its base a, $f(x) = \log_a x$ is either growth or decay:
If $a > 1$, then $f(x) = \log_a x$ is a growth function.
If $0 < a < 1$, then the function $f(x) = \log_a x$ is a decay function.

There are two types of logarithms distinguished from all other logarithms, the common logarithm and the natural logarithm. The *common logarithm* has base 10, $\log_{10} x = \log x$; the *natural logarithm* has base e, $\log_e x = \ln x$. (Euler's number $e \approx 2.71828$; see more about e and some of its properties in chapter 9.) The major properties of the logarithmic function follow directly from its definition as well:

$$a^{\log_a b} = b,$$
$$\log_a (m \cdot n) = \log_a m + \log_a n, \quad m > 0, n > 0,$$
$$\log_a (m \div n) = \log_a m - \log_a n, \quad m > 0, n > 0,$$
$$\log_a b^m = m \cdot \log_a b,$$
$$\log_a b = \frac{\log_c b}{\log_c a}, \quad \text{for any number } c > 0.$$

As the consequence of the last property, $\log_a b = \frac{1}{\log_b a}$.

A few examples:

$$\log_3 81 = 4, \text{ because } 3^4 = 81,$$
$$\log 1000 = \log 10^3 = 3, \text{ because } 10^3 = 1000,$$
$$\log_3 54 = \log_3 (27 \cdot 2) = \log_3 27 + \log_3 2 = 3 + \log_3 2,$$
$$\log_2 32 - \log_2 8 = \log_2 \frac{32}{8} = \log_2 4 = 2,$$
$$\ln \sqrt{e} = \ln e^{0.5} = 0.5 \cdot \ln e = 0.5 \cdot 1 = 0.5.$$

The properties of exponential and logarithmic functions are broadly used in the solution of exponential and logarithmic equations. By an exponential equation we will understand an equation of the form $a^{f(x)} = a^{g(x)}$, when $a > 0$, $a \neq 1$. The solution of those equations is always reduced to the equation $f(x) = g(x)$. If $g(x) = b$ (a constant), the equation becomes $f(x) = b$.

The main idea behind the solution of the most exponential equations is getting all the terms with the exponents to be expressed as powers with the common base and then transforming the equation to algebraic form.

Problem 1. Solve the equation

$$0.6^x \cdot \left(\frac{25}{9}\right)^{x^2-12} = \left(\frac{27}{125}\right)^3.$$

Solution. To solve this equation, we can express all the terms as exponents with the common base of $\frac{3}{5}$ and make a few simplifications to transform it to an algebraic equation.

$$\left(\frac{3}{5}\right)^x \cdot \left(\frac{5}{3}\right)^{2(x^2-12)} = \left(\frac{3}{5}\right)^{3\cdot3},$$

$$\left(\frac{3}{5}\right)^x \cdot \left(\frac{3}{5}\right)^{-2(x^2-12)} = \left(\frac{3}{5}\right)^9,$$

$$\left(\frac{3}{5}\right)^{x-2x^2+24} = \left(\frac{3}{5}\right)^9.$$

Since the bases are equal, we can equate the exponents and get

$$x - 2x^2 + 24 = 9,$$
$$2x^2 - x - 15 = 0,$$
$$D = 1 + 120 = 121.$$

$x = \frac{1\pm\sqrt{121}}{4}$, and the roots are $x = 3$, $x = -2.5$.

Answer: $-2.5, 3$.

Problem 2. Solve the equation

$$\sqrt[x]{64} - \sqrt[x]{2^{3x+3}} + 12 = 0.$$

Solution. Obviously, $x \neq 0$. The equation can be rewritten in exponential form as

$$2^{\frac{6}{x}} - 2^{\frac{3x+3}{x}} + 12 = 0.$$

Since

$$2^{\frac{3x+3}{x}} = 2^{3+\frac{3}{x}} = 2^3 \cdot 2^{\frac{3}{x}},$$

we can rewrite the equation as

$$(2^{\frac{3}{x}})^2 - 8 \cdot 2^{\frac{3}{x}} + 12 = 0.$$

Substituting $y = 2^{\frac{3}{x}}$ $(y > 0)$ gives the quadratic equation

$$y^2 - 8y + 12 = 0,$$
$$D = 64 - 48 = 16,$$
$$y = \frac{8\pm\sqrt{16}}{2},$$

so $y_1 = 6$, $y_2 = 2$. Therefore, $2^{\frac{3}{x}} = 6$ or $2^{\frac{3}{x}} = 2$.

From the first equation, $\frac{3}{x} = \log_2 6$, $x = \frac{3}{\log_2 6} = 3\log_6 2$.

From the second equation, $\frac{3}{x} = 1$, from which $x = 3$.

Answer: $3, 3\log_6 2$.

Problem 3. Solve the equation

$$27^x - 13 \cdot 9^x + 13 \cdot 3^{x+1} - 27 = 0.$$

Solution. Since $27 = 3^3, 9 = 3^2$, and $3^{x+1} = 3^x \cdot 3$ the equation can be rewritten as

$$3^{3x} - 13 \cdot 3^{2x} + 39 \cdot 3^x - 27 = 0.$$

Substituting $y = 3^x$ $(y > 0)$ gives

$$y^3 - 13y^2 + 39y - 27 = 0.$$

This cubic equation can be solved by regrouping its terms and further factorization.

$$y^3 - 12y^2 - y^2 + 27y + 12y - 27 = 0,$$
$$(y^3 - y^2) - (12y^2 - 12y) + (27y - 27) = 0,$$
$$y^2(y - 1) - 12y(y - 1) + 27(y - 1) = 0,$$
$$(y - 1)(y^2 - 12y + 27) = 0.$$

Factoring the quadratic $y^2 - 12y + 27$ using Viète's formulas gives

$$y^2 - 12y + 27 = (y - 9)(y - 3).$$

Therefore, the equation can be rewritten as

$$(y - 1)(y - 9)(y - 3) = 0.$$

The last equation is reduced to the three simple equations $y - 1 = 0$, $y - 9 = 0$, or $y - 3 = 0$. Therefore, $y = 1$, $y = 9$, or $y = 3$. It follows that

$$3^x = 1, \text{ then } x = 0;$$
$$3^x = 9, \text{ then } x = 2;$$
$$3^x = 3, \text{ then } x = 1.$$

Answer: 0, 1, 2.

Problem 4. Solve the equation

$$9^x + 6^x = 2^{2x+1}.$$

Solution. In the previous equations it was not hard to see the common base for each term in the equation, but in this case all three bases are different. First, let's rewrite it after slightly modifying the bases as

$$3^{2x} + 2^x \cdot 3^x = 2 \cdot 2^{2x}.$$

Here the idea is to divide both sides of the equation by 2^{2x} (we don't have to worry about losing any potential roots since $2^{2x} > 0$ for any real x). The good question to ask is how did you think of this trick? Why do we divide by this specific expression? If you look carefully, you may realize that this decision would lead to a quadratic equation with one variable, which is easily solvable. We will go over this technique in detail in later chapters when we talk about homogenous equations. The equation is transformed to

$$\frac{3^{2x}}{2^{2x}} + \frac{3^x}{2^x} - 2 = 0,$$

$$\left(\frac{3^x}{2^x}\right)^2 + \frac{3^x}{2^x} - 2 = 0.$$

Substituting $y = \frac{3^x}{2^x}$ ($y > 0$) gives the quadratic equation $y^2 + y - 2 = 0$, from which by using Viète's formulas we get $y_1 = 1$, $y_2 = -2$. Substituting the value $y_1 = 1$ for $\frac{3^x}{2^x}$ yields $\frac{3^x}{2^x} = 1$, or $\left(\frac{3}{2}\right)^x = 1$ and $x = 0$. Since $\frac{3^x}{2^x} > 0$, the value $y_2 = -2$ does not satisfy the equation. Hence, the original equation has only one solution, 0.

Answer: 0.

Problem 5. Given that $\tan \alpha = 3^x$, $\tan \beta = 3^{-x}$, and $\alpha - \beta = \frac{\pi}{6}$. Find x.

Solution. Recall from chapter 6 the formula for the tangent of the difference of two angles:

$$\tan(\alpha - \beta) = \frac{\tan \alpha - \tan \beta}{1 + \tan \alpha \tan \beta}.$$

On the other hand, it is given that $\tan(\alpha - \beta) = \tan \frac{\pi}{6} = \frac{1}{\sqrt{3}}$. It follows that $\frac{\tan \alpha - \tan \beta}{1 + \tan \alpha \tan \beta} = \frac{1}{\sqrt{3}}$, or substituting the given values $\tan \alpha = 3^x$ and $\tan \beta = 3^{-x}$, we obtain

$$\frac{3^x - 3^{-x}}{1 + 3^x \cdot 3^{-x}} = \frac{1}{\sqrt{3}},$$

$$\frac{3^x - 3^{-x}}{1 + 1} = \frac{1}{\sqrt{3}},$$

$$\frac{3^x - 3^{-x}}{2} = \frac{1}{\sqrt{3}}.$$

Substituting $y = 3^x$ ($y > 0$) gives $y - \frac{1}{y} - \frac{2}{\sqrt{3}} = 0$, or equivalently,

$$\sqrt{3}y^2 - 2y - \sqrt{3} = 0,$$

$$D = 4 + 12 = 16.$$

$$y = \frac{2 \pm \sqrt{16}}{2\sqrt{3}},$$

so $y_1 = \frac{3}{\sqrt{3}}$, $y_2 = -\frac{1}{\sqrt{3}}$.

The second solution does not satisfy the condition $y > 0$, so we reject it. If $y = \frac{3}{\sqrt{3}} = \sqrt{3} = 3^{\frac{1}{2}}$, then $3^x = 3^{\frac{1}{2}}$, which yields $x = \frac{1}{2}$.

Answer: $\frac{1}{2}$.

Problem 6. Solve the equation

$$1 + a + a^2 + \cdots + a^x = (1+a)(1+a^2)(1+a^4)(1+a^8)(1+a^{16})(1+a^{32}),$$

where a is some natural number.

Solution. Let's consider two choices for a.

Choice 1. $a = 1$. Then the left-hand side becomes

$$1 + a + a^2 + \cdots + a^x = 1 + 1 + \cdots + 1 = x + 1$$

(a is added $(x + 1)$ times).

The right-hand side becomes

$$(1+a)(1+a^2)(1+a^4)(1+a^8)(1+a^{16})(1+a^{32}) = 2 \cdot 2 \cdot 2 \cdot 2 \cdot 2 \cdot 2 = 2^6.$$

Therefore, $x + 1 = 2^6$, which yields $x = 63$.

Choice 2. $a \neq 1$. By multiplying both sides of the equation by $(1 - a)$, on the left-hand side we get

$$(1 + a + a^2 + \cdots + a^x)(1 - a) = 1 - a^{x+1}. \qquad (1)$$

On the right-hand side the adjacent factors will produce a difference of squares six times.

$$(1 - a)(1 + a) = 1 - a^2,$$
$$(1 - a^2)(1 + a^2) = 1 - a^4, \ldots, (1 - a^{32})(1 + a^{32}) = 1 - a^{64}.$$

Therefore,

$$(1 - a)(1 + a)(1 + a^2)(1 + a^4)(1 + a^8)(1 + a^{16})(1 + a^{32}) = 1 - a^{64}. \qquad (2)$$

From (1) and (2) it follows that

$$1 - a^{x+1} = 1 - a^{64},$$
$$a^{x+1} = a^{64},$$
$$x + 1 = 64,$$
$$x = 63.$$

We see that in each case the solution is exactly the same, $x = 63$.

I recommend remembering the trick used in this problem—multiplying both sides of the equation by the same number not equal to 0. It might be useful when you try to get rid of multiple exponents as was done in the problem above. By applying the formula for the difference of squares a few times, we were able to significantly simplify the equation. This technique is similar to adding and subtracting the same number when you try to complete a perfect square and then factor polynomials. There is no universal rule about when to apply it. In every problem you need to make a judgment about how useful and helpful it is in a specific case.

Problem 7. Solve the equation

$$(\sin^{2x} 18° + \cos^{2x} 36°) \cdot 4^x - 2 = 0.$$

Solution. As we calculated earlier, $\sin 18° = \frac{\sqrt{5}-1}{4}$. Using it in the formula for the cosine of a double angle gives

$$\cos 36° = \cos^2 18° - \sin^2 18° = 1 - 2\sin^2 18° = 1 - 2\left(\frac{\sqrt{5}-1}{4}\right)^2$$

$$= 1 - 2 \cdot \frac{5 - 2\sqrt{5} + 1}{16} = \frac{8 - 6 + 2\sqrt{5}}{8} = \frac{\sqrt{5}+1}{4}.$$

Substituting the values for $\sin 18°$ and $\cos 36°$ in the equation and observing that

$$\left(\frac{\sqrt{5} \pm 1}{2}\right)^2 = \frac{5 + 1 \pm 2\sqrt{5}}{4} = \frac{6 \pm 2\sqrt{5}}{4} = \frac{3 \pm \sqrt{5}}{2}$$

yields

$$(\sin^{2x} 18° + \cos^{2x} 36°) \cdot 4^x - 2 = \left(\left(\frac{\sqrt{5}-1}{4}\right)^{2x} + \left(\frac{\sqrt{5}+1}{4}\right)^{2x}\right) \cdot 4^x - 2$$

$$= \frac{(\sqrt{5}-1)^{2x}}{4^x} + \frac{(\sqrt{5}+1)^{2x}}{4^x} - 2$$

$$= \left(\frac{\sqrt{5}-1}{2}\right)^{2x} + \left(\frac{\sqrt{5}+1}{2}\right)^{2x} - 2$$

$$= \left(\frac{3-\sqrt{5}}{2}\right)^x + \left(\frac{3+\sqrt{5}}{2}\right)^x - 2.$$

Hence, the original equation can be though of as

$$\left(\frac{3-\sqrt{5}}{2}\right)^x + \left(\frac{3+\sqrt{5}}{2}\right)^x - 2 = 0.$$

Note that

$$\left(\frac{3-\sqrt{5}}{2}\right) \cdot \left(\frac{3+\sqrt{5}}{2}\right) = \frac{9-5}{4} = 1.$$

Therefore,

$$\frac{3-\sqrt{5}}{2} = \frac{1}{\frac{3+\sqrt{5}}{2}}.$$

Substituting $\left(\frac{3-\sqrt{5}}{2}\right)^x = y$ $(y \neq 0)$ gives $\left(\frac{3+\sqrt{5}}{2}\right)^x = \frac{1}{y}$. It follows that $y + \frac{1}{y} - 2 = 0.$ $y^2 - 2y + 1 = 0$, or equivalently, $(y-1)^2 = 0$, from which $y = 1$. Thus, $\left(\frac{3-\sqrt{5}}{2}\right)^x = 1$, giving $x = 0$.

Answer: 0.

Solving logarithmic equations usually requires using properties of logarithms. Once you have used the properties to condense any logarithm expressions in the equation, you can solve the problem by changing the logarithmic equation into an exponential or polynomial equation. In the solution of the following logarithmic equations we will explore several different techniques.

1 Logarithm's definition

No special tricks are to be employed in this case. Just use the definition of a logarithm to convert the equation into exponential form. Next, use properties of exponential functions to solve the resulting equation.

Problem 8. Solve the equation

$$\log_2(9^x + 3^{2x-1} - 2^{x+0.5}) = 3.5 + x.$$

Solution. Using the definition of a logarithm gives

$$9^x + 3^{2x-1} - 2^{x+0.5} = 2^{3.5+x},$$

$$3^{2x} + \frac{1}{3} \cdot 3^{2x} = \sqrt{2} \cdot 2^x + 8\sqrt{2} \cdot 2^x,$$

$$3^{2x}\underbrace{\left(1 + \frac{1}{3}\right)}_{\frac{4}{3}} = 2^x\underbrace{\left(\sqrt{2} + 8\sqrt{2}\right)}_{9\sqrt{2}}.$$

Dividing both sides by 2^x gives

$$\frac{3^{2x}}{2^x} = \frac{9\sqrt{2}}{\frac{4}{3}},$$

$$\frac{9^x}{2^x} = \frac{27\sqrt{2}}{4},$$

$$\left(\frac{9}{2}\right)^x = \left(\frac{9}{2}\right)^{\frac{3}{2}},$$

$$x = \frac{3}{2}.$$

It is easier to verify the solution by substituting $\frac{3}{2}$ into the original equation, instead of finding the equation's domain. Calculating the left-hand side gives

$$\log_2\left(9^{\frac{3}{2}} + 3^{2\cdot\frac{3}{2}-1} - 2^{\frac{3}{2}+0.5}\right) = \log_2(27+9-4) = \log_2 32 = 5.$$

Calculating the right-hand side gives $3.5+1.5 = 5$. We see that $5 = 5$.

Answer: $\frac{3}{2}$.

2 Taking the logarithm of the same base of each side of an equation

Problem 9. Solve the equation

$$x^{\frac{\log x+5}{3}} = 10^{5+\log x}.$$

Solution. The domain of this equation is all positive real numbers, $x > 0$. If two positive quantities are equal, then the logarithms of each side with any base must be equal as well. The decision of how to select the base of the logarithm depends on a problem's conditions. In this case it makes sense to apply the common logarithm:

$$\log x^{\frac{\log x+5}{3}} = \log 10^{5+\log x}.$$

Using the property $\log_a b^m = m \cdot \log_a b$, we can rewrite the equation as

$$\frac{\log x+5}{3} \cdot \log x = (5+\log x) \cdot \underbrace{\log 10}_{=1},$$

$$\frac{\log x+5}{3} \cdot \log x = 5+\log x.$$

Multiplying both sides by 3 gives

$$(\log x+5)\log x = 15+3\log x.$$

Introducing the new variable $y = \log x$ gives

$$(y+5)y = 15+3y,$$
$$y^2+2y-15 = 0.$$

Solving the quadratic equation by Viète's formulas gives $y_1 = -5$, $y_2 = 3$. Hence, $\log x = -5$, from which $x = 10^{-5}$ or $\log x = 3$, from which $x = 10^3$. Both roots are positive numbers and belong to the equation's domain.

Answer: 10^{-5}, 10^3.

3   Converting logarithmic equations to exponential equations

The goal is to simplify an equation by using the properties of logarithms to a single logarithm term with a further conversion into exponential form or just dropping the logarithms of the same base on both sides of the equation.

Problem 10. Solve the equation

$$\log(5-x)+2\log\sqrt{3-x} = 1.$$

Solution. To start, we need to determine the domain of this equation:

$$5-x > 0,$$
$$3-x > 0.$$

The domain will be all real numbers less than 3, $x < 3$. Using the properties $\log_a b^m = m \cdot \log_a b$ and $\log_a(m \cdot n) = \log_a m + \log_a n$ and substituting 1 in the right-hand side for log 10 gives

$$\log(5-x)+\log(\sqrt{3-x})^2 = 1,$$
$$\log(5-x)+\log(3-x) = \log 10,$$
$$\log((5-x)(3-x)) = \log 10.$$

At this point we can drop the common logarithms and get

$$(5-x)(3-x) = 10,$$
$$x^2 - 8x + 5 = 0,$$
$$D = 64 - 20 = 44.$$

$x = \frac{8\pm\sqrt{44}}{2}$. Then $x = 4+\sqrt{11}$ or $x = 4-\sqrt{11}$. We reject the first root since it does not belong to the domain of the equation, because $4+\sqrt{11} > 3$. The second root is in the domain; therefore, the equation has only the solution $4-\sqrt{11}$.

Answer: $4-\sqrt{11}$.

Problem 11. Solve the equation

$$\log(3^x - 2^{4-x}) = 2 + \frac{1}{4}\log 16 - \frac{1}{2}x\log 4.$$

Solution. The domain of the equation is determined from the solutions of the inequality $3^x - 2^{4-x} > 0$. Since $2^{4-x} = \frac{16}{2^x}$, it can be rewritten as $3^x - \frac{16}{2^x} > 0$ or equivalently as $\frac{6^x-16}{2^x} > 0$. The denominator on the left-hand side of the inequality is positive number for any real x, $2^x > 0$. Therefore, the domain of the equation consists of all real numbers satisfying the inequality $6^x - 16 > 0$, from which

$x > \log_6 16$. Note that $\frac{1}{4}\log 16 = \frac{1}{4}\log 2^4 = \log 2$ and $\frac{1}{2}x\log 4 = \frac{1}{2}x\log 2^2 = x\log 2 = \log 2^x$. The equation can be rewritten as

$$\log(3^x - 2^{4-x}) = 2 + \log 2 - \log 2^x,$$

or equivalently as

$$\log(3^x - 2^{4-x}) - \log 2 + \log 2^x = 2.$$

Applying properties of the sum and the difference of logarithms gives

$$\log \frac{(3^x - 2^{4-x}) \cdot 2^x}{2} = 2.$$

Applying the definition of logarithm yields

$$\frac{(3^x - 2^{4-x}) \cdot 2^x}{2} = 10^2,$$

$$(3^x - 2^{4-x}) \cdot 2^x = 200,$$

$$6^x - 2^{4-x} \cdot 2^x = 200,$$

$$6^x - 2^4 = 200,$$

$$6^x = 216,$$

$$x = 3.$$

Obviously, $3 > \log_6 16$, so it belongs to the equation's domain.

Answer: 3.

4 Application of the property $a^{\log_a b} = b$

The reason we consider this property's application in a separate section is, first of all, because it is a very powerful property; secondly, it is a universal tool for getting rid of logarithms as soon as you have the base of the exponent equal to the logarithm's base at the exponent's degree.

Problem 12. Solve the equation

$$6^{(\log_6 x)^2} + x^{\log_6 x} = 12.$$

Solution. First, determining the domain, we see that x must be a positive number, $x > 0$. Rewrite the equation as

$$\underbrace{\left(6^{\log_6 x}\right)}_{=x}{}^{\log_6 x} + x^{\log_6 x} = 12,$$

$$x^{\log_6 x} + x^{\log_6 x} = 12,$$

$$2x^{\log_6 x} = 12,$$

$$x^{\log_6 x} = 6.$$

Now finding the logarithm with the base 6 of both sides of the equation yields

$$\log_6 x \log_6 x = \log_6 6,$$
$$(\log_6 x)^2 = 1.$$

Then $\log_6 x = 1$ or $\log_6 x = -1$.

Solving each equation, we obtain that $x = 6$ or $x = \frac{1}{6}$. Both roots are positive and belong to the equation's domain.

Answer: 6, $\frac{1}{6}$.

Before proceeding to the next problem, let's prove one more compelling property of logarithms:

$$a^{\log_c b} = b^{\log_c a}.$$

Letting $a^{\log_c b} = m$, then finding the logarithm with base c from both sides of the equality gives

$$\log_c a^{\log_c b} = \log_c m,$$
$$\log_c b \cdot \log_c a = \log_c m,$$
$$\log_c b \cdot \log_c a = \log_c a^{\log_c b},$$
$$\log_c b \cdot \log_c a = \log_c b \cdot \log_c a,$$
$$\log_c b \cdot \log_c a = \log_c a \cdot \log_c b,$$
$$\log_c a^{\log_c b} = \log_c b^{\log_c a},$$

so $a^{\log_c b} = b^{\log_c a}$, which is the desired result.

Problem 13. Solve the equation

$$5^{\log x} = 50 - x^{\log 5}.$$

Solution. Finding the equation's domain, we see that $x > 0$. As we just proved, $x^{\log 5} = 5^{\log x}$, so

$$5^{\log x} = 50 - 5^{\log x},$$
$$2 \cdot 5^{\log x} = 50,$$
$$5^{\log x} = 25,$$

or equivalently,

$$5^{\log x} = 5^2,$$

which leads to

$$\log x = 2,$$
$$x = 100.$$

It belongs to the equation's domain.

Answer: 100.

5 Transformation to the common base

In the solutions of the problems below the goal will be to use properties of logarithms to convert all logarithms in an equation to logarithms of the same base. Then one may apply substitution to simplify the equation and transform it to a manageable form. While changing the logarithm's base formula, it is possible to lose some potential solutions. Therefore, it is important to check the solutions before finishing the problem.

Problem 14. Solve the equation

$$20 \log_{4x} \sqrt{x} + 7 \log_{16x} x^3 - 3 \log_{\frac{x}{2}} x^2 = 0.$$

Solution. The domain of the equation is all positive real numbers, $x > 0$. As we can see, all the logarithms in this equation have different bases. However, the expressions under each logarithm might be presented as a power of x. The key to the solution will be to make a conversion from the original equation to an equation with logarithms with the same bases. As a result of the conversion to logarithms with base x, we narrow the domain of the original equation for the value of $x = 1$, which means we may lose that value as a possible solution. Substituting 1 for x in the equation yields

$$20 \log_4 \sqrt{1} + 7 \log_{16} 1^3 - 3 \log_{\frac{1}{2}} 1^2 = 20 \cdot 0 + 7 \cdot 0 - 3 \cdot 0 = 0,$$

$0 = 0$ is true. Thus, it is verified that $x = 1$ is a solution of the equation. Now, as we already found one of the solutions, proceed with the modification of the equation

$$20 \log_{4x} \sqrt{x} + 7 \log_{16x} x^3 - 3 \log_{\frac{x}{2}} x^2 = 0,$$

$$20 \cdot \frac{1}{2} \log_{4x} x + 7 \cdot 3 \log_{16x} x - 3 \cdot 2 \log_{\frac{x}{2}} x = 0,$$

$$10 \log_{4x} x + 21 \log_{16x} x - 6 \log_{\frac{x}{2}} x = 0,$$

$$\frac{10}{\log_x (4x)} + \frac{21}{\log_x (16x)} - \frac{6}{\log_x \left(\frac{x}{2}\right)} = 0,$$

$$\frac{10}{\log_x x + \log_x 4} + \frac{21}{\log_x x + \log_x 16} - \frac{6}{\log_x x + \log_x 2^{-1}} = 0,$$

$$\frac{10}{1 + 2 \log_x 2} + \frac{21}{1 + 4 \log_x 2} - \frac{6}{1 - \log_x 2} = 0.$$

Substituting $y = \log_x 2$ gives

$$\frac{10}{1 + 2y} + \frac{21}{1 + 4y} - \frac{6}{1 - y} = 0.$$

After finding the common denominator and combining like terms, we get the quadratic equation

$$26y^2 - 3y - 5 = 0,$$

where $y \neq -\frac{1}{2},\ y \neq -\frac{1}{4},\ y \neq 1.$

$$D = 9 + 520 = 529.$$

$$y = \frac{3 \pm \sqrt{529}}{52},$$

so, $y = \frac{1}{2}$ or $y = -\frac{5}{13}$. Then, $\log_x 2 = \frac{1}{2}$, which yields $x = 4$; or $\log_x 2 = -\frac{5}{13}$, which yields $x = 2^{-\frac{13}{5}} = \frac{1}{2^{\frac{13}{5}}} = \frac{1}{4\sqrt[5]{8}}$. All the roots belong to the equation's domain.

Answer: $1, 4, \frac{1}{4\sqrt[5]{8}}$.

The final topic of this chapter is the discussion of solutions of the equations in the form $f(x)^{\varphi(x)} = f(x)^{g(x)}$, where $f(x)$, $\varphi(x)$, and $g(x)$ are some functions of a variable x. The solutions of these equations come to exploring two cases:

1) $f(x) = 1$,

2) $\varphi(x) = g(x)$.

Problem 15. Solve the equation

$$(x - 1)^{3x+1} = (x - 1)^{2x+4}.$$

Solution. If the base $x - 1 = 1$, from which $x = 2$, then $1^{3x+1} = 1^{2x+4}$ is true. Therefore, the first solution of the equation is $x = 2$. To find the other solutions, equate the exponents and consider the equation

$$3x + 1 = 2x + 4,$$

$$x = 3.$$

Answer: 2, 3.

Problem 16. Solve the equation

$$x^{x^2+1} = x^5.$$

Solution. $x = 1$ satisfies the equation. The other solutions will be derived by equating the exponents, which gives the equation

$$x^2 + 1 = 5, \text{ or } x^2 = 4.$$

$$x = \pm 2.$$

Answer: 1, −2, 2.

Problem 17. Solve the equation

$$x^{\sqrt{x-2}} = x^2.$$

Solution. First, let's determine the domain of this equation. Solving the inequality $x - 2 \geq 0$ yields $x \geq 2$. Clearly, in this case, $x = 1$ will not be a solution of the equation, because 1 does not belong to the domain of the equation. The next step is to solve the equation by equating the exponents: $\sqrt{x-2} = 2$. Squaring each side gives $x - 2 = 4$, from which $x = 6$.

Answer: 6.

Problem 18. Solve the equation

$$(\tan x)^{\cos^2 x} = (\cot x)^{\sin x}.$$

Solution. The equation can be rewritten in such a way that the powers have the same base on each side:

$$(\tan x)^{\cos^2 x} = (\tan x)^{-\sin x}.$$

Now we follow the standard procedure and explore two cases:

1) The base on each side equals 1, $\tan x = 1$. Then $x = \frac{\pi}{4} + \pi k$, $k \in Z$.

2) The exponents are equal to each other:

$$\cos^2 x = -\sin x,$$
$$1 - \sin^2 x = -\sin x,$$
$$\sin^2 x - \sin x - 1 = 0.$$

Substituting $y = \sin x$ gives $y^2 - y - 1 = 0$.

$$D = 1 + 4 = 5,$$
$$y = \frac{1 \pm \sqrt{5}}{2}.$$

Therefore,

$$y = \frac{1 + \sqrt{5}}{2} \text{ or } y = \frac{1 - \sqrt{5}}{2}.$$

Finally, we have to solve two equations for x:

$$\sin x = \frac{1 + \sqrt{5}}{2} \text{ or } \sin x = \frac{1 - \sqrt{5}}{2}.$$

The first equation has no solutions because $\frac{1+\sqrt{5}}{2} > 1$ and it is beyond the range of the sine function (recall that $|\sin x| \leq 1$). We leave it to readers to verify that $\left|\frac{1-\sqrt{5}}{2}\right| < 1$ and therefore the second equation has solutions $x = (-1)^n \arcsin \frac{1-\sqrt{5}}{2} + \pi n$, $n \in Z$.

Answer: $x = \frac{\pi}{4} + \pi k$, $k \in Z$, and $x = (-1)^n \arcsin \frac{1-\sqrt{5}}{2} + \pi n$, $n \in Z$.

Exercises

Problem 19. Solve the equation $3^{2x+4} + 45 \cdot 6^x - 9 \cdot 2^{2x+2} = 0$.

Problem 20. Solve the equation $2^{\sin^2 x} + 4 \cdot 2^{\cos^2 x} = 6$.

Problem 21. Solve the equation $2 + 3 \log_5 2 - x = \log_5(3^x - 5^{2-x})$.

Problem 22. Solve the equation $\log(81 \cdot \sqrt[3]{3^{x^2-8x}}) = 0$.

Problem 23. Solve the equation $3 \log_x 4 + 2 \log_{4x} 4 + 3 \log_{16x} 4 = 0$.

Problem 24. Solve the equation $(x + 1)^{x^2+3x} = (x + 1)^{10x-12}$.

Problem 25. Solve the equation $(x^2 - x - 1)^{x^2-1} = 1$.

Chapter 9

Classic Inequalities

One of the very efficient and powerful techniques often used in the solution of many non-standard equations, including multivariable equations, which will be explored in the next chapter, is the application of classic inequalities. Most often they prove useful in the "assessment-evaluation" problems, whenever we try to evaluate the maximum or the minimum value of an expression, which allows us to obtain elegant and efficient solutions. In this chapter we will show a few examples of non-routine problems offered on various mathematical contests in the past, the solutions of which are derived using classic inequalities.

But first, we have to introduce the classic inequalities standing behind the suggested methods.

The Arithmetic Mean–Geometric Mean (AM–GM) inequality

The arithmetic mean of any n nonnegative real numbers is greater than or equal to their geometric mean. The two means are equal if and only if all the numbers are equal:

$$\frac{a_1 + a_2 + \cdots + a_n}{n} \geq \sqrt[n]{a_1 \cdot a_2 \cdot \ldots \cdot a_n}.$$

This inequality is also called *Cauchy's Inequality* after the prominent French mathematician Augustin-Louis Cauchy (1789–1857).

Proof. The theorem can be proved by induction. If $n = 2$, we need to prove that $\frac{a_1+a_2}{2} \geq \sqrt{a_1 \cdot a_2}$. The square of a real number is a nonnegative number. Therefore, $(\sqrt{a_1} - \sqrt{a_2})^2 \geq 0$, or equivalently, $a_1 + a_2 - 2\sqrt{a_1 a_2} \geq 0$, which leads to $\frac{a_1+a_2}{2} \geq \sqrt{a_1 \cdot a_2}$. The equality holds only when the difference is equal to 0, or $a_1 = a_2$. The desired statement for two numbers is proved. Assume now that the inequality is true for $n = m$. Let's prove that with that assumption, it also has to be true for the number $2m$. Indeed,

$$
\begin{aligned}
\frac{a_1 + a_2 + \cdots + a_{2m}}{2m} &= \frac{\frac{a_1+a_2}{2} + \frac{a_3+a_4}{2} + \cdots + \frac{a_{2m-1}+a_{2m}}{2}}{m} \\
&\geq \sqrt[m]{\frac{a_1 + a_2}{2} \cdot \frac{a_3 + a_4}{2} \cdots \cdots \frac{a_{2m-1} + a_{2m}}{2}} \\
&\geq \sqrt[m]{\sqrt{a_1 \cdot a_2} \cdot \sqrt{a_3 \cdot a_4} \cdots \cdots \sqrt{a_{2m-1} \cdot a_{2m}}} \\
&= \sqrt[2m]{a_1 \cdot a_2 \cdots \cdots a_{2m}}.
\end{aligned}
$$

As the equality holds for $n = 2$, it will be true for $n = 4, n = 8, \ldots, n = 2^p$ for any natural number p. Let's now consider any natural number $n \neq 2^p$. There must exist some natural number s such that $n + s = 2^p$. According to our induction assumption and just proved statement, we get that

$$
\frac{a_1 + a_2 + \cdots + a_n + a_{n+1} + \cdots + a_{n+s}}{n + s} \geq \sqrt[n+s]{a_1 \cdot a_2 \cdots \cdots a_{n+s}}.
$$

If in the last inequality we set $a_{n+1} = \cdots = a_{n+s} = \frac{a_1+a_2+\cdots+a_n}{n}$ then

$$
\frac{a_1 + a_2 + \cdots + a_n + \frac{(a_1+a_2+\cdots+a_n)s}{n}}{n + s} \geq \sqrt[n+s]{a_1 \cdot a_2 \cdots \cdots a_n \cdot \left(\frac{a_1 + a_2 + \cdots + a_n}{n}\right)^s}.
$$

Let's denote

$$
A_n = \frac{a_1 + a_2 + \cdots + a_n}{n}. \tag{*}
$$

It follows that

$$
\begin{aligned}
&\frac{a_1 + a_2 + \cdots + a_n + \frac{(a_1+a_2+\cdots+a_n)s}{n}}{n + s} \\
&= \frac{a_1(n + s) + a_2(n + s) + \cdots + a_n(n + s)}{n(n + s)} = \frac{a_1 + a_2 + \cdots + a_n}{n} = A_n.
\end{aligned}
$$

Therefore,

$$
A_n \geq \sqrt[n+s]{a_1 \cdot a_2 \cdots \cdots a_n \cdot \left(\frac{a_1 + a_2 + \cdots + a_n}{n}\right)^s} = \sqrt[n+s]{a_1 \cdot a_2 \cdots \cdots a_n \cdot (A_n)^s}.
$$

Raising both sides of the last inequality to the power $(n+s)$ leads to $A_n^{n+s} \geq a_1 \cdot a_2 \cdot \ldots \cdot a_n \cdot (A_n)^s$. Dividing both sides by $(A_n)^s$ $((A_n)^s > 0)$ gives $A_n^n \geq a_1 \cdot a_2 \cdot \ldots \cdot a_n$. Substituting into the last inequality the expression of A_n from the equality (∗) and finding the nth root of both sides, we arrive at the desired result

$$\frac{a_1 + a_2 + \cdots + a_n}{n} \geq \sqrt[n]{a_1 \cdot a_2 \cdot \ldots \cdot a_n}.$$

Therefore, by mathematical induction the inequality holds for all natural numbers n. Clearly, if all the numbers a_n are equal, the inequality becomes an equality. Assume now that some two numbers are not equal. For example, $a_1 \neq a_2$. If that is the case, it's not hard to show that the equality will not hold. Indeed,

$$\frac{a_1 + a_2 + \cdots + a_n}{n} = \frac{\frac{a_1+a_2}{2} + \frac{a_1+a_2}{2} + a_3 + \cdots + a_n}{n}$$

$$\geq \sqrt[n]{\left(\frac{a_1 + a_2}{2}\right)^2 \cdot a_3 \cdot \ldots \cdot a_n}.$$

Since $a_1 \neq a_2$, then $\frac{a_1+a_2}{2} > \sqrt{a_1 \cdot a_2}$ and so, it follows that

$$\sqrt[n]{\left(\frac{a_1 + a_2}{2}\right)^2 \cdot a_3 \cdot \ldots \cdot a_n} > \sqrt[n]{a_1 \cdot a_2 \cdot \ldots \cdot a_n},$$

which yields

$$\frac{a_1 + a_2 + \cdots + a_n}{n} > \sqrt[n]{a_1 \cdot a_2 \cdot \ldots \cdot a_n}.$$

So, if not all the numbers a_n are equal to each other, the equality will not hold. In other words, the equality holds only when $a_1 = a_2 = \cdots = a_n$, as was to be proved.   □

Cauchy-Bunyakovsky inequality

For any real numbers the following inequality is true:

$$(a_1 b_1 + a_2 b_2 + \cdots + a_n b_n)^2 \leq (a_1^2 + a_2^2 + \cdots + a_n^2)(b_1^2 + b_2^2 + \cdots + b_n^2).$$

Equality holds only when numbers a_k and b_k are multiples of each other. This classic inequality is named after Augustin-Louis Cauchy and the Ukrainian mathematician Viktor Bunyakovsky (1804–1889). It also bears the name the *Cauchy–Schwartz inequality*. The inequality for sums was published by Cauchy in 1821. Viktor Bunyakovsky proved it for integrals in 1859. The modern proof of the integral inequality was given by German mathematician Karl Hermann Amandus Schwarz (1843–1921) in 1888.

Proof. If all the numbers a_i equal 0, $a_i = 0$ $(i = 1, \ldots, n)$, the inequality becomes $0 \leq 0$, which is obviously true. Assume now that at least one of the numbers a_i does not equal 0, i.e., $a_1^2 + a_2^2 + \cdots + a_n^2 > 0$. We will introduce an auxiliary quadratic polynomial with variable x

$$(a_1 x + b_1)^2 + (a_2 x + b_2)^2 + \cdots + (a_n x + b_n)^2.$$

It can be rewritten as

$$ax^2 + 2bx + c = (a_1 x + b_1)^2 + (a_2 x + b_2)^2 + \cdots + (a_n x + b_n)^2, \qquad (1)$$

where

$$\begin{aligned} a &= a_1^2 + a_2^2 + \cdots + a_n^2, \\ b &= a_1 b_1 + a_2 b_2 + \cdots + a_n b_n, \quad \text{and} \\ c &= b_1^2 + b_2^2 + \cdots + b_n^2. \end{aligned} \qquad (2)$$

The right-hand side of the equality (1) is greater than or equal to 0 as the sum of nonnegative numbers. Thus, the left-hand side has to be greater than or equal to 0 as well, $ax^2 + bx + c \geq 0$. That will be possible only when the discriminant of the quadratic trinomial is less than or equal to 0. This implies that $D = 4b^2 - 4ac \leq 0$, from which, $b^2 \leq ac$. Substituting the expressions for the coefficients a, b, and c from formulas (2) into the last inequality yields

$$(a_1 b_1 + a_2 b_2 + \cdots + a_n b_n)^2 \leq (a_1^2 + a_2^2 + \cdots + a_n^2)(b_1^2 + b_2^2 + \cdots + b_n^2),$$

which is what had to be proved.

Equality is possible only when $D = 0$. The quadratic trinomial then will have two equal roots $x_1 = x_2$ and so,

$$ax_1^2 + 2bx_1 + c = (a_1 x_1 + b_1)^2 + (a_2 x_1 + b_2)^2 + \cdots + (a_n x_1 + b_n)^2 = 0.$$

This equality holds only when each addend on the left-hand side equals 0 (because, being a square, each addend is a nonnegative number): $a_k x_1 + b_k = 0$ $(k = 1, \ldots, n)$. The last equality can be rewritten as $\frac{b_k}{a_k} = -x_1$. The number $x_1 \neq 0$ is the root of the quadratic trinomial, so the numbers a_k and b_k are proportional with the ratio $(-x_1)$, which proves the fact that equality holds only when the numbers a_k and b_k are multiples of each other. $\square$

I want to emphasize the non-standard method applied in this proof, the technique of the introduction of an auxiliary polynomial, which worked well for us a few times before. In all the previous chapters our main goal was to find the roots of an equation. However, sometimes to solve a problem, you need to do the opposite task, to construct a polynomial, whose roots are the numbers that we are given. A similar problem, problem 9, was considered in chapter 2. This powerful method proved to be effective again; it helped not just significantly simplify the solution of a problem, but made it look elegant and beautiful.

Bernoulli's inequality is named after the Swiss mathematician Jacob Bernoulli (1654–1705).

For any real number $a \geq -1$ and any natural number n, $(1+a)^n \geq 1+na$.

Proof. We need to consider two choices: when the right-hand side of the inequality is a negative number and when it is a positive number. If $1+na < 0$, then the inequality holds. Indeed, it is given that $a \geq -1$, thus, $(1+a)^n \geq 0 > 1+na$. Let's now work with $1+na > 0$. Consider the set of numbers: $\underbrace{1, 1, \ldots, 1}_{n-1}, 1+na$

(number 1 is selected $n-1$ times). According to the AM-GM inequality,

$$\frac{1+1+\cdots+1+(1+na)}{n} \geq \sqrt[n]{(1+na)\cdot 1\cdot 1\cdot\ldots\cdot 1},$$

or equivalently,

$$\frac{n+na}{n} \geq \sqrt[n]{(1+na)},$$

which yields $(1+a)^n \geq 1+na$, and we arrived at the desired result.

Bernoulli's inequality can be extended for real exponents.

For any real numbers a and b, such that $a > -1$, the following inequalities hold true:

$(1+a)^b \leq 1+ab$, when $0 \leq b \leq 1$,

$(1+a)^b \geq 1+ab$, when $b \in\ ]-\infty; 0] \cup [1; +\infty[$.

In some texts Bernoulli's inequality is used in an alternative form:

$(1-a)^b \geq 1-ab$, for $b \geq 1$ and $0 \leq a \leq 1$,

$(1-a)^b \leq 1-ab$, for $0 \leq b \leq 1$ and $0 \leq a \leq 1$.

Finally, we introduce one more classic inequality relationship for the *harmonic mean*, *quadratic mean*, and their relationships with the AM-GM inequality. The *harmonic mean* H of the positive numbers $a_1, a_2, \ldots, a_n$ is defined as

$$H = \frac{n}{\frac{1}{a_1} + \frac{1}{a_2} + \cdots + \frac{1}{a_n}}.$$

The *Root mean square* (abbreviated RMS), also known as the *quadratic mean*, is defined as the square root of the arithmetic mean of the squares of a set of numbers

$$RMS = \sqrt{\frac{a_1^2 + a_2^2 + \cdots + a_n^2}{n}}.$$

The following inequalities are true for any positive numbers a_k ($k = 1, 2, \ldots, n$) and any natural number n:

$$\frac{n}{\frac{1}{a_1} + \frac{1}{a_2} + \cdots + \frac{1}{a_n}} \leq \sqrt[n]{a_1 \cdot a_2 \cdot \ldots \cdot a_n} \leq \frac{a_1 + a_2 + \cdots + a_n}{n} \leq \sqrt{\frac{a_1^2 + a_2^2 + \cdots + a_n^2}{n}}.$$

We will prove the inequalities for two positive numbers and we invite readers to prove the general case on their own. In the case of two positive numbers a and b

$$\frac{2}{\frac{1}{a}+\frac{1}{b}} \le \sqrt{ab} \le \frac{a+b}{2} \le \sqrt{\frac{a^2+b^2}{2}}.$$

Proof. First, we will show that $\frac{2}{\frac{1}{a}+\frac{1}{b}} \le \sqrt{ab}$. Rewrite the inequality as $\frac{2ab}{b+a} \le \sqrt{ab}$ and find the difference

$$\sqrt{ab} - \frac{2ab}{b+a} = \frac{b\sqrt{ab}+a\sqrt{ab}-2ab}{b+a} = \frac{\sqrt{ab}(b+a-2\sqrt{ab})}{b+a}$$

$$= \frac{\sqrt{ab}(\sqrt{b}-\sqrt{a})^2}{b+a}.$$

Clearly, the last expression is greater than 0, $\frac{\sqrt{ab}(\sqrt{b}-\sqrt{a})^2}{b+a} \ge 0$, because $\sqrt{ab} > 0$, $(\sqrt{b}-\sqrt{a})^2 \ge 0$, and $b+a > 0$. Therefore, $\sqrt{ab} - \frac{2ab}{b+a} \ge 0$ or equivalently, $\frac{2ab}{b+a} \le \sqrt{ab}$, which is the desired result.

The last step is to prove that $\frac{a+b}{2} \le \sqrt{\frac{a^2+b^2}{2}}$. It is equivalent to proving that $\left(\frac{a+b}{2}\right)^2 \le \frac{a^2+b^2}{2}$. Once again, we will consider the difference

$$\frac{a^2+b^2}{2} - \left(\frac{a+b}{2}\right)^2 = \frac{2a^2+2b^2-a^2-2ab-b^2}{4} = \frac{a^2+b^2-2ab}{4}$$

$$= \frac{(a-b)^2}{4} \ge 0.$$

Hence, the statement $\left(\frac{a+b}{2}\right)^2 \le \frac{a^2+b^2}{2}$ holds true.

Since the middle inequality is true, as proved above for the AM-GM inequality, the proof for all four inequalities is completed. □

This property has an interesting geometrical interpretation:

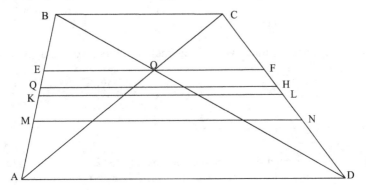

Let ABCD be the trapezoid with the bases $AD = a$ and $BC = b$. The following statements are true:

1. The length of the middle-line KL equals the arithmetic mean of a and b:
   $KL = \frac{a+b}{2}$.

2. The length of the segment QH equals the geometric mean of a and b:
   $QH = \sqrt{ab}$. The segment QH is parallel to the bases and divides the trapezoid ABCD into two similar trapezoids BCHQ and QHDA.

3. The length of the segment EF equals the harmonic mean of a and b:
   $EF = \frac{2}{\frac{1}{a}+\frac{1}{b}}$. Segment EF is parallel to the bases and passes through the point of intersection of the diagonals of the trapezoid ABCD.

4. The length of the segment MN equals the quadratic mean of a and b:
   $MN = \sqrt{\frac{a^2+b^2}{2}}$. MN is parallel to the bases and splits the trapezoid ABCD into two trapezoids of equal areas.

Though the first three properties are easy to prove, the last one presents a nice challenge. You may find one of the solutions to this in the chapter "Areas of similar polygons" in my book *Geometrical Kaleidoscope*, Dover Publications, 2017.

Now that we are armed with the classic inequalities, let's illustrate their applications in problem solving.

Problem 1. Prove that $a \sin x + b \cos x \leq \sqrt{a^2 + b^2}$.

Proof. According to Cauchy-Bunyakovsky inequality,

$$(a \sin x + b \cos x)^2 \leq (a^2 + b^2)(\sin^2 x + \cos^2 x) = (a^2 + b^2) \cdot 1 = a^2 + b^2.$$

Therefore, we proved an even more powerful statement than was required (compare this to the solution offered in chapter 7):

$$-\sqrt{a^2 + b^2} \leq a \sin x + b \cos x \leq \sqrt{a^2 + b^2}.$$

Problem 2. Prove that for any positive numbers a_i $(i = 1, 2, \ldots, n)$ and any natural number n the following inequality is true

$$(a_1 + a_2 + \cdots + a_n)\left(\frac{1}{a_1} + \frac{1}{a_2} + \cdots + \frac{1}{a_n}\right) \geq n^2.$$

Proof. We will demonstrate two different methods. First, we will apply the AM-GM inequality twice:

$$a_1 + a_2 + \cdots + a_n \geq n \cdot \sqrt[n]{a_1 \cdot a_2 \cdot \ldots \cdot a_n}. \tag{1}$$

$$\frac{1}{a_1} + \frac{1}{a_2} + \cdots + \frac{1}{a_n} \geq n \cdot \sqrt[n]{\frac{1}{a_1} \cdot \frac{1}{a_2} \cdot \ldots \cdot \frac{1}{a_n}}. \tag{2}$$

All the numbers on each side of the inequalities (1) and (2) are positive. Therefore, by multiplying these inequalities, the inequality sign in the resulting inequality will not change:

$$(a_1 + a_2 + \cdots + a_n)\left(\frac{1}{a_1} + \frac{1}{a_2} + \cdots + \frac{1}{a_n}\right)$$

$$\geq n^2 \cdot \sqrt[n]{a_1 \cdot a_2 \cdots a_n} \cdot \underbrace{\sqrt[n]{\frac{1}{a_1} \cdot \frac{1}{a_2} \cdots \frac{1}{a_n}}}_{1} = n^2,$$

which completes the proof.

The second method is to utilize Cauchy-Bunyakovsky inequality for the two sets of numbers $\sqrt{a_1}, \sqrt{a_2}, \ldots, \sqrt{a_n}$ and $\frac{1}{\sqrt{a_1}}, \frac{1}{\sqrt{a_2}}, \ldots, \frac{1}{\sqrt{a_n}}$:

$$\left((\sqrt{a_1})^2 + (\sqrt{a_2})^2 + \cdots + (\sqrt{a_n})^2\right) \cdot \left(\left(\frac{1}{\sqrt{a_1}}\right)^2 + \left(\frac{1}{\sqrt{a_2}}\right)^2 + \cdots + \left(\frac{1}{\sqrt{a_n}}\right)^2\right)$$

$$\geq \left(\sqrt{a_1} \cdot \frac{1}{\sqrt{a_1}} + \sqrt{a_2} \cdot \frac{1}{\sqrt{a_2}} + \cdots + \sqrt{a_n} \cdot \frac{1}{\sqrt{a_n}}\right)^2$$

$$= \left(\underbrace{1 + 1 + \cdots + 1}_{=n}\right)^2 = n^2.$$

Clearly, $(\sqrt{a_i})^2 = a_i$ and $\left(\frac{1}{\sqrt{a_i}}\right)^2 = \frac{1}{a_i}$ for $i = 1, 2, \ldots, n$. Thus, we arrive at our desired result of

$$(a_1 + a_2 + \cdots + a_n)\left(\frac{1}{a_1} + \frac{1}{a_2} + \cdots + \frac{1}{a_n}\right) \geq n^2.$$

Problem 3. Find the minimum value of the expression $x^4 + y^4 + z^4 - 4xyz$, where x, y, and z are real numbers.

Solution. Applying the AM-GM inequality to four terms x^4, y^4, z^4, and 1 (all are nonnegative numbers) gives

$$x^4 + y^4 + z^4 + 1 \geq 4\sqrt[4]{x^4 \cdot y^4 \cdot z^4 \cdot 1} = 4xyz.$$

Therefore, $x^4 + y^4 + z^4 - 4xyz \geq -1$. Equality will be achieved when $x^4 = y^4 = z^4 = 1$ and $xyz \geq 0$. The minimum value of the given expression equals -1. It is attained for any combination of numbers $(x, y, z) = (1, 1, 1), (1, -1, -1), (-1, 1, -1), (-1, -1, 1)$.

Problem 4. Which is larger, 99! or 50^{99}?

Solution. By definition, $n!$ it is the product of all numbers from 1 to n. Thus, $99! = 1 \cdot 2 \cdot 3 \cdot \ldots \cdot 97 \cdot 98 \cdot 99$. We can rewrite it by grouping the products of the numbers equidistant from both ends, $99! = (1 \cdot 99) \cdot (2 \cdot 98) \cdot \ldots \cdot (49 \cdot 51) \cdot 50$. Applying the AM-GM inequality, we get

$$99! = (1 \cdot 99) \cdot (2 \cdot 98) \cdot \ldots \cdot (49 \cdot 51) \cdot 50$$
$$< \left(\frac{1+99}{2}\right)^2 \cdot \left(\frac{2+98}{2}\right)^2 \cdot \ldots \cdot \left(\frac{49+51}{2}\right)^2 \cdot 50$$
$$= 50^2 \cdot 50^2 \cdot \ldots \cdot 50^2 \cdot 50 = 50^{2 \cdot 49 + 1} = 50^{99}.$$

Therefore, we conclude that $99! < 50^{99}$.

Problem 5. The polynomial $P(x) = x^n + a_1 x^{n-1} + \cdots + a_{n-1} x + 1$ with the nonnegative coefficients $a_1, a_2, \ldots, a_{n-1}$ has n real roots. Prove that $P(2) \geq 3^n$.

Solution. Since it is given that all the coefficients of $P(x)$ are nonnegative numbers, then none of its n roots can be positive. Denoting those negative roots as $x_1, \ldots, x_n$, let's consider the opposite numbers $y_1 = -x_1, \ldots, y_n = -x_n$. The polynomial can be expressed then as $P(x) = (x + y_1) \cdot (x + y_2) \cdot \ldots \cdot (x + y_n)$, where $y_i > 0$, $i = 1, 2, \ldots, n$. Applying the AM-GM inequality n times, $2 + y_i = 1 + 1 + y_i \geq 3\sqrt[3]{1 \cdot 1 \cdot y_i} = 3\sqrt[3]{y_i}$, for $i = 1, 2, \ldots, n$ and observing that by Viète's formulas for the polynomial $P(x)$, $y_1 \cdot y_2 \cdot \ldots \cdot y_n = 1$ (see chapter 2), we obtain

$$P(2) = (2 + y_1) \cdot \ldots \cdot (2 + y_n) \geq 3\sqrt[3]{y_1} \cdot \ldots \cdot 3\sqrt[3]{y_n}$$
$$= 3^n \cdot \sqrt[3]{y_1 \cdot \ldots \cdot y_n} = 3^n \cdot 1 = 3^n,$$

as required to be proved.

As we get acquainted with some applications of classic inequalities to problem solving, let's turn to the main subject of the book, the equations.

Problem 6. Solve the equation

$$\frac{1}{\cos x} + \cos^2 x + \sin x = 3\sqrt[3]{\frac{1}{2}\sin 2x},$$

when $0 \leq x < \frac{\pi}{2}$.

Solution. Most likely, the straightforward approach to this equation would lead nowhere. Even though it's easy to convert all the trigonometric functions to the function of one variable x (if you use, for example, the identity for the sine of a double angle on the right-hand side), it is not clear what to do next. So, it's not a bad idea to analyze the left-hand side by using the AM-GM inequality and transform the sum on the left to a product. Recalling that $\sin 2x = 2\cos x \cdot \sin x$, we get

$$\frac{1}{\cos x} + \cos^2 x + \sin x \geq 3\sqrt[3]{\frac{1}{\cos x} \cdot \cos^2 x \cdot \sin x}$$

$$= 3\sqrt[3]{\cos x \cdot \sin x} = 3\sqrt[3]{\frac{1}{2}\sin 2x},$$

which represents the expression on the right-hand side of the given equation. The equality will be possible if and only if all three addends on the left-hand side are equal

$$\frac{1}{\cos x} = \cos^2 x,$$

$$\frac{1}{\cos x} = \sin x,$$

$$\cos^2 x = \sin x.$$

Consider the second equation, $\frac{1}{\cos x} = \sin x$, and modify it to $\sin x \cos x = 1$. Multiplying both sides by 2 gives $2\sin x \cos x = 2$, or equivalently, $\sin 2x = 2$. This equation has no solutions because the range of the sine function consists of all real numbers not exceeding 1 in absolute value. Therefore, we may ignore the first and the third equation in our system and conclude that the original equation has no solutions.

Problem 7. Solve the equation

$$(\sin 54° - \cos 72°)^x + 4^{x^2} + \sqrt{2} = 3\sqrt[3]{5^{-\log_5 2^{x-2x^2-\frac{1}{2}}}}.$$

Solution. Before making a plan for the solution, let's simplify each side of the equation as much as possible. We start with the left-hand side.

Applying the cofunction identities and the double-angle identity $\cos 2\alpha = 1 - 2\sin^2\alpha$ gives

$$\sin 54° - \cos 72° = \sin(90° - 36°) - \cos(90° - 18°)$$

$$= \cos 36° - \sin 18° = 1 - 2\sin^2 18° - \sin 18°. \qquad (1)$$

Using the results from chapter 6, where we calculated the value of $\sin 18°$ as $\sin\frac{\pi}{10} = \sin 18° = \frac{\sqrt{5}-1}{4}$ and substituting it into (1) gives

$$\sin 54° - \cos 72° = 1 - 2\cdot\left(\frac{\sqrt{5}-1}{4}\right)^2 - \frac{\sqrt{5}-1}{4}$$

$$= 1 - 2\cdot\frac{6-2\sqrt{5}}{16} - \frac{\sqrt{5}-1}{4} = 1 - \frac{3-\sqrt{5}}{4} - \frac{\sqrt{5}-1}{4}$$

$$= \frac{4-3+\sqrt{5}-\sqrt{5}+1}{4} = \frac{2}{4} = \frac{1}{2} = 2^{-1}.$$

The right-hand side is simplified by using a property of logarithms

$$3\sqrt[3]{5^{-\log_5 2^{x-2x^2-\frac{1}{2}}}} = 3\sqrt[3]{5^{\log_5 2^{-\left(x-2x^2-\frac{1}{2}\right)}}} = 3\sqrt[3]{2^{2x^2-x+\frac{1}{2}}}.$$

Thus, the original equation can be rewritten as

$$2^{-x} + 2^{2x^2} + 2^{\frac{1}{2}} = 3\sqrt[3]{2^{2x^2-x+\frac{1}{2}}}.$$

Each number on the left-hand side is positive; therefore we may apply the AM-GM inequality transforming the sum on the left to a product. We have

$$2^{-x} + 2^{2x^2} + 2^{\frac{1}{2}} \geq 3\sqrt[3]{2^{-x} \cdot 2^{2x^2} \cdot 2^{\frac{1}{2}}} = 3\sqrt[3]{2^{2x^2-x+\frac{1}{2}}}.$$

Note that $3\sqrt[3]{2^{2x^2-x+\frac{1}{2}}}$ is the given expression on the right-hand side of the original equation. The equality is possible only when $2^{-x} = 2^{2x^2} = 2^{\frac{1}{2}}$, which leads to the following system of equations:

$$
\begin{aligned}
-x &= 2x^2, \\
-x &= \tfrac{1}{2}, \qquad\text{or equivalently} \\
2x^2 &= \tfrac{1}{2}.
\end{aligned}
\qquad
\begin{aligned}
x(2x+1) &= 0, \\
x &= -\tfrac{1}{2}, \\
x^2 &= \tfrac{1}{4}.
\end{aligned}
$$

The solutions of the first equation are 0 and $-\tfrac{1}{2}$.
The solution of the second equation is $-\tfrac{1}{2}$.
The solutions of the third equation are $\tfrac{1}{2}$ and $-\tfrac{1}{2}$.

Only the common solution of $-\tfrac{1}{2}$ satisfies all three equations at the same time. The domain of the original equation is all real numbers. Therefore, $-\tfrac{1}{2}$ is the solution of the original equation.

Answer: $-\tfrac{1}{2}$.

The following two problems present examples of more difficult challenges offered on national math Olympiads.

Problem 8.   (Bulgarian national math Olympiad, 1984). The polynomial $P(x) = ax^n - ax^{n-1} + c_2 x^{n-2} + \cdots + c_{n-2}x^2 - n^2bx + b$ has exactly n positive roots. Prove that they are all equal.

Solution. It is given that $P(x)$ has n positive roots. Therefore, its degree cannot be less than n, which implies $a \neq 0$. Denoting the roots of $P(x)$ as $x_1, x_2, \ldots, x_n$ and applying Viète's formulas, we get

$$x_1 + x_2 + \cdots + x_n = -\frac{a}{-a} = 1,$$

$$\sum_{i=1}^{n} x_1 \cdot x_2 \cdot \ldots \cdot x_{i-1} \cdot x_{i+1} \cdot \ldots \cdot x_n = (-1)^n n^2 \cdot \frac{b}{a},$$

$$x_1 \cdot x_2 \cdot \ldots \cdot x_n = (-1)^n \cdot \frac{b}{a}.$$

From the above formulas, clearly, $b \neq 0$. Let's observe that n^2 can be represented as

$$n^2 = 1 \cdot \frac{(-1)^n \cdot n^2 \cdot \frac{b}{a}}{(-1)^n \cdot \frac{b}{a}}. \tag{1}$$

Substituting into the fraction $\frac{(-1)^n \cdot n^2 \cdot \frac{b}{a}}{(-1)^n \cdot \frac{b}{a}}$ the values from Viète's formulas expressions, it can be rewritten as

$$\frac{(-1)^n \cdot n^2 \cdot \frac{b}{a}}{(-1)^n \cdot \frac{b}{a}} = \frac{x_2 \cdot x_3 \cdot \ldots \cdot x_n + x_1 \cdot x_3 \cdot \ldots \cdot x_n + \cdots + x_1 \cdot x_2 \cdot \ldots \cdot x_{n-1}}{x_1 \cdot x_2 \cdot \ldots \cdot x_n}$$

$$= \left(\frac{1}{x_1} + \frac{1}{x_2} + \cdots + \frac{1}{x_n} \right).$$

Also note that from the first of Viète's formulas, 1 can be expressed as $1 = x_1 + x_2 + \cdots + x_n$. Furthermore, applying the AM-GM inequality to (1) gives

$$n^2 = 1 \cdot \frac{(-1)^n \cdot n^2 \cdot \frac{b}{a}}{(-1)^n \cdot \frac{b}{a}} = (x_1 + x_2 + \cdots + x_n) \left(\frac{1}{x_1} + \frac{1}{x_2} + \cdots + \frac{1}{x_n} \right)$$

$$\geq (n \cdot \sqrt[n]{x_1 \cdot x_2 \cdot \ldots \cdot x_n}) \left(n \cdot \sqrt[n]{\frac{1}{x_1} \cdot \frac{1}{x_2} \cdot \ldots \cdot \frac{1}{x_n}} \right)$$

$$= n^2 \cdot \sqrt[n]{\left(x_1 \cdot \frac{1}{x_1} \right) \left(x_2 \cdot \frac{1}{x_2} \right) \cdot \ldots \cdot \left(x_n \cdot \frac{1}{x_n} \right)} = n^2.$$

This is possible only for $x_1 = x_2 = \cdots = x_n = \frac{1}{n}$, which is the result we need.

Problem 9. (Canadian national math Olympiad, 1982, modified). Find the positive solutions of the equation $\frac{1-x^a}{1-x} = (1+x)^{a-1}$ for $x \neq 1$ and $0 < a < 1$.

Solution. Instead of attempting to find positive solutions of the equation, we will prove they do not exist. Let $b = 1 - a$. From the given conditions, $0 < b < 1$. First, we will investigate the solutions of the equation for x such that $0 < x < 1$. Applying

Bernoulli's inequality for the exponent b, $0 < b < 1$, we get $(1+x)^b \leq 1+bx$. Consequently,

$$(1+x)^{-b} \geq \frac{1}{1+bx}. \tag{1}$$

Applying an alternative form of Bernoulli's inequality to the expression $(1+b)x^b$ for $0 < x < 1$ (which implies $0 < 1-x < 1$) gives

$$(1+b)x^b = (1+b)(1-(1-x))^b \leq (1+b)(1-b(1-x))$$
$$= 1-b+bx+b-b^2+b^2x = 1+bx+b^2(x-1) < 1+bx$$

(because $(x-1) < 0$ for $0 < x < 1$).

The last inequality can be rewritten as $1+bx > (1+b)x^b$. Now, going back to the substitution $b = 1-a$ for any given values of x and a, we obtain $1+bx > (1+b)x^{1-a}$, or equivalently, $1+bx > \frac{(1+b)x}{x^a}$. Multiplying both sides of the last inequality by the positive number x^a leads to the inequality

$$(1+bx)\cdot x^a > (1+b)x \text{ or equivalently, } x^a + bx^{a+1} > (1+b)x. \tag{2}$$

Let's now consider the difference $(1+x)^{a-1} - \frac{1-x^a}{1-x}$. Using inequality (1) gives

$$(1+x)^{a-1} - \frac{1-x^a}{1-x} = (1+x)^{-b} - \frac{1-x^a}{1-x} \geq \frac{1}{1+bx} - \frac{1-x^a}{1-x}$$
$$= \frac{1-x+(x^a-1)(1+bx)}{(1+bx)(1-x)}$$
$$= \frac{1-x+x^a-1+bx^{a+1}-bx}{(1+bx)(1-x)}$$
$$= \frac{x^a+bx^{a+1}-(1+b)x}{(1+bx)(1-x)} > 0.$$

From (2), the numerator of the last fraction is a positive number, $x^a + bx^{a+1} - (1+b)x > 0$. Clearly, for the considered values of x, $0 < x < 1$, the denominator is also a positive number, $(1+bx)(1-x) > 0$. Therefore, the fraction is positive as well. We arrive at the conclusion that $\frac{1-x^a}{1-x} < (1+x)^{a-1}$, which implies that the original equation has no solutions when $0 < x < 1$.

The case when $x > 1$ converts to the previous case by substituting $x = \frac{1}{t}$. It follows that $0 < t < 1$ and, utilizing the results from above, we obtain

$$\frac{1-x^a}{1-x} = \frac{1-\frac{1}{t^a}}{1-\frac{1}{t}} = \frac{t^a-1}{(t-1)\cdot t^{a-1}} = \frac{1-t^a}{1-t}\cdot t^{1-a}$$
$$= t^b\cdot\frac{1-t^a}{1-t} < t^b\cdot(1+t)^{a-1} = t^b\cdot(1+t)^{-b} = \frac{(1+t)^{-b}}{t^{-b}}$$
$$= \left(\frac{1}{t}+1\right)^{-b} = (1+x)^{-b} = (1+x)^{a-1}.$$

We see again that $\frac{1-x^a}{1-x} < (1+x)^{a-1}$. So, for $x > 1$ we get to the same result as for $0 < x < 1$, that the equation has no solutions. Finally, combining the outcomes of both scenarios for x, we conclude that the original equation has no positive solutions under the given conditions for $0 < a < 1$.

Our main focus in this book is practically approaching various types of equations and the methods of their solution. The classic inequalities considered in this chapter present a fruitful and efficient technique in assessing the range of the terms comprising some equations. Such analysis allows making conclusions regarding the solvability of an equation and gives a clear path to finding its roots. However, as we demonstrated in a few problems here, the applications of various classic inequalities are not only restricted to that goal and engender many fine extensions. You may find numerous algebraic problems that are connected to inequalities, specifically the problems related to bounded sequences, convergent sequences, and the limit of convergent sequences. For example, the remarkable outcome related to Bernoulli's inequality was the discovery of one of the proofs of the number e's existence. Sometimes it is called *Euler's number* after the prominent Swiss mathematician Leonhard Euler (1707–1783) who was the first to introduce the choice of the symbol e. The constant was discovered by Jacob Bernoulli while studying compound interest calculations. He was evaluating the value of

$$\lim_{n \to \infty} \left(1 + \frac{1}{n}\right)^n,$$

which turned out to exist and equal e, one of the most amazing and exceptional constants in mathematics. We encountered number e as the base of the natural logarithm in the previous chapter. Similar to π, e is irrational (it can't be represented as a ratio of two integers) and transcendental (it is not a root of any non-zero polynomial with rational coefficients). Let's recall here Euler's formula $e^{ix} = \cos x + i \sin x$ (where $i = \sqrt{-1}$), which was called "the most remarkable formula in mathematics" by the Nobel Prize winner American physicist Richard Feynman (1918–1988). The formula establishes the fundamental relationship between the trigonometric functions and the complex exponential function. Using Euler's formula, it is easy to prove de Moivre's formula, mentioned in chapter 6,

$$(\cos x + i \sin x)^n = \cos(nx) + i \sin(nx).$$

The number e plays a significant role across mathematics and especially in calculus. Its history and properties became the subject matter of numerous articles in math magazines and educational books. To acquaint readers with a few of those properties, see the examples below.

In this chapter we provided an overview of several well-known inequalities important in our equations world exploration. There are many more that we will leave to the reader to discover. We encourage the reader to dive into the amazing world of various (including geometric) inequalities, which might become the start of another great mathematical adventure. Meanwhile, we will continue our odyssey, in which the next stop will be at the investigation of multivariable equations. One of the useful techniques applied in their solutions is the utilization of the classic inequalities covered in this chapter.

Follow along. More revelations are coming ahead.

Euler's number e

$$e \approx 2.718281828459045\ldots$$

$$(\cos x + i \sin x)^n = (e^{ix})^n = e^{ixn} = \cos(nx) + i \sin(nx)$$

$$\left(1 + \frac{1}{x}\right)^x < e < \left(1 + \frac{1}{x}\right)^{x+1}$$

$$\lim_{n \to \infty} \left(1 + \frac{1}{n}\right)^n = e$$

$$\lim_{n \to \infty} \frac{n}{\sqrt[n]{n!}} = e, \text{ where } n! = 1 \cdot 2 \cdot 3 \cdot \ldots \cdot n$$

$$e^x = 1 + \frac{x}{1!} + \frac{x^2}{2!} + \frac{x^3}{3!} + \cdots, \quad -\infty < x < \infty \text{ or}$$

$$e = 1 + \frac{1}{1!} + \frac{1}{2!} + \frac{1}{3!} + \cdots = 1 + \frac{1}{1} + \frac{1}{1 \cdot 2} + \frac{1}{1 \cdot 2 \cdot 3} + \cdots$$

$$e = 2 + \cfrac{1}{1 + \cfrac{1}{2 + \cfrac{1}{1 + \cfrac{1}{1 + \cfrac{1}{4 + \cfrac{1}{1 + \cfrac{1}{1 + \cdots}}}}}}}$$

$$e = 1 + \cfrac{1}{0 + \cfrac{1}{1 + \cfrac{1}{1 + \cfrac{1}{2 + \cfrac{1}{1 + \cfrac{1}{1 + \cfrac{1}{4 + \cdots}}}}}}}$$

Chapter 10

Diophantine Equations, Multivariable Equations

Up until now, the problems we have exhibited encompassed various types of equations with one variable. How challenging and exciting would it be to solve an equation that contains multiple variables?

I remember, many years ago, the confusion and bewilderment of my students after failing to solve the following problem offered at the city mathematical contest:

Find the natural solutions of the equation $x + \frac{1}{y+\frac{1}{z}} = \frac{10}{7}$.

The most difficult thing about this problem is the fact that it has three (!) variables in one equation. Every well-known trick for the solutions of one-variable equations will fail in this case. However, the problem is not that complicated. It suffices to notice that you may rewrite the number $\frac{10}{7}$ on the right-hand side as

$$\frac{10}{7} = 1\frac{3}{7} = 1 + \frac{1}{\frac{7}{3}} = 1 + \frac{1}{2 + \frac{1}{3}}.$$

The critical condition here is that all solutions have to be natural numbers. The equality is attainable only when the respective whole numbers are equal on both sides of the equation. Therefore, the only solution exists when $x = 1, y = 2, z = 3$.

Multivariable equations are always challenging due to their non-standard nature. To a large extent they are solved by trial. Usually the common perception is that those equations are difficult or even impossible to solve because of the lack of information that is provided. We will make an argument against this and demonstrate a few useful techniques of how to overcome some of these tricky challenges.

The most famous of the multivariable equations are the *Diophantine equations*. A Diophantine equation for integer-valued (meaning that the solutions are to be integer numbers) variables $x, y, \ldots, z$ is an equation $P(x, y, \ldots, z) = 0$, where P is a polynomial in the given variables with integer coefficients.

The prominent Greek mathematician Diophantus of Alexandria was the first to study problems having fewer equations than unknown variables solvable in integer numbers. He lived in the 3rd century A.D. but his exact date of birth is unknown. The age at which he died, however, can be figured out by solving the problem written as his epitaph. (See problem 29, chapter 1.) In his series of books *Arithmetica*, he gave integer solutions of various linear, quadratic, and cubic equations, and of systems of equations with integer coefficients in two or more variables. In his honor the solutions of such equations are called *diophantine* and the equations with multiple integer-valued variables are called *Diophantine equations*.

Unfortunately, seven out of the 13 *Arithmetica* books have never been found. Despite this, the remaining six volumes delivered invaluable results of his work, which made huge contributions to the field of mathematics, and even became the inspiration and reference for many great mathematicians from Pierre Fermat (1601–1665) and René Descartes (1596–1650) to David Hilbert (1862–1943).

The famous Fermat's Last Theorem *"The equation $a^n + b^n = c^n$ has no positive integer solutions for any natural number n greater than 2"* became the mystery and the biggest challenge for many great mathematicians for three and a half centuries. It is probably the most popular and well-known math problem even to those not familiar with math whatsoever. A lot of novels and movies in popular culture were devoted to this historical problem. It was finally solved in 1995 by the British mathematician Sir Andrew John Wiles.

The problem of finding the solutions of a multivariable equation is often difficult. It becomes even more complicated for a Diophantine equation because of the restriction of seeking only integer solutions. There is no uniform algorithm or technique figuring out the solutions of such equations or even for determining if those solutions exist at all. In 1900, the prominent German mathematician David Hilbert formulated his famous 23 unsolved problems (at that time), which, in his opinion, should determine the main directions in mathematics' development in the new millennium. The problems have received significant attention, and even today work on the remaining unsolved problems is still considered to be of the greatest importance.

One of his fundamental problems, problem 10, is the only "decision problem" on that list; it raised a question regarding the solvability of Diophantine equations:

Given a Diophantine equation with any number of unknown quantities and with rational integral numerical coefficients: To devise a process according to which it can be determined in a finite number of operations whether the equation is solvable in rational integers.

The negative answer to this question was given by the Soviet mathematician Yuri Matiyashevich in 1970 (completing the work of American mathematicians Martin Davis, Hilary Putnam, and Julia Robinson). According to his theorem, there is no algorithm or uniform method to solving any Diophantine equation; the same is true for other types of equations with multiple variables. However, some of those equations do have solutions and are often offered on various math contests and Olympiads. In this chapter we will examine a few useful techniques and hints of how to handle those intriguing challenges.

To start, let's consider an interesting "magic" trick of guessing someone's birthday. You can impress your friend by figuring out his/her exact date of birth by knowing only one number, which is the sum of the products of the day of birth by 12 and the month of birth by 31. Assume your friend's birthday is June 16. Then your friend has to multiply 16 by 12 and add the result to the product of 6 by 31, $16 \cdot 12 + 6 \cdot 31 = 378$. Once the number 378 is revealed to you, your task is to tell your friend's exact date of birth. Denoting by x the day and by y the month of the friend's birthday gives the Diophantine equation $12x + 31y = 378$ with the restrictions on x and y such that $1 \le x \le 31$ and $1 \le y \le 12$. Expressing x in terms of y from the equation and introducing one more variable t gives $x = \frac{378-31y}{12} = 31 - 2y + \frac{6-7y}{12} = 31 - 2y + t$ where $t = \frac{6-7y}{12}$ and since x is a natural number, t has to be an integer. From the last equality we express y in terms of t as $y = \frac{6-12t}{7} = -t + \frac{6-5t}{7} = -t + n$, where $n = \frac{6-5t}{7}$ and n also has to be some integer. Solving now for t, we get that $t = \frac{6-7n}{5} = 1 - n + \frac{1-2n}{5} = 1 - n + p$, where $p = \frac{1-2n}{5}$. Expressing n in terms of p, we get $n = \frac{1-5p}{2} = -2p + \frac{1-p}{2} = -2p + m$, where $m = \frac{1-p}{2}$ and m is some integer. Finally, from the last equality, $p = 1 - 2m$.

To complete the solution, we need to back-substitute the $p = 1 - 2m$ and after a few steps we obtain the expressions for y and x:

$$y = -t + n = -1 + 2n - p = -1 - 5p + 2m = 12m - 6, \qquad (1)$$
$$x = 31 - 2\underbrace{(12m - 6)}_{y} + \underbrace{(1 - n + p)}_{t} = 47 - 31m. \qquad (2)$$

Recalling the restrictions on x and y and using (1) and (2) yields

$$1 \le 12m - 6 \le 12,$$
$$1 \le -31m + 47 \le 31.$$

Solving the first inequality gives $\frac{7}{12} \le m \le 1\frac{1}{2}$.
Solving the second inequality gives $\frac{16}{31} \le m \le 1\frac{15}{31}$.

Since m has to be an integer, the only possible solution of the system of inequalities is $m = 1$. Therefore, $y = 12m - 6 = 6$ and $x = -31m + 47 = 16$. Your friend's date of birth is correctly determined as June 16th.

It's interesting to note that this problem always has a unique solution for a specific date of birth. Indeed, assume that the equation $12x + 31y = n$ (n is the number revealed to you after the calculations, it was equal to 378 in our example) has two distinct solutions, pairs of numbers (x_1, y_1) and (x_2, y_2). These numbers have to satisfy the equation, so it follows that

$$12x_1 + 31y_1 = n,$$
$$12x_2 + 31y_2 = n.$$

Subtracting the second equation from the first gives $12(x_1 - x_2) + 31(y_1 - y_2) = 0$, or equivalently, $12(x_1 - x_2) = -31(y_1 - y_2)$. Since all our solutions are natural numbers, it follows from the last equality that the difference $(x_1 - x_2)$ has to be divisible by 31. Recalling that each of the numbers x_1 and x_2 cannot exceed 31 (the maximum number of days in a month), the only time when their difference is divisible by 31 is when $x_1 = x_2$, which contradicts our assumption. So, you may play this trick on any of your new friends without being afraid of failure. You will always "guess" the correct date of birth, because the Diophantine equation has only one solution for any specifically calculated number $12x + 31y = n$.

Evidently, solving this equation without having restrictions on values of x and y, we can use the formulas $x = -31m + 47$ and $y = 12m - 6$ for generating infinitely many integer solutions. Substituting integer values for m will produce integer solutions for x and y.

Generally, a linear Diophantine equation either has no solutions, or an infinite number of solutions. Depending on further restrictions imposed on the variables, there may be a finite number of solutions.

The linear Diophantine equation has the form $ax + by = c$, where the constants a, b, and c are integers, and x and y are the variables. Such an equation has integer solutions if and only if c is a multiple of the greatest common divisor of a and b. This statement is not difficult to prove, and we leave its justification to the reader.

Let's assume now that a and b are relatively prime numbers and the pair of integers (x_0, y_0) is a solution of the given equation. Hence, $ax + by = ax_0 + by_0 = c$, which leads to

$$x = x_0 - \frac{b}{a}(y - y_0). \qquad (*)$$

If a and b are relatively prime numbers, then the number $(y - y_0)$ must be divisible by a. Therefore, there must exist an integer t such that $(y - y_0) = at$. Expressing y using the last equality, we get $y = y_0 + at$. Substituting the value of y into the equality $(*)$ gives

$$x = x_0 - \frac{b}{a}(y - y_0) = x_0 - \frac{b}{a} \cdot at = x_0 - bt.$$

Let's summarize the results: if you manage to find one pair of integer solutions of the Diophantine linear equation $ax + by = c$, then all other solutions can be expressed as

$$x = x_0 - bt,$$
$$y = y_0 + at, \quad t \in Z.$$

For example, the equation $4x + 6y = 7$ has no integer solutions, because you may rewrite it as $2(2x + 3y) = 7$ and it is clear that it is impossible to find integers x and y to make an even number on the left-hand side be equal to an odd number on the right-hand side.

In another example, in the equation $2x + 5y = 4$, $a = 2$ and $b = 5$ are relatively prime numbers. The pair $x_0 = -3$ and $y_0 = 2$ satisfies the equation. Then, there is an infinite number of solutions expressed as

$$x = -3 - 5t,$$
$$y = 2 + 2t, \quad t \in Z.$$

As you can see, the critical condition for the solvability of a Diophantine equation $ax + by = c$ is the requirement that c is a multiple of the greatest common divisor of a and b. The problem of finding the greatest common divisor of two numbers, GCD, is efficiently solved by the algorithm, described in Euclid's work *Elements*. The prominent ancient Greek mathematician Euclid of Alexandria (who lived mid-4th century BCE—mid-3rd century BCE), commonly regarded as the "father of geometry," took a geometrical approach in finding the biggest number that divides two given numbers: *"If you have two distances, AB and CD, and you always take away the smaller from the bigger, you will end up with a distance that measures both of them."* The main ideas behind the algorithm that bears Euclid's name are the following:

If b is a divisor of a, then their greatest common divisor is b, $GCD(a, b) = b$;
If $a = bq + r$, for integers q and r, then $GCD(a, b) = GCD(b, r)$.
The algorithm's formal description is the following:

1. Divide a by b and get the remainder r. If $r = 0$, use b as the GCD of a and b.

2. If $r \neq 0$, divide b by r and get the remainder q. If $q = 0$, use r as the GCD of b and r. If $q \neq 0$, repeat the steps until the remainder equals 0.

Since at each step the remainder decreases by at least 1, eventually it must be 0. A formal proof would involve mathematical induction. We suggest readers prove this statement on their own.

Denoting the quotients in Euclid's algorithm by $q_1, q_2, \ldots, q_n$, we can express the $\frac{a}{b}$ as the continued fraction

$$\frac{a}{b} = q_1 + \cfrac{1}{q_2 + \cfrac{1}{q_3 + \cdots + \cfrac{1}{q_n}}}$$

Let's demonstrate how the algorithm works by solving an interesting problem that provides a vivid geometrical interpretation of it.

Given a rectangle with sides 1368 and 532, cut from it several squares with a side length of 532 until you get the remaining rectangle with one of the sides of a length less than 532. From the new rectangle cut squares with sides equal in length to its smaller side as many times as it is possible. You repeat this process up to a point when the last square is cut. What is the length of the side of the last square?

Solution. We start with finding the GCD of two numbers $a = 1368$ and $b = 532$.

$1368 = 532 \cdot 2 + 304$,   GCD$(1368, 532) = $ GCD$(532, 304)$,   2   squares   with side 532.

$532 = 304 \cdot 1 + 228$, GCD$(532, 304) = $ GCD$(304, 228)$, 1 square with side 304.

$304 = 228 \cdot 1 + 76$, GCD$(304, 228) = $ GCD$(228, 76)$, 1 square with side 228.

$228 = 76 \cdot 3 + 0$, GCD$(228, 76) = 76$, 3 squares with side 76.

Therefore, GCD$(1368, 532) = 76$ and the last square's side has the length of 76. We got seven squares of four different sizes.

There is an interesting variation of this problem, if you pose a little more complex question: find natural numbers a and b such that it is possible to cut a rectangle of size $a \times b$ (a—length, b—width) into squares of n different sizes for the given natural number n. We invite readers to find a solution to this problem before continuing to the next chapter (should it be fun to get it on your own?), as we will get back to this problem while discussing the amazing properties of the Fibonacci numbers.

The Euclidean algorithm has many practical applications, one of which is used in the solution of Diophantine equations.

The logic behind the steps we were making in solving the "guessing birthday" trick was in finding expressions with integer coefficients for the variables x and y in terms of some integer m. The same goal can be achieved by the Euclidian algorithm. Indeed, using the Euclidean algorithm on two numbers 12 and 31 from the equation $12x + 31y = 378$ we solved, gives

$$31 = 12 \cdot 2 + 7$$
$$12 = 7 \cdot 1 + 5$$
$$7 = 5 \cdot 1 + 2$$

$$5 = 2 \cdot 2 + 1$$
$$2 = 1 \cdot 2 + 0.$$

The numbers in the GCD calculation 31, 12, 7, 5, 2, 1 are the same numbers that represented denominators and coefficients of the variables in the numerators in the fractions when we were solving the equation. If we reverse the steps, we will find the solution to the equation $12x + 31y = 1$:

$$1 = 5 - 2 \cdot 2 = 5 - 2 \cdot (7 - 5 \cdot 1) = -2 \cdot 7 + 3 \cdot 5 = -2 \cdot 7 + 3 \cdot (12 - 7 \cdot 1)$$
$$= 3 \cdot 12 - 5 \cdot 7 = 3 \cdot 12 - 5 \cdot (31 - 12 \cdot 2) = -5 \cdot 31 + 13 \cdot 12.$$

So, $x_0 = 13$, $y_0 = -5$. Hence, all other solutions are expressed as

$$x = 13 - 31t,$$
$$y = -5 + 12t, \quad t \in Z.$$

Since the original equation to solve was $12x + 31y = 378$, if we multiply x_0 and y_0 by 378, then the left side will increase by a factor of 378. So, by increasing the right side by the same factor we will get a pair (x, y) that satisfies our original equation. Accordingly, a solution is $x = 13 \cdot 378 = 4{,}914$ and $y = -5 \cdot 378 = -1890$. Therefore, the general solution to the original equation is

$$x = 4{,}914 - 31t,$$
$$y = -1890 + 12t, \quad t \in Z.$$

Recalling the restrictions on x and y, $1 \leq x \leq 31$ and $1 \leq y \leq 12$, we get two inequalities

$$1 \leq 4{,}914 - 31t \leq 31,$$
$$1 \leq -1890 + 12t \leq 12.$$

Solving the first inequality yields $157\frac{16}{31} \leq t \leq 158\frac{15}{31}$.

Solving the second inequality yields $157\frac{7}{12} \leq t \leq 158\frac{1}{2}$.

Since t has to be an integer, the only possible solution of the system of inequalities is $t = 158$. Therefore, $x = 4{,}914 - 31 \cdot 158 = 16$ and $y = -1890 + 12 \cdot 158 = 6$, which is the same result as we got before, the birthday is June 16.

After our introduction to linear Diophantine equations, we will present the reader with one practical application of these equations in the determining of the common or extraneous solutions in trigonometric equations. In some equations you need to figure out the common solutions to get to the answer. In other problems you need to eliminate repetitive solutions or get rid of extraneous solutions. Either way, the suggested method is very useful and helpful. Before we proceed, I want to point out that in the problems below it is much easier to get solutions to the Diophantine equations using the "guess and check" technique. However, keep in mind that for solutions of more complicated equations you always have in your arsenal the Euclidean algorithm.

Problem 1. Solve the equation

$$\cos 6x + \sin \frac{5x}{2} = 2.$$

Solution. Recalling the range of the functions sine and cosine yields

$$|\cos 6x| \le 1 \quad \text{and} \quad \left|\sin \frac{5x}{2}\right| \le 1.$$

Therefore, the equality will be attainable only when each addend on the left-hand side of the equation equals 1:

$$\cos 6x = 1,$$
$$\sin \frac{5x}{2} = 1.$$

The solutions of each equation are:

$$x = \frac{\pi k}{3}, \quad k \in Z,$$
$$x = \frac{\pi}{5}(4n + 1), \quad n \in Z.$$

Do we have common solutions? If yes, how can we find them? The answers to these questions are derived from the solutions to the following Diophantine equation:

$$\frac{\pi k}{3} = \frac{\pi}{5}(4n + 1),$$

or equivalently,

$$5k - 12n = 3.$$

It's not hard to see that the pair $k_0 = 3$ and $n_0 = 1$ satisfies the equation. Then all the other solutions for k and n are determined by the formulas

$$k = 3 + 12t,$$
$$n = 1 + 5t, \quad t \in Z.$$

So, to find common solutions, we just need to use one of the above formulas for k or for n,

$$x = \frac{\pi k}{3} = \frac{\pi(3 + 12t)}{3} = \pi(1 + 4t), \quad t \in Z.$$

Answer: $x = \pi(1 + 4t), t \in Z.$

Problem 2. Solve the equation

$$\tan 2x \cdot \tan 7x = 1.$$

Solution. Using the definition of tangent, rewrite the equation as

$$\frac{\sin 2x}{\cos 2x} \cdot \frac{\sin 7x}{\cos 7x} - 1 = 0,$$

or equivalently,

$$\frac{\sin 2x \sin 7x - \cos 2x \cos 7x}{\cos 2x \cos 7x} = 0.$$

Applying the formula for the cosine of the sum of two angles, the numerator can be rewritten as $(-\cos 9x)$ and so the equation becomes

$$\frac{-\cos 9x}{\cos 2x \cos 7x} = 0,$$

from which

$$\cos 9x = 0,$$
$$\cos 2x \neq 0,$$
$$\cos 7x \neq 0.$$

The solutions of the first equation are $x = \frac{\pi}{18}(2n + 1)$, $n \in Z$. We have to exclude from those solutions the values $x = \frac{\pi}{4}(2k + 1)$, $k \in Z$, for which $\cos 2x = 0$ and $x = \frac{\pi}{14}(2p + 1)$, $p \in Z$, for which $\cos 7x = 0$. Consider the Diophantine equation

$$\frac{\pi}{18}(2n + 1) = \frac{\pi}{4}(2k + 1),$$
$$4n + 2 = 18k + 9,$$
$$4n - 18k = 7.$$

The last equation has no integer solutions. Therefore, there are no values of x to be excluded from the solutions of the original equation related to the restriction $\cos 2x \neq 0$.

Consider the second Diophantine equation

$$\frac{\pi}{18}(2n + 1) = \frac{\pi}{14}(2p + 1),$$
$$14n + 7 = 18p + 9,$$
$$14n - 18p = 2,$$
$$7n - 9p = 1.$$

The feasible solutions are $n_0 = 4$, $p_0 = 3$. Thus, the general solutions to be excluded can be expressed as

$$n = 4 + 9t, \quad t \in Z, \quad \text{or}$$
$$p = 3 + 7t, \quad t \in Z.$$

Answer: $x = \frac{\pi}{18}(2n + 1), n \in Z; n \neq 4 + 9t, t \in Z.$

Problem 3. Solve the equation

$$\cos x \cdot \cos 2x \cdot \cos 4x \cdot \cos 8x = \frac{1}{16}.$$

Solution. To solve the equation multiply both sides by $16 \sin x$. This trick would allow utilizing the formula for the sine of a double angle four times on the left-hand side transforming the original equation to the equation $\sin 16x = \sin x$. Indeed, after multiplying both sides by $16 \sin x$, the left-hand side transforms to

$$8 \cdot (2 \sin x \cos x) \cdot \cos 2x \cdot \cos 4x \cdot \cos 8x = 4 \cdot (2 \sin 2x \cdot \cos 2x) \cdot \cos 4x \cdot \cos 8x$$
$$= 2 \cdot (2 \sin 4x \cdot \cos 4x) \cdot \cos 8x$$
$$= 2 \sin 8x \cdot \cos 8x$$
$$= \sin 16x.$$

The right-hand side becomes

$$\frac{1}{16} \cdot 16 \sin x = \sin x.$$

It follows that
$$\sin 16x = \sin x.$$

That's an easy equation to solve and we should be happy at this point. But we forgot to mention one thing—we can multiply both sides of the equation by the same number and assume nothing changes only when we multiply by a non-zero number. Thus, all the values of x for which $\sin x = 0$ must be eliminated from the solutions of the equation (it is easy to verify that x for which $\sin x = 0$ does not satisfy the equation, so we do not lose any solutions). In other words, as a result of our manipulations, we get to the following system:

$$\sin 16x - \sin x = 0,$$
$$\sin x \neq 0.$$

Applying the identity for the difference of sines of two angles gives

$$2 \sin \frac{15x}{2} \cdot \cos \frac{17x}{2} = 0,$$

It follows that either $\sin \frac{15x}{2} = 0$ or $\cos \frac{17x}{2} = 0$. This leads to the following systems:

$$x = \frac{2\pi k}{15}, \quad n \in Z,$$
$$x \neq \pi p, \quad p \in Z,$$

or

$$x = \frac{\pi}{17}(2n + 1), \quad n \in Z,$$
$$x \neq \pi p, \quad p \in Z.$$

To get the solutions of the original equation, we have to eliminate the numbers πp, $p \in Z$ from each set of the solutions for x.

First, consider the Diophantine equation $\frac{2\pi k}{15} = \pi p$, or equivalently, $2k - 15p = 0$. The obvious solution is $k_0 = 15$, $p_0 = 2$. Then the other integer solutions are $k = 15 + 15t = 15m, m \in Z$, and $p = 2 + 2t' = 2m, m \in Z$.

Second, $\frac{\pi}{17}(2n + 1) = \pi p$, or equivalently, $2n - 17p = -1$. The solution will be $n_0 = 8$, $p_0 = 1$. Thus, the general solutions are $n = 8 + 17t, t \in Z$, or $p = 1 + 2t, t \in Z$. Those are the numbers to be eliminated from the solutions.

Answer: $x = \frac{2\pi k}{15}, k \neq 15m, k \in Z, m \in Z; x = \frac{\pi}{17}(2n + 1), n \neq 8 + 17t, n \in Z, t \in Z.$

Let's turn now to other multivariable equations and investigate some useful and efficient methods of their solutions. In many of the following problems you would need to use not only one of the suggested techniques, but various combinations of them. Also keep in mind that in solving Diophantine equations we considered only integer solutions, while now we will be seeking all real, not just integer, solutions, unless there are specifically indicated restrictions on the set of solutions.

Completing the square

Problem 4. Solve the equation

$$6x^2 + 2y^2 + z^2 + 6xy - 2xz - 2yz - 2x + 1 = 0.$$

Solution. The idea behind this method is reflected in its name. Try to complete the square or maybe a few squares in the equation and then analyze the final product. Sometimes this is an easy task; sometimes it needs a fair number of manipulations, as is the case in this problem. Let's regroup the terms and rewrite the equation:

$$(x^2 - 2x + 1) + (4x^2 + 4xy + y^2) + (x^2 + y^2 + z^2 + 2xy - 2xz - 2yz) = 0,$$
$$(x - 1)^2 + (2x + y)^2 + (x + y - z)^2 = 0.$$

Since the square of a real number is always nonnegative, the left-hand side of the last equation, as the sum of three squares, has to be a nonnegative number.

Therefore, equality to 0 will be possible only when each addend equals 0 at the same time:

$$x - 1 = 0,$$
$$2x + y = 0,$$
$$x + y - z = 0.$$

From the first equation, $x = 1$. Substituting this value into the second equation yields $y = -2$. Finally, substituting values of x and y into the third equation gives $z = -1$.

Answer: $(1, -2, -1)$.

One might ask how to figure out the best way to combine the terms in order to complete the squares. Well, there is no universal rule. Carefully analyzing the given conditions in the problem should provide some hints. In the last problem we obviously made the first two combinations and put all the remaining terms aside. It happened that all the remaining terms formed a perfect square as well. Not every time will you be this lucky, however, even the negative results (they are always around!) will get you closer to the final destination.

Problem 5. Solve the equation

$$\log_2^2(x + y) + \log_2^2(xy) + 1 = 2\log_2(x + y).$$

Solution. Rewriting the equation and regrouping its terms gives

$$(\log_2^2(x + y) - 2\log_2(x + y) + 1) + \log_2^2(xy) = 0,$$

or equivalently,

$$(\log_2(x + y) - 1)^2 + \log_2^2(xy) = 0.$$

The left-hand side of the last equation is the sum of two squares. Thus, it has to be a nonnegative number. Equality to 0 is possible only when each square on the left-hand side equals 0:

$$(\log_2(x + y) - 1)^2 = 0,$$
$$\log_2^2(xy) = 0.$$

$$\log_2(x + y) = 1,$$
$$\log_2(xy) = 0.$$

$$x + y = 2,$$
$$xy = 1.$$

Expressing y from the first equation in terms of x as $y = 2 - x$ and substituting it in the second equation gives $x(2 - x) = 1$, or equivalently, $x^2 - 2x + 1 = 0$, or $(x - 1)^2 = 0$, from which $x = 1$. Substituting this value into $y = 2 - x$ gives $y = 1$. Both are positive numbers and belong to the equation's domain. The same result will be achieved by applying directly Viète's formulas to x and y as to the roots of a quadratic equation with the coefficients -2 and 1. We will discuss this technique in greater detail in chapter 12.

Answer: $x = y = 1$.

Function properties analysis

We have already encountered this technique in previous chapters. Determining the range of the functions comprising each side of the equation and calculating the maximum or minimum values, you may draw interesting conclusions leading to the solution of the equation.

Problem 6. Solve the equation

$$(3 - \sin x)(4 - \sin^{-2} x) = 12 + \cos^2 y.$$

Solution. The range of the functions sine and cosine is all real numbers not exceeding 1: $|\sin x| \le 1$, $|\cos y| \le 1$. It follows that the left-hand side of this equation cannot exceed 12, while the right-hand side is greater than or equal to 12.
  Indeed, $3 - \sin x \le 3 + 1 = 4$; $4 - \sin^{-2} x \le 4 - 1 = 3$. It follows $(3 - \sin x)(4 - \sin^{-2} x) \le 3 \cdot 4 = 12$.
  On the other hand, $12 + \cos^2 y \ge 12 + 0 = 12$.
  The equality then will be attainable only when the following equalities simultaneously are true:

$$3 - \sin x = 4,$$
$$4 - \sin^{-2} x = 3,$$
$$12 + \cos^2 y = 12.$$

Solving this system of equations gives

$$\sin x = -1,$$
$$\sin^2 x = 1,$$
$$\cos^2 y = 0.$$

Common solutions are $x = -\frac{\pi}{2} + 2\pi k$, $k \in Z$; $y = \frac{\pi}{2} + \pi n$, $n \in Z$.

Answer: $x = -\frac{\pi}{2} + 2\pi k$, $k \in Z$; $y = \frac{\pi}{2} + \pi n$, $n \in Z$.

Problem 7. Solve the equation

$$1 - 2x - x^2 = \tan^2(x + y) + \cot^2(x + y).$$

Solution. At first glance, the problem looks very confusing and maybe even unsolvable. The way to approach it is to compare the range of the left-hand side and the range of the right-hand side of the equation. That comparison might provide some insight for the next steps to take.

First, let's do some preliminary work based on the method of completing the square. The goal is to complete (if possible, of course) the square on each side. Subtracting 2 from both sides of the equation gives

$$1 - 2x - x^2 - 2 = \tan^2(x+y) + \cot^2(x+y) - 2,$$
$$-1 - 2x - x^2 = \tan^2(x+y) - 2 + \cot^2(x+y),$$
$$-(1+x)^2 = (\tan(x+y) - \cot(x+y))^2.$$

At this point we clearly see that the range of the left-hand side of the last equation consists of all nonpositive numbers, while the range of the right-hand side consists of all nonnegative numbers. The equality is possible only when each side equals 0. Therefore, the solution of the equation comes to the solution of the system of equations

$$1 + x = 0,$$
$$\tan(x+y) - \cot(x+y) = 0.$$

From the first equation $x = -1$. Substituting this value in the second equation yields

$$\tan(y-1) - \cot(y-1) = 0.$$

Introducing a new variable $z = \tan(y-1)$ gives

$$z - \frac{1}{z} = 0,$$
$$z^2 - 1 = 0,$$
$$z = \pm 1,$$

thus $\tan(y-1) = 1$, or $\tan(y-1) = -1$.

If $\tan(y-1) = 1$, then $y - 1 = \frac{\pi}{4} + \pi k, k \in Z$, or $y = 1 + \frac{\pi}{4} + \pi k, k \in Z$.
If $\tan(y-1) = -1$, then $y - 1 = -\frac{\pi}{4} + \pi m, m \in Z$, or $y = 1 - \frac{\pi}{4} + \pi m$, $m \in Z$.

These two answers can be expressed by one formula as

$$y = 1 + \frac{\pi}{4} + \frac{\pi n}{2}, \quad n \in Z.$$

Answer: $\left(-1; 1 + \frac{\pi}{4} + \frac{\pi n}{2}, n \in Z\right)$.

Problem 8. Solve the equation

$$(\cos^2 x + \cos^{-2} x)(1 + \tan^2 2y)(3 + \sin 3z) = 4.$$

Solution. Let's examine and analyze each factor on the left-hand side.

First, note that $\cos^2 x + \cos^{-2} x \geq 2$. Indeed, $a + \frac{1}{a} \geq 2$ for any positive number a, as it was already proved in chapter 7.

Secondly, $\tan^2 2y \geq 0$, thus $1 + \tan^2 2y \geq 1$.

Finally, since $|\sin 3z| \leq 1$, the minimum value of $\sin 3z$ that could be attained is -1. Then $3 + \sin 3z \geq 2$.

The conclusion from the above observations is that the left-hand side of the equation exceeds or equals $2 \cdot 1 \cdot 2 = 4$. The equality will be possible only when each factor on the left-hand side has a minimum value:

$$\cos^2 x + \cos^{-2} x = 2,$$
$$1 + \tan^2 2y = 1,$$
$$3 + \sin 3z = 2.$$

Solving the first equation gives

$$\cos^2 x + \cos^{-2} x - 2 = 0,$$
$$\left(\cos x - \frac{1}{\cos x}\right)^2 = 0,$$
$$\cos^2 x - 1 = 0,$$
$$\cos x = \pm 1,$$

then $x = \pi k, k \in Z$.

Solving the second equation gives

$$1 + \tan^2 2y = 1,$$
$$\tan^2 2y = 0, \text{ or } \tan 2y = 0$$

and then $2y = \pi m, m \in Z, y = \frac{\pi m}{2}, m \in Z$.

Solving the third equation gives

$$3 + \sin 3z = 2,$$
$$\sin 3z = -1,$$

then $3z = -\frac{\pi}{2} + 2\pi n, n \in Z$ and $z = -\frac{\pi}{6} + \frac{2\pi n}{3}, n \in Z$.

Answer: $\left(\pi k, k \in Z; \frac{\pi m}{2}, m \in Z; -\frac{\pi}{6} + \frac{2\pi n}{3}, n \in Z\right)$.

Problem 9. Solve the equation

$$x\sqrt{1 - y^2} + y\sqrt{1 - x^2} = 3.$$

Solution. The domain of this equation is determined from the solutions of the following inequalities:

$$1 - y^2 \geq 0,$$
$$1 - x^2 \geq 0.$$

The domain is all real numbers x and y such that $|x| \leq 1$ and $|y| \leq 1$. For these values of x and y there must exist angles α and β such that $\sin \alpha = x$ and $\sin \beta = y$. So,

$$\sqrt{1 - y^2} = \sqrt{1 - \sin^2 \beta} = \sqrt{\cos^2 \beta} = \cos \beta$$

and

$$\sqrt{1 - x^2} = \sqrt{1 - \sin^2 \alpha} = \sqrt{\cos^2 \alpha} = \cos \alpha.$$

After substituting these expressions for x and y into the original equation, it's not hard to see that on the left-hand side of the equation we obtain the sine of the sum of the angles α and β:

$$\sin \alpha \cos \beta + \sin \beta \cos \alpha = \sin(\alpha + \beta).$$

The range of the sine function is all real numbers not exceeding 1 in absolute value. The maximum value of the left-hand side is 1. Therefore, it is impossible that $\sin(\alpha + \beta) = 3$. The original equation has no solutions.

Note that, generally speaking, $\sqrt{m^2} = |m|$, so $\sqrt{1 - y^2} = |\cos \beta|$ and $\sqrt{1 - x^2} = |\cos \alpha|$, but dropping an absolute value sign in this case without properly considering two outcomes did not affect the solution analysis. Anyway, we would get the sine of a sum or difference of two angles and come to the same conclusion regarding the solvability of the equation.

Problem 10. Solve the equation

$$2(x^2 - 2x + 3)(y^2 - 3y + 4) = 7.$$

Solution. Each of the factors on the left-hand side is a quadratic trinomial with the first coefficient 1, which is a positive number. Recalling the properties of a quadratic function, it follows that the minimum value of each factor will be attained at the vertex of the parabolas (they open upward). Let's consider the function $f(x) = x^2 - 2x + 3$. It has a minimum when $x = \frac{-b}{2a} = \frac{2}{2} = 1$. For the second function, $f(y) = y^2 - 3y + 4$, its minimum value is attained when $y = \frac{-b}{2a} = \frac{3}{2}$. Therefore, the left-hand side of the given equation exceeds or equals

$$2 \cdot (1^2 - 2 \cdot 1 + 3)\left(\left(\frac{3}{2}\right)^2 - 3 \cdot \frac{3}{2} + 4 \right) = 2 \cdot 2 \cdot \frac{7}{4} = 7.$$

Equality is attained when $x = 1$ and $y = \frac{3}{2}$. Hence, the only possible solutions of the given equation are $x = 1$, $y = \frac{3}{2}$.

Answer: $x = 1$, $y = \frac{3}{2}$.

Quadratic equations method

When you are dealing with a quadratic polynomial containing multiple variables, you can ignore the other variables and treat it as a quadratic polynomial with only one variable. This gives you the ability to express one unknown in terms of the other unknown(s) and might lead to a very short and elegant solution.

Problem 11. Solve the equation

$$x^2 - 2xy + 2y^2 - 2x - 2y + 5 = 0.$$

Solution. Let's consider the equation as a quadratic equation in x with coefficients depending on y. Basically, this is the idea behind "ignoring" the other variable(s) in the explanation of the method given above. We rewrite the equation as

$$x^2 - 2(y+1)x + (2y^2 - 2y + 5) = 0.$$

This equation has solutions only when the discriminant is a nonnegative number. Finding D, or rather $\frac{D}{4}$, which is easier here, gives

$$\frac{D}{4} = (y+1)^2 - (2y^2 - 2y + 5) = y^2 + 2y + 1 - 2y^2 + 2y - 5$$
$$= -y^2 + 4y - 4 = -(y-2)^2 \le 0.$$

The original equation will be solvable if and only if $\frac{D}{4} = 0$, from which we obtain $y = 2$. It follows that

$$x = \frac{y + 1 \pm \sqrt{0}}{1} = 3.$$

Answer: $(3, 2)$.

Classic inequalities application

Next, we will turn to the techniques outlined in the previous chapter and apply them to the solutions of multivariable equations to follow.

Problem 12. Solve the equation

$$\sqrt{5(x^2 + 2yz)} + \sqrt{6(y^2 + 2xz)} + \sqrt{5(z^2 + 2xy)} = 4(x + y + z).$$

Solution. Noticing that the sum of three nonnegative square roots has to be nonnegative yields that the number on the right-hand side must be nonnegative as well.

Thus, it follows that $x + y + z \geq 0$. From the Cauchy-Bunyakovsky inequality for the three nonnegative numbers

$$\sqrt{a_1} \cdot \sqrt{b_1} + \sqrt{a_2} \cdot \sqrt{b_2} + \sqrt{a_3} \cdot \sqrt{b_3}$$
$$\leq \sqrt{(\sqrt{a_1})^2 + (\sqrt{a_2})^2 + (\sqrt{a_3})^2} \cdot \sqrt{(\sqrt{b_1})^2 + (\sqrt{b_2})^2 + (\sqrt{b_3})^2}$$
$$= \sqrt{a_1 + a_2 + a_3} \cdot \sqrt{b_1 + b_2 + b_3}.$$

Applying the Cauchy-Bunyakovsky inequality to the equation's left-hand side yields

$$\sqrt{5(x^2 + 2yz)} + \sqrt{6(y^2 + 2xz)} + \sqrt{5(z^2 + 2xy)}$$
$$= \sqrt{5} \cdot \sqrt{x^2 + 2yz} + \sqrt{6} \cdot \sqrt{y^2 + 2xz} + \sqrt{5} \cdot \sqrt{z^2 + 2xy}$$
$$\leq \sqrt{5 + 6 + 5} \cdot \sqrt{x^2 + 2yz + y^2 + 2xz + z^2 + 2xy}$$
$$= \sqrt{16} \cdot \sqrt{(x + y + z)^2} = 4|x + y + z| = 4(x + y + z).$$

We can drop the absolute value sign because $x + y + z \geq 0$, as it was established at the very beginning of the solution. Recalling that equality in the Cauchy-Bunyakovsky inequality is attained only when the respective numbers are in proportion leads to

$$\frac{\sqrt{x^2 + 2yz}}{\sqrt{5}} = \frac{\sqrt{y^2 + 2xz}}{\sqrt{6}} = \frac{\sqrt{z^2 + 2xy}}{\sqrt{5}}.$$

Hence, our equation is reduced to the following system of equations

$$\frac{\sqrt{x^2 + 2yz}}{\sqrt{5}} = \frac{\sqrt{y^2 + 2xz}}{\sqrt{6}},$$
$$\frac{\sqrt{x^2 + 2yz}}{\sqrt{5}} = \frac{\sqrt{z^2 + 2xy}}{\sqrt{5}},$$

or equivalently,

$$6(x^2 + 2yz) = 5(y^2 + 2xz),$$
$$x^2 + 2yz = z^2 + 2xy.$$

Let's rewrite and examine the second equation of the system.

$$x^2 + 2yz - z^2 - 2xy = 0,$$
$$(x^2 - z^2) - (2xy - 2yz) = 0,$$
$$(x - z)(x + z) - 2y(x - z) = 0,$$
$$(x - z)(x + z - 2y) = 0.$$

It follows that $x = z$ or $y = \frac{x+z}{2}$. Substituting those expressions in the first equation of the system gives the following two equations:

$$6(x^2 + 2yx) = 5(y^2 + 2x^2) \quad \text{or} \quad 6(x^2 + (x+z)z) = 5\left(\left(\frac{x+z}{2}\right)^2 + 2xz\right).$$

If $x = z$, then $6(x^2 + 2yx) = 5(y^2 + 2x^2)$, or equivalently $5y^2 - 12yx + 4x^2 = 0$.
Solve the last equation as a quadratic in y.

$$D = 144x^2 - 80x^2 = 64x^2.$$

$$y = \frac{12x \pm \sqrt{64x^2}}{10},$$

which gives $y_1 = 2x$ or $y_2 = \frac{2}{5}x$.

So, in this case we get that any triple of numbers $(k, 2k, k)$ or $\left(m, \frac{2}{5}m, m\right)$ is a solution of the original equation for any nonnegative real numbers k and m. In other words, the original equation has indefinitely many solutions: $(k, 2k, k)$ and $\left(m, \frac{2}{5}m, m\right)$, where $k \geq 0$, $m \geq 0$.

The second root is $y = \frac{x+z}{2}$. Then the first equation of the system transforms into

$$6(x^2 + (x+z)z) = 5\left(\left(\frac{x+z}{2}\right)^2 + 2xz\right).$$

This simplifies to $19x^2 - 26xz + 19z^2 = 0$. The discriminant of this quadratic equation for x is nonpositive:

$$D = 676z^2 - 1444z^2 = -768z^2 \leq 0.$$

Therefore, no real solutions exist when $z \neq 0$.

If $z = 0$, then $D = 0$ and respectively, $x = 0$ and $y = \frac{x+z}{2} = 0$.

We did not find any new solutions, because the solution $(0, 0, 0)$ belongs to the already found solutions, expressed above as $(k, 2k, k)$ or $\left(m, \frac{2}{5}m, m\right)$.

Answer: The solutions of the equation are triples $(k, 2k, k)$ or $\left(m, \frac{2}{5}m, m\right)$ for any real numbers $k \geq 0$, $m \geq 0$.

Problem 13. Solve the equation

$$\sqrt{3x - 2y} + \sqrt{3y - 2z} + \sqrt{3z - 2x} = \frac{9\sqrt{(3x - 2y)(3y - 2z)(3z - 2x)}}{\sqrt{(3x - 2y)(3y - 2z)} + \sqrt{(3y - 2z)(3z - 2x)} + \sqrt{(3x - 2y)(3z - 2x)}}.$$

Solution. Modifying the right-hand side of the equation by dividing its numerator and denominator by $\sqrt{(3x-2y)(3y-2z)(3z-2x)}$ leads to

$$\frac{9\sqrt{(3x-2y)(3y-2z)(3z-2x)}}{\sqrt{(3x-2y)(3y-2z)}+\sqrt{(3y-2z)(3z-2x)}+\sqrt{(3x-2y)(3z-2x)}}$$

$$=\frac{9}{\frac{\sqrt{(3x-2y)(3y-2z)}+\sqrt{(3y-2z)(3z-2x)}+\sqrt{(3x-2y)(3z-2x)}}{\sqrt{(3x-2y)(3y-2z)(3z-2x)}}}$$

$$=\frac{9}{\frac{1}{\sqrt{3z-2x}}+\frac{1}{\sqrt{3x-2y}}+\frac{1}{\sqrt{3y-2z}}}.$$

The original equation can be rewritten as

$$(\sqrt{3x-2y}+\sqrt{3y-2z}+\sqrt{3z-2x})\left(\frac{1}{\sqrt{3z-2x}}+\frac{1}{\sqrt{3x-2y}}+\frac{1}{\sqrt{3y-2z}}\right)=9.$$

The next step will be to apply the property proved in problem 2 from chapter 9:

$$(a_1+a_2+a_3)\left(\frac{1}{a_1}+\frac{1}{a_2}+\frac{1}{a_3}\right)\geq 3^2,$$

where each number is positive with equality if and only if all the numbers are equal, $a_1=a_2=a_3$.

According to this inequality, the left-hand side of the equation has to be greater than or equal to 9. Equality is attained only when all three addends are equal to each other: $\sqrt{3x-2y}=\sqrt{3y-2z}=\sqrt{3z-2x}$, which can be written as the system of equations:

$$3x-2y=3y-2z,$$
$$3z-2x=3y-2z.$$

Solving each equation for x gives

$$x=\frac{5y-2z}{3},$$
$$x=\frac{5z-3y}{2}.$$

It follows that $\frac{5y-2z}{3}=\frac{5z-3y}{2}$, which yields $10y-4z=15z-9y$, or equivalently, $y=z$. After substituting this expression into the first equation of the system, we obtain $x=y=z$.

Answer: The equation has indefinitely many solutions: (k,k,k), for any real positive numbers k (the expression under each square root has to be a nonnegative number, excluding 0 in the denominator of the fraction).

Homogeneous equations

In many multivariable equations the key to the solution is finding some relationships between the variables. If you somehow manage to express one variable in terms of the other, you would either simplify the original equation or at least be able to reduce it to a few simpler equations. As we just saw in the problems above, by expressing multiple variables in terms of each other, we managed not only to find the solutions, but also explained the solutions' origin. Surprisingly, you may apply this technique while solving single-variable equations as well. In some cases, you translate the original equation with one variable into a multivariable equation. It significantly simplifies the original equation and provides a path to its solution by expressing the new variables in terms of each other. So-called *homogeneous equations* are a great example of this technique. We will call the equation $P(x_1, x_2, \ldots, x_n) = 0$ a *homogeneous equation* of degree m, if P is a polynomial in n variables in which all the monomials are of the same degree m. The degree of a monomial is the sum of the degrees of its factors. For example, the equation $x^4 + 2yz^2x - 6y^4 + z^3x = 0$ is a homogeneous equation with three variables of the fourth degree because the sum of the degrees of all the variables in each monomial equals 4. Consider a homogeneous equation of degree n with two variables x and y,

$$a_n x^n + a_{n-1}x^{n-1}y + \cdots + a_1 xy^{n-1} + a_0 y^n = 0 \quad \text{(when } a_n \neq 0\text{)}.$$

The pair $(0, 0)$ is a solution of this equation. Assume $a_n \neq 0$ and $x \neq 0$. Then if $y = 0$, the equation will have no solutions. If we divide both sides by x^n and introduce a new variable $z = \frac{y}{x}$, the original equation will be translated into an equation with one variable z, $a_n + a_{n-1}z + \cdots + a_1 z^{n-1} + a_0 z^n = 0$. This property of homogeneous equations is useful in solving certain equations.

Problem 14. Solve the equation

$$(x^2 - x + 1)^2 + (x + 1)^2 = 2(x^3 + 1).$$

Solution. Using the formula for the sum of cubes on the right-hand side, we rewrite the equation as

$$(x^2 - x + 1)^2 + (x + 1)^2 = 2(x + 1)(x^2 - x + 1).$$

It leads to a decision of how to introduce new variables:

$$u = x^2 - x + 1,$$
$$v = x + 1.$$

Hence, we get a homogeneous equation for the variables u and v:

$$u^2 + v^2 = 2uv.$$

Clearly $x = -1$ is not a solution of the original equation ($(1+1+1)^3 + 0 \neq 2 \cdot 0$, $27 = 0$ is wrong). Therefore, $v \neq 0$ and we can divide both sides of the last equation by v^2:

$$\left(\frac{u}{v}\right)^2 - 2\frac{u}{v} + 1 = 0.$$

The left-hand side is a perfect square. So,

$$\left(\frac{u}{v} - 1\right)^2 = 0 \quad \text{then} \quad \frac{u}{v} = 1,$$

which leads to the following equation for x

$$\frac{x^2 - x + 1}{x + 1} = 1,$$
$$x^2 - x + 1 = x + 1,$$
$$x^2 - 2x = 0,$$

or equivalently, $x(x - 2) = 0$, from which $x = 0$ or $x = 2$.

Answer: 0, 2.

Problem 15. Solve the equation

$$6\sqrt[3]{x-3} + \sqrt[3]{x-2} = 5\sqrt[6]{(x-2)(x-3)}.$$

Solution. Let $u = \sqrt[6]{x-3}$ and $v = \sqrt[6]{x-2}$. It follows that $6u^2 + v^2 = 5uv$. It's easy to verify that $x = 2$ is not a solution of the original equation, so we can divide both sides by v^2 ($v \neq 0$):

$$6\left(\frac{u}{v}\right)^2 + 1 - 5\frac{u}{v} = 0.$$

Introducing the new variable $y = \frac{u}{v}$ leads to the equation $6y^2 - 5y + 1 = 0$.

$$D = 25 - 24 = 1,$$
$$y = \frac{5 \pm 1}{12},$$

from which $y = \frac{1}{2}$ or $y = \frac{1}{3}$. Substituting these values back for $\frac{u}{v}$ gives $\frac{u}{v} = \frac{1}{2}$ or $\frac{u}{v} = \frac{1}{3}$. So, $v = 2u$ or $v = 3u$.

Finally, we get two equations to solve for x:

$$\sqrt[6]{x-2} = 2\sqrt[6]{x-3} \quad \text{or} \quad \sqrt[6]{x-2} = 3\sqrt[6]{x-3}.$$

Solving the first equation yields $x - 2 = 64(x - 3)$, or equivalently, $63x = 190$, from which $x = \frac{190}{63} = 3\frac{1}{63}$.

Solving the second equation yields $x - 2 = 729(x - 3)$, or equivalently, $728x = 2185$, from which $x = \frac{2185}{728} = 3\frac{1}{728}$.

The domain of the original equation is determined from the solutions of the inequality $(x - 2)(x - 3) \geq 0$. So, $x \in \,]-\infty, 2] \cup [3, +\infty[$. Clearly, each solution $3\frac{1}{63}$ and $3\frac{1}{728}$ belong to the domain.

Answer: $3\frac{1}{63}, 3\frac{1}{728}$.

Problem 16. Solve the equation

$$9^x + 6^x = 2^{2x+1}.$$

This problem was offered as problem 4 in chapter 8.

As a matter of fact, it's a typical homogeneous equation. Solving the problem in chapter 8, we applied the methods considered for homogeneous equations, but did not fully explain how we made the decision to divide both sides of the equation by 2^{2x}. It should now be much clearer if we introduce the new variables $u = 3^x$ and $v = 2^x$. The equation translates into $u^2 + uv = 2v^2$. By dividing both sides of the last equation by v^2, we get an equation with one variable, and the rest of the solution becomes pretty straightforward. You may return to chapter 8 and go over the solution one more time, viewing the equation now as homogeneous.

The technique can be extended to trigonometric equations of the type

$$a_0 \sin^n x + a_1 \sin^{n-1} x \cos x + a_2 \sin^{n-2} x \cos^2 x + \cdots + a_n \cos^n x = 0, \quad (*)$$

where the coefficients $a_0, a_1, \ldots, a_n$ are real numbers. Such an equation is called a homogeneous equation of degree n for $\sin x$ and $\cos x$. If $a_0 = 0$, then all the solutions of the equation $\cos x = 0$ (the numbers $\frac{\pi}{2} + \pi k, k \in Z$) satisfy $(*)$. If $a_0 \neq 0$, then the numbers $\frac{\pi}{2} + \pi k, k \in Z$ will not be solutions of $(*)$, because for $x = \frac{\pi}{2} + \pi k, k \in Z, \cos x = 0, \sin x = \pm 1$ and the left-hand side of $(*)$ becomes $\pm a_0 \neq 0$. So, now we will be looking for the solutions of $(*)$ for all x such that $x \neq \frac{\pi}{2} + \pi k, k \in Z$. Since in this case $\cos x \neq 0$, dividing both sides of $(*)$ by $\cos^n x$ gives

$$a_0 \tan^n x + a_1 \tan^{n-1} x + \cdots + a_{n-1} \tan x + a_n = 0.$$

The last equation is equivalent to $(*)$ when $a_0 \neq 0$ and is solvable as an algebraic equation of one variable $\tan x$.

If readers want to see more about homogeneous equations, they may read a very good article "Homogeneous equations" by L. Ryzhkov and Y. Ionin in the currently defunct magazine *Quantum*, May/June 1998 issue. It has a great collection of problems on homogeneous equations giving explanations and detailed solutions.

The Cartesian coordinate system and vector algebra method

We will close this chapter with an intriguing alternative to the conventional and traditional methods for equations solving attributed to the Cartesian coordinate system and vector algebra techniques. The unique and very powerful Cartesian coordinate system method was first introduced by the outstanding French mathematician and philosopher René Descartes (1596–1650), whose name has been mentioned before. One of the key figures in the scientific revolution of medieval times, René Descartes made tremendous contributions to the field of mathematics. His devotion to mathematics is best described in his own words: ". . . In my opinion, all things in nature occur mathematically."

He is considered the father of analytic geometry, the very first and the most complete system linking together algebra and geometry; he was the first to consider algebra as "universal mathematics" and viewed it as the symbolic logic allowing the description of logical principles symbolically through variables and operations with them. His work laid the foundation for the later development of calculus by Isaac Newton and Gottfried Leibniz. Many historians believe that his ideas had great influence on the development of Isaac Newton's views of algebraic methods and specifically in the treatment of independent variables.

We expect readers to be familiar with the basic definitions and properties of vector algebra in two- and three-dimensional space. Let's make a brief visit to some of them. As we know, any point in a Cartesian coordinate system is defined by its coordinates. Having the coordinates of the points $A(x_A, y_A)$ and $B(x_B, y_B)$ in two-dimensional space, the distance between them is calculated by the formula

$$AB = \sqrt{(x_B - x_A)^2 + (y_B - y_A)^2}.$$

A vector as a geometric subject is defined by its magnitude or length and its direction. In the Cartesian coordinate system, a vector is identified by the coordinates of its initial and terminal point, that is, a vector $\overrightarrow{AB}$ has coordinates (or components) $(x_B - x_A, y_B - y_A)$. The length of a vector $\overrightarrow{AB}$ is determined in a similar way as the length of a segment AB. The analogous definitions are applicable for vectors in three-dimensional space. Since each point in three-dimensional space has three coordinates, then the length of the segment AB, where $A(x_A, y_A, z_A)$ and $B(x_B, y_B, z_B)$, is calculated as

$$AB = \sqrt{(x_B - x_A)^2 + (y_B - y_A)^2 + (z_B - z_A)^2}.$$

The *vector scalar product* (or *dot product*) equals the product of the vectors' lengths by the cosine of the angle they form. If two vectors are given by the coordinates $\vec{a} = (x, y)$ and $\vec{b} = (m, n)$, and γ is the angle they form, then their scalar product is calculated by the formula $\vec{a} \cdot \vec{b} = \|\vec{a}\| \cdot \|\vec{b}\| \cdot \cos \gamma$, or in a coordinate form, $\vec{a} \cdot \vec{b} = xm + yn$. In solving many problems utilization of the vector scalar product definition provides an elegant and easy solution. It allows us to establish relationships between variables leading to important conclusions regarding an equation's solvability, as it will be demonstrated in the following problem 17.

Problem 17. Solve the equation

$$\sqrt{x^2 + y^2} = \frac{x + 4y - 1}{\sqrt{17}}.$$

Solution. The equation can be rewritten as

$$\sqrt{17} \cdot \sqrt{x^2 + y^2} = x + 4y - 1.$$

Consider two vectors $\vec{a} = (x, y)$ and $\vec{b} = (1, 4)$. Their lengths are

$$\|\vec{a}\| = \sqrt{x^2 + y^2} \quad \text{and} \quad \|\vec{b}\| = \sqrt{1^2 + 4^2} = \sqrt{17}.$$

The scalar product of the vectors in a coordinate form is

$$\vec{a} \cdot \vec{b} = x \cdot 1 + y \cdot 4 = x + 4y.$$

On the other hand, if they form an angle γ, then the vectors' scalar product equals

$$\vec{a} \cdot \vec{b} = \|\vec{a}\| \cdot \|\vec{b}\| \cdot \cos \gamma = \sqrt{x^2 + y^2} \cdot \sqrt{17} \cdot \cos \gamma.$$

No matter what the value of γ is, $|\cos \gamma| \leq 1$. Therefore, $\|\vec{a}\| \cdot \|\vec{b}\| \geq \vec{a} \cdot \vec{b}$, which means that $\sqrt{x^2 + y^2} \cdot \sqrt{17} \geq x + 4y > x + 4y - 1$. From the last inequality we deduce that the original equation has no solutions.

Two vectors are called collinear if they lie on the same line or on parallel lines. They have either the same or opposite direction and we can express one of the two vectors in terms of the other by multiplying by some number k: if two vectors $\vec{m}$ and $\vec{n}$ are collinear, then there is a number k such as $\vec{m} = k \cdot \vec{n}$. The converse statement is correct as well, that is, if there exists a number k such as $\vec{m} = k \cdot \vec{n}$, then the vectors $\vec{m}$ and $\vec{n}$ are collinear. In a coordinate form this means that the vectors' coordinates are proportional. If the vectors have the same direction, k is a positive number; when the vectors have opposite directions, k is a negative number. The collinearity of vectors can be used to prove a lot of things in geometry and in vector algebra. For example, if it is given that three points are on the same straight line, the vectors passing through these points must be collinear. An application of these statements is illustrated in the following problem 18.

Problem 18. Solve the equation

$$\sqrt{x^2 + (y - 1)^2 + (z + 3)^2} + \sqrt{(x - 2)^2 + (y - 4)^2 + (z - 3)^2} = 7.$$

Solution. In this problem we extend the Cartesian coordinate system method to three dimensions. Consider points with their coordinates, $A(0, 1, -3)$, $B(2, 4, 3)$, and $C(x, y, z)$. The distance in a three-dimensional space between the pairs of points is expressed in Cartesian coordinates as

$$AC = \sqrt{x^2 + (y-1)^2 + (z+3)^2},$$

$$BC = \sqrt{(x-2)^2 + (y-4)^2 + (z-3)^2},$$

$$AB = \sqrt{(2-0)^2 + (4-1)^2 + (3-(-3))^2} = \sqrt{4+9+36} = \sqrt{49} = 7.$$

With the conditions given, we see that $AC + BC = AB$, which implies that C belongs to segment AB. Hence, the three vectors $\overrightarrow{AC}$, $\overrightarrow{CB}$, and $\overrightarrow{AB}$ are collinear and therefore their respective coordinates must be proportional. Since the vectors $\overrightarrow{AC}$ and $\overrightarrow{AB}$ are defined by their coordinates as $\overrightarrow{AC} = (x, y-1, z+3)$ and $\overrightarrow{AB} = (2, 3, 6)$ the collinearity of vectors $\overrightarrow{AC}$ and $\overrightarrow{AB}$ can be expressed in a coordinate form as

$$\frac{x}{2} = \frac{y-1}{3} = \frac{z+3}{6}.$$

Introducing k as the ratio $k = \frac{x}{2} = \frac{y-1}{3} = \frac{z+3}{6}$, we obtain that $x = 2k$, $y = 3k+1$, and $z = 6k - 3$. C lies between A and B, so the ratio k has to satisfy the conditions $0 \le k \le 1$. Therefore, we can conclude that the equation has infinitely many solutions expressed as triples $(2k, 3k+1, 6k-3)$, where k satisfies the condition $0 \le k \le 1$.

There are a number of other methods for solving multivariable equations and there are many pursuits one can take for further explorations. In this chapter we gave those that we believe are most commonly used. In the next chapter we are going to extend our studies of multivariable equations and cover an interesting and compelling but less well-known technique for their solutions involving linear homogeneous recurrences and their properties. Meanwhile, see a few exercises to have some practice with the methods discussed earlier.

Exercises

Problem 19. Solve the equation $2 + y^2 - 2y = 1 - \cos^2 \frac{x}{y}$.

Problem 20. Solve the equation $\left(2 + \frac{1}{\cos^2 x}\right)(4 - 2\cos^4 x) = 1 + 5\sin 3y$.

Problem 21. Solve the equation $2^{|x|} - \cos y + \ln(1 + x^2 + |y|) = 0$.

Problem 22. Solve the equation

$$\sqrt{2x-y} + \sqrt{2z-x} + \sqrt{2y-z} = \frac{\sqrt{81(2x-y)(2z-x)(2y-z)}}{\sqrt{(2x-y)(2z-x)} + \sqrt{(2x-y)(2y-z)} + \sqrt{(2z-x)(2y-z)}}.$$

Problem 23. Solve the equation

$$\sqrt{(1+u)^2 + (v-3)^2} + \sqrt{(u+2)^2 + 9v^2} = |2v+3|.$$

Problem 24. Solve the equation $x^2 + xy - \sqrt{3}x + y^2 - \sqrt{3}y + 1 = 0$.

Problem 25. Solve the equation $(x^2 - 4x + 5)(y^2 - 6y + 10) = 1$.

Chapter 11

Search for Natural Solutions

Mathematics is the gate and key to the sciences
Roger Bacon.

From the conversation of two friends:
"What was the bravest thing you have ever done in your life?"
"I raised my hand once in an algebra class."

As funny as it sounds, unfortunately sometimes the common perception of algebra and math is that they are scary and intimidating. Many students are afraid to raise their hand and ask a question. On the other hand, kids who really love math and are eager to learn hit on interesting questions and find challenging problems intriguing. Education then becomes mutually beneficial: as a teacher you help your students and at the same time you learn from them and along with them.

"May I ask you for help with an equation I can't solve? I got it from one of the math journals I subscribe to" inquired one of the best students in a class, with a friendly expression on her face. It was early in September of the second year of my teaching career. I inherited the class from a very good teacher who took maternity leave. The kids missed her and were not happy having the new person. Many of them had solid backgrounds and expressed strong interest in math. By asking that question, the girl was trying to achieve two goals: first, to get the problem solved (she admitted later that she spent a lot of time without any result) and second, probably even more importantly, to test the new young teacher.

It was a non-standard Diophantine equation, which did not look familiar to me at all. After a few sleepless nights I came back to the girl with the solution. Some scientists believe that under stress our capabilities increase and sometimes people produce great results in a very short time just because of being afraid of failure. I think in my case it was a valid proof of that theory. I have no idea how it was solved that quickly; most likely, it was because I understood how important it was to prove myself to the kids from the very first days of my teaching. It was the first step in establishing the mutual trust and respect. The problem itself (let's call it the Main Problem) turned out to be very interesting, not just because of its complexity, but mostly because of the methods used in its solution and their application to other related problems. Moreover, it unexpectedly brought up the discussion of second order linear homogeneous recurrences and their properties, which in turn led to the famous *Fibonacci sequence*, and its connections with the *golden ratio*. We'll consider the Main Problem's analysis here as the essential continuation of the methods outlined in the previous chapter.

Main Problem. *Given the equation* $4x^k + (x+1)^2 = y^2$. *Prove that*

1) *There are no natural solutions when* $k = 1$.

2) *There are at least two natural solutions when* $k = 2$.

3) *There exists an infinite number of natural solutions when* $k = 2$.

There are a few different approaches to the solution of the problem. We will go over the solution involving exploring infinite sequences of natural numbers that satisfy specific conditions for their terms. Before proceeding directly to the solution of the problem, we have to cover some introductory important explanations.

Many real-life problems deal with infinite sequences and their properties. One of the most difficult problems related to infinite sequences is to determine (if possible, of course) an explicit rule that gives its nth term as a function of the term's position number n in the sequence. There is another way to define a sequence: by a recursive rule. A recursive rule gives the beginning term(s) of a sequence and an equation (recursive equation) connecting the nth term with one or more preceding terms. For example, some of the well-known recursive equations for arithmetic and geometric sequences respectively are:

$a_n = a_{n-1} + d$, where d is the common difference.
$a_n = r \cdot a_{n-1}$, where r is the common ratio.

When each term in a sequence a_n is expressed as linear combination of k preceding terms, then the relationship is called a linear homogeneous recurrence relation of degree k and $a_n = c_1 a_{n-1} + c_2 a_{n-2} + \cdots + c_k a_{n-k}$, where coefficients $c_1, c_2, \ldots, c_k$ are real numbers and $c_k \neq 0$. The transition from the recurrence formula to the explicit formula $a_n = r^n$ is achieved if and only if $r^n = c_1 r^{n-1} + c_2 r^{n-2} + \cdots + c_k r^{n-k}$. Dividing both sides of the last equality by r^{n-k} and collecting terms gives $r^k - c_1 r^{k-1} - c_2 r^{k-2} - \cdots - c_{k-1} r - c_k = 0$, which is called the *characteristic equation* of the recurrence relation.

The polynomial $r^k - c_1 r^{k-1} - c_2 r^{k-2} - \cdots - c_{k-1} r - c_k$ is called the *char-acteristic polynomial* or *auxiliary polynomial*. Its roots are called the characteristic roots of the recursive relation and can be used to find the transition rule to give an explicit formula for the sequence. If those roots exist, then the transition from recursive to explicit rule will be accomplished. The following theorem gives necessary and sufficient conditions:

Let $c_1, c_2, \ldots, c_k$ be real coefficients in the characteristic equation $r^k - c_1 r^{k-1} - c_2 r^{k-2} - \cdots - c_{k-1} r - c_k = 0$, which has k distinct roots $r_1, r_2, \ldots, r_k$. A sequence $\{a_n\}$ is a solution of the recurrence relation

$$a_n = c_1 a_{n-1} + c_2 a_{n-2} + \cdots + c_k a_{n-k}$$

if and only if

$$a_n = \gamma_1 r_1^n + \gamma_2 r_2^n + \cdots + \gamma_k r_k^n \quad \text{for } n = 0, 1, 2, \ldots,$$

where $\gamma_1, \gamma_2, \ldots, \gamma_k$ are real numbers.

We are not going to prove this theorem and cover in detail the general case of linear homogeneous recurrence relations of degree k here, but instead we will concentrate on the investigation of second order linear homogeneous recurrences. A second order linear homogeneous recurrence is defined as a recurrence of the form $a_n = c_1 a_{n-1} + c_2 a_{n-2}$, where c_1 and c_2 are some real numbers.

During the solution of the Main Problem we will come across a second order linear homogeneous recurrence for the specifically built sequence of numbers n_i. The goal will be to find the explicit formula and then analyze the result to get to the final solution.

Let us first introduce two important lemmas that will be used in all the following considerations.

Lemma 1. *Given two sequences $\{a_n\}$ and $\{b_n\}$ such that the nth term of each sequence satisfies the recursive rule $r_n = \alpha r_{n-1} + \beta r_{n-2}$, where α and β are some real numbers. Then, for any real numbers c_1 and c_2 the sequence $\{c_1 a_n + c_2 b_n\}$ will also satisfy the same recursive rule.*

Proof. Let's write the given recursive rule for $\{a_n\}$ and $\{b_n\}$:

$$a_n = \alpha a_{n-1} + \beta a_{n-2},$$
$$b_n = \alpha b_{n-1} + \beta b_{n-2}.$$

It follows that, for any real numbers c_1 and c_2

$$\begin{aligned} c_1 a_n + c_2 b_n &= c_1(\alpha a_{n-1} + \beta a_{n-2}) + c_2(\alpha b_{n-1} + \beta b_{n-2}) \\ &= c_1 \alpha a_{n-1} + c_1 \beta a_{n-2} + c_2 \alpha b_{n-1} + c_2 \beta b_{n-2} \\ &= \alpha(c_1 a_{n-1} + c_2 b_{n-1}) + \beta(c_1 a_{n-2} + c_2 b_{n-2}). \end{aligned}$$

We see that the lemma's statement is correct. $\qquad\square$

Lemma 2. *Let α and β be real numbers and suppose $r^2 = \alpha r + \beta$ is the characteristic equation of a second order linear homogeneous recurrence that has two distinct roots r_1 and r_2. The sequence $\{a_n\}$ is a solution of the recurrence relation*

$$a_n = \alpha a_{n-1} + \beta a_{n-2}$$

if and only if

$$a_n = cr_1^n + dr_2^n \quad for\ n = 0, 1, 2, \ldots,$$

where c and d are real numbers dependent upon the initial conditions for the recursive formula.

Proof. First, notice that the values of the first and second term, a_1 and a_2, will define a unique sequence. Those values will be used to determine the coefficients c and d. Next, let's solve the equation $r^2 - \alpha r - \beta = 0$.
$r_{1,2} = \frac{\alpha \pm \sqrt{\alpha^2 + 4\beta}}{2}$. Selecting the root $r_1 = \frac{\alpha + \sqrt{\alpha^2 + 4\beta}}{2}$, we will build the sequence $\left\{ a_n = c \cdot \left(\frac{\alpha + \sqrt{\alpha^2 + 4\beta}}{2} \right)^n \right\}$:

$$a_1 = c \cdot \frac{\alpha + \sqrt{\alpha^2 + 4\beta}}{2}, a_2 = c \cdot \left(\frac{\alpha + \sqrt{\alpha^2 + 4\beta}}{2} \right)^2, \ldots, a_n = c \cdot \left(\frac{\alpha + \sqrt{\alpha^2 + 4\beta}}{2} \right)^n.$$

Our goal is to prove that for this sequence the following recursive rule is true: $a_n = \alpha \cdot a_{n-1} + \beta \cdot a_{n-2}$, or substituting the value of r_1, we have to prove that

$$c \cdot \left(\frac{\alpha + \sqrt{\alpha^2 + 4\beta}}{2} \right)^n = c\alpha \left(\frac{\alpha + \sqrt{\alpha^2 + 4\beta}}{2} \right)^{n-1} + c\beta \left(\frac{\alpha + \sqrt{\alpha^2 + 4\beta}}{2} \right)^{n-2}.$$

Consider the right-hand side of the last equality. Factoring $c \left(\frac{\alpha}{2} + \sqrt{\left(\frac{\alpha}{2} \right)^2 + \beta} \right)^{n-2}$ out of each term, modifies it to:

$$\alpha \cdot a_{n-1} + \beta \cdot a_{n-2} = c\alpha \left(\frac{\alpha}{2} + \sqrt{\left(\frac{\alpha}{2} \right)^2 + \beta} \right)^{n-1} + c\beta \left(\frac{\alpha}{2} + \sqrt{\left(\frac{\alpha}{2} \right)^2 + \beta} \right)^{n-2}$$

$$= \underbrace{c \left(\frac{\alpha}{2} + \sqrt{\left(\frac{\alpha}{2} \right)^2 + \beta} \right)^{n-2}}_{common\ factor} \cdot \left(\alpha \cdot \left(\frac{\alpha}{2} + \sqrt{\left(\frac{\alpha}{2} \right)^2 + \beta} \right) + \beta \right)$$

$$= c \left(\frac{\alpha}{2} + \sqrt{\left(\frac{\alpha}{2} \right)^2 + \beta} \right)^{n-2} \cdot \underbrace{\left(\frac{\alpha^2}{2} + \alpha \cdot \sqrt{\left(\frac{\alpha}{2} \right)^2 + \beta} + \beta \right)}_{perfect\ square}. \qquad (*)$$

The second factor in the last expression is a perfect square. Indeed,

$$\frac{\alpha^2}{2} + \alpha \cdot \sqrt{\left(\frac{\alpha}{2}\right)^2 + \beta} + \beta = \frac{\alpha^2}{4} + 2 \cdot \frac{\alpha}{2} \cdot \sqrt{\left(\frac{\alpha}{2}\right)^2 + \beta} + \frac{\alpha^2}{4} + \beta$$

$$= \left(\frac{\alpha}{2} + \sqrt{\left(\frac{\alpha}{2}\right)^2 + \beta}\right)^2.$$

Substituting this into (∗) and using the property of exponential functions $x^y \cdot x^z = x^{y+z}$ gives

$$c \left(\frac{\alpha}{2} + \sqrt{\left(\frac{\alpha}{2}\right)^2 + \beta}\right)^{n-2} \cdot \left(\frac{\alpha}{2} + \sqrt{\left(\frac{\alpha}{2}\right)^2 + \beta}\right)^2 = c \left(\frac{\alpha}{2} + \sqrt{\left(\frac{\alpha}{2}\right)^2 + \beta}\right)^{n-2+2}$$

$$= c \left(\frac{\alpha + \sqrt{\alpha^2 + 4\beta}}{2}\right)^n$$

$$= a_n,$$

which completes the proof.   □

If instead of r_1 we select the second root $r_2 = \frac{\alpha - \sqrt{\alpha^2 + 4\beta}}{2}$ and the coefficient d, the proof would be identical. Using Lemma 1, we can state now that the sequence $a_n = cr_1^n + dr_2^n$ satisfies the recursive rule $a_n = \alpha a_{n-1} + \beta a_{n-2}$. Numbers c and d are to be selected in such a way that after their substitution into the recursive formula for a_1 and a_2, the solution of the system of two equations produces the given values for the first and the second term of the sequence in each specific case.

Let's now prove the converse statement by using mathematical induction. It is given that $a_1 = cr_1 + dr_2, a_2 = cr_1^2 + dr_2^2$, and $a_n = (r_1 + r_2) \cdot a_{n-1} - r_1 r_2 \cdot a_{n-2}$. We need to prove that the explicit formula for nth term of this sequence is $a_n = cr_1^n + dr_2^n$. Let's check it for $n = 3$:

$$a_3 = (r_1 + r_2) \cdot a_2 - r_1 r_2 \cdot a_1 = (r_1 + r_2) \cdot (cr_1^2 + dr_2^2) - r_1 r_2 \cdot (cr_1 + dr_2)$$

$$= cr_1^3 + cr_2 r_1^2 + dr_1 r_2^2 + dr_2^3 - cr_2 r_1^2 - dr_1 r_2^2 = cr_1^3 + dr_2^3.$$

As you see, it works. Next, assuming that the equality has already been proved for $n \le k$, let's make sure it's true for $n = k + 1$. So, we assume that if $a_k = (r_1 + r_2)a_{k-1} - r_1 r_2 \cdot a_{k-2}$, then $a_k = cr_1^k + dr_2^k$ is true. We need to show that under this assumption the following statement will also be true:

$$a_{k+1} = cr_1^{k+1} + dr_2^{k+1}.$$

Consider and modify a_{k+1},

$$a_{k+1} = (r_1 + r_2) \cdot a_k - r_1 r_2 \cdot a_{k-1}$$
$$= (r_1 + r_2)(cr_1^k + dr_2^k) - r_1 r_2(cr_1^{k-1} + dr_2^{k-1})$$
$$= cr_1^{k+1} + dr_1 r_2^k + cr_2 r_1^k + dr_2^{k+1} - cr_2 r_1^k - dr_1 r_2^k$$
$$= cr_1^{k+1} + dr_2^{k+1}.$$

Using the induction assumption, the statement has to hold for all natural n, which completes the proof.

Lemmas 1 and 2 provide the rules for translating from recursive to explicit formulas (and vice versa) for second degree linear homogeneous recurrences. Before we come back to our Main Problem, let's go over some examples to become more familiar with the method and its application.

Problem 1. Two sequences are given $\{x_n\}$ and $\{y_n\}$ by the recursive rules $x_1 = 1$, $x_2 = 2$, $x_n = 4x_{n-1} - 3x_{n-2}$ and $y_1 = \frac{1}{2}$, $y_2 = \frac{3}{2}$, $y_n = 7y_{n-1} - 12y_{n-2}$. Prove that for any natural number n: $x_n - y_n = \frac{1}{2}$.

Proof. To solve the problem, one can translate from recursive rules for nth term to explicit rules. Solving by Viète's formulas the quadratic equation $r^2 - 4r + 3 = 0$ gives the roots, $r_1 = 1, r_2 = 3$. It follows that, for the first sequence its explicit formula is $x_n = a \cdot 1^n + b \cdot 3^n$. We just need to select a and b to satisfy the conditions $x_1 = 1$, $x_2 = 2$. The solutions of the following system of the linear equations will provide the desired values:

$$a + 3b = 1,$$
$$a + 9b = 2.$$

Subtracting the first equation from the second equation gives $b = \frac{1}{6}$. Substituting $b = \frac{1}{6}$ into the first equation gives $a = \frac{1}{2}$. Therefore, the explicit formula for the first sequence is

$$x_n = \frac{1}{2} \cdot 1^n + \frac{1}{6} \cdot 3^n = \frac{1}{2} + \frac{1}{6} \cdot 3^n.$$

Similarly, we find the explicit formula for the second sequence. Solving the equation $r^2 - 7r + 12 = 0$ gives the roots $r_1 = 3$, $r_2 = 4$. It follows that $y_n = c \cdot 3^n + d \cdot 4^n$. To satisfy the conditions for $y_1 = \frac{1}{2}$ and $y_2 = \frac{3}{2}$, coefficients c and d can be calculated from the following system of equations:

$$3c + 4d = \frac{1}{2},$$
$$9c + 16d = \frac{3}{2}.$$

Multiplying the first equation by 3 and subtracting it from the second equation gives $d = 0$. Substituting $d = 0$ into the first equation gives $c = \frac{1}{6}$. Therefore,

$y_n = \frac{1}{6} \cdot 3^n$. The last step is to find the difference

$$x_n - y_n = \frac{1}{2} + \frac{1}{6} \cdot 3^n - \frac{1}{6} \cdot 3^n = \frac{1}{2},$$

which is the desired result. Our proof is completed.   □

Problem 2. Prove that each term in the sequence $a_n = \frac{(2+\sqrt{3})^n - (2-\sqrt{3})^n}{2\sqrt{3}}$ is an integer for any natural number n.

Solution. Using Viète's formulas to find the coefficients of a quadratic equation that has the roots $r_1 = 2 + \sqrt{3}$ and $r_2 = 2 - \sqrt{3}$ gives

$$r_1 + r_2 = 2 + \sqrt{3} + 2 - \sqrt{3} = 4,$$
$$r_1 \cdot r_2 = (2 + \sqrt{3}) \cdot (2 - \sqrt{3}) = 4 - 3 = 1.$$

Based on Lemma 2, the recursive formula for the given sequence is

$a_n = 4a_{n-1} - a_{n-2}$ and

$$a_1 = \frac{(2+\sqrt{3})^1 - (2-\sqrt{3})^1}{2\sqrt{3}} = \frac{2+\sqrt{3}-2+\sqrt{3}}{2\sqrt{3}} = 1,$$

$$a_2 = \frac{(2+\sqrt{3})^2 - (2-\sqrt{3})^2}{2\sqrt{3}} = \frac{4+4\sqrt{3}+3-4+4\sqrt{3}-3}{2\sqrt{3}} = \frac{8\sqrt{3}}{2\sqrt{3}} = 4.$$

Therefore, the first two terms in the sequence are integers and all the other terms, starting with the third, satisfy $a_n = 4a_{n-1} - a_{n-2}$. So, each term is the result of some simple arithmetic operations with integers. Thus, all the terms in this sequence are integers, which is what had to be proved.

Problem 3. Given the quadratic equation $ax^2 + bx + c = 0$, $a \neq 0$. Let $S_n = \alpha^n + \beta^n$, where α and β are the roots of the equation. Establish a relationship between S_n, S_{n+1}, and S_{n+2}.

Solution. Rewrite the equation as

$$x^2 + \frac{b}{a}x + \frac{c}{a} = 0.$$

By Viète's formulas,

$$\alpha + \beta = -\frac{b}{a}, \quad \alpha \cdot \beta = \frac{c}{a}.$$

According to Lemma 2, we get

$$S_{n+2} = -\frac{b}{a} \cdot S_{n+1} - \frac{c}{a} \cdot S_n.$$

Problem 4. Show that the number $(7 + \sqrt{48})^{135} + (7 - \sqrt{48})^{135}$ is an integer and is divisible by 14.

Solution. Considering the sequence $a_n = (7+\sqrt{48})^n + (7-\sqrt{48})^n$, let's find its recursive formula. The numbers $7+\sqrt{48}$ and $7-\sqrt{48}$ are the roots of the quadratic equation $r^2 - 14r + 1 = 0$, because $7+\sqrt{48}+7-\sqrt{48} = 14$ and $(7+\sqrt{48})(7-\sqrt{48}) = 49 - 48 = 1$. Therefore, the recursive formula can be expressed as $a_n = 14a_{n-1} - a_{n-2}$. Calculating the values of a_1 and a_2, we see that $a_1 = 7+\sqrt{48}+7-\sqrt{48} = 14$, $a_2 = (7+\sqrt{48})^2 + (7-\sqrt{48})^2 = 49 + 14\sqrt{48} + 48 + 49 - 14\sqrt{48} + 48 = 194$. Therefore, the first two terms in the sequence are integers and all the other terms, starting with the third, satisfy $a_n = 14a_{n-1} - a_{n-2}$. So, each term is the result of some simple arithmetic operations with integers. Thus, all the terms in this sequence are integers. The number $(7+\sqrt{48})^{135} + (7-\sqrt{48})^{135}$ represents the 135th term in this sequence, so it has to be an integer as well. To prove that it is divisible by 14, it suffices to apply property 3 of division of a sum of two powers with a natural odd exponent by the sum of their bases, as studied in chapter 4. Since the sum of the bases equals 14, it is divisible by 14. Thus, the sum of the powers must be divisible by 14 as well.

Problem 5. Given the sequence $x_n = \left(\frac{3+\sqrt{5}}{2}\right)^n + \left(\frac{3-\sqrt{5}}{2}\right)^n - 2$, where n is a natural number. Prove that every term in this sequence is a natural number of the form $5t^2$ for even terms and of the form p^2 for odd terms, where t and p are some integers ($t \in Z, p \in Z$).

Proof. First of all, let's make two important observations:

$$\left(\frac{1\pm\sqrt{5}}{2}\right)^2 = \frac{1\pm2\sqrt{5}+5}{4} = \frac{3\pm\sqrt{5}}{2} \quad \text{and}$$

$$\left(\frac{\sqrt{5}+1}{2}\right)^n \cdot \left(\frac{\sqrt{5}-1}{2}\right)^n = \left(\frac{5-1}{4}\right)^n = 1.$$

Therefore, after completing a perfect square, the given expression can be rewritten as

$$\left(\frac{3+\sqrt{5}}{2}\right)^n + \left(\frac{3-\sqrt{5}}{2}\right)^n - 2$$

$$= \left(\frac{1+\sqrt{5}}{2}\right)^{2n} + \left(\frac{\sqrt{5}-1}{2}\right)^{2n} - 2\cdot\left(\frac{1+\sqrt{5}}{2}\right)^n \cdot \left(\frac{\sqrt{5}-1}{2}\right)^n$$

$$= \left(\left(\frac{1+\sqrt{5}}{2}\right)^n - \left(\frac{\sqrt{5}-1}{2}\right)^n\right)^2.$$

Let us now investigate the properties of the sequence

$$a_n = \left(\frac{1+\sqrt{5}}{2}\right)^n - \left(\frac{\sqrt{5}-1}{2}\right)^n.$$

As in the previous problems, relying on Lemma 2 we will translate from an explicit formula to a recursive formula. In order to do that, we need to find the coefficients of the quadratic equation with roots $\frac{1+\sqrt{5}}{2}$ and $\frac{\sqrt{5}-1}{2}$:

$$r_1 + r_2 = \frac{1+\sqrt{5}}{2} + \frac{\sqrt{5}-1}{2} = \sqrt{5},$$

$$r_1 \cdot r_2 = \frac{1+\sqrt{5}}{2} \cdot \frac{\sqrt{5}-1}{2} = \frac{(\sqrt{5}+1)(\sqrt{5}-1)}{4} = \frac{5-1}{4} = 1.$$

Therefore, $a_n = \sqrt{5}a_{n-1} - a_{n-2}$.

If $n = 1$, then $a_1 = \frac{1+\sqrt{5}}{2} - \frac{\sqrt{5}-1}{2} = 1$; if $n = 2$, then $a_2 = \left(\frac{1+\sqrt{5}}{2}\right)^2 - \left(\frac{\sqrt{5}-1}{2}\right)^2 = \frac{1+2\sqrt{5}+5}{4} - \frac{1-2\sqrt{5}+5}{4} = \frac{4\sqrt{5}}{4} = \sqrt{5}$.

We will use mathematical induction to prove that every odd term, starting from a_3 ($a_3 = \sqrt{5} \cdot \sqrt{5} - 1 = 4$), in this sequence is an integer and every even term has the form $\sqrt{5}m$, where m is some natural number.

Suppose that the statement is correct for $n = k$, that is, a_{2k-1} is an integer number and $a_{2k} = \sqrt{5}m$, $m \in N$. Then for every odd term $a_{2k+1} = \sqrt{5}a_{2k} - a_{2k-1} = \sqrt{5} \cdot \sqrt{5}m - a_{2k-1} = 5m - a_{2k-1}$. $5m$ is a natural number and a_{2k-1} is an integer due to the induction assumption. Therefore, their difference has to be an integer number. So, we proved that every odd term is an integer.

Consider now an even term

$$a_{2k+2} = \sqrt{5}a_{2k+1} - a_{2k} = \sqrt{5}a_{2k+1} - \sqrt{5}m = \sqrt{5}(a_{2k+1} - m).$$

We proved already that a_{2k+1} is an integer. Therefore, $(a_{2k+1} - m)$ must be an integer as well and if we let $a_{2k+1} - m = t$, $t \in Z$, then $a_{2k+2} = \sqrt{5}t$, $t \in Z$. So, by induction our statement holds for all natural n.

Now, having justified that every odd term in the sequence is an integer and every even term has the form $\sqrt{5}t$, $t \in Z$, recall that our discussion was held for the sequence $a_n = \left(\frac{1+\sqrt{5}}{2}\right)^n - \left(\frac{\sqrt{5}-1}{2}\right)^n$. If we now go back to the originally given sequence and observe that every term in it is a square of the respective term in the sequence above,

$$x_n = \left(\frac{3+\sqrt{5}}{2}\right)^n + \left(\frac{3-\sqrt{5}}{2}\right)^n - 2 = \left(\left(\frac{1+\sqrt{5}}{2}\right)^n - \left(\frac{\sqrt{5}-1}{2}\right)^n\right)^2 = a_n^2,$$

we conclude that the statement of the problem is true. Indeed, every term in the original sequence, being the square of an integer, must be a natural number; also, every even term is a square of $\sqrt{5}t$, which is $5t^2$, $t \in Z$, and every odd term is a square of some integer, or p^2, $p \in Z$. The proof is completed.   □

Armed with the two important lemmas and practical application of the methods, we are now ready to proceed to the Main Problem's solution.

Main Problem Solution. If $k = 1$, the equation becomes $4x + (x + 1)^2 = y^2$. It can be rewritten as $x^2 + 6x + 1 = y^2$, or equivalently, after completing a square on the left-hand side, as

$$(x + 3)^2 - 8 = y^2. \tag{1}$$

Let's analyze it.

Since x and y have to be natural solutions, then $x > 0$, $y > 0$ and from the original equation we see that $y^2 = 4x + (x + 1)^2 > (x + 1)^2$. Thus, $y > x + 1$. On the other hand, $y^2 = (x + 3)^2 - 8 < (x + 3)^2$, which yields $y < x + 3$. Combining these two conclusions yields $x + 1 < y < x + 3$. Hence, to be a natural number and satisfy the last double-sided inequality, y must equal to $x + 2$, $y = x + 2$. Substituting this expression into (1) and solving for x gives

$$(x + 3)^2 - 8 = (x + 2)^2,$$
$$x^2 + 6x + 9 - 8 = x^2 + 4x + 4,$$
$$2x = 3.$$

It follows $x = 1.5$, which is not a natural solution. Since $x = 1.5$ is the unique solution of the last equation, the first statement of the problem is correct and the original equation does not have any natural solutions when $k = 1$.

Let's now turn to the second part of the problem. Assume $k = 2$. The equation becomes $4x^2 + (x + 1)^2 = y^2$, or using the difference of squares formula, it can be rewritten as

$$(y - 2x)(y + 2x) = (x + 1)^2. \tag{2}$$

The expression on the right-hand side of the equation is a square. Remember, we are looking for the natural solutions, so $x \neq 0$. It follows that the factors on the left-hand side cannot be equal to each other. Assume $(y - 2x)$ and $(y + 2x)$ are relatively prime numbers. If they are not relatively prime, then the number on the right-hand side should be divisible by their common factor and by dividing both sides of the equation by such a number the equation (2) can always be rewritten as having relatively prime factors on the left-hand side. Hence, (2) will hold true for natural numbers x and y only when $(y - 2x)$ and $(y + 2x)$ each represent some square of a natural number (if two relatively prime numbers are not equal and their product equals a square of some third number, then each of them must be the square of some number):

$$y - 2x = n^2,$$
$$y + 2x = m^2.$$

If we manage to identify at least two pairs of natural numbers m and n such that the system has natural solutions, our goal will be accomplished.

Subtracting the first equation from the second gives

$$x = \frac{m^2 - n^2}{4}. \tag{3}$$

Adding the equations gives

$$y = \frac{m^2 + n^2}{2}. \tag{4}$$

Let's now show how to select m and n to get to the natural solutions of the system.

Rewriting (2) as $n^2 \cdot m^2 = (x+1)^2$ gives $x + 1 = mn$, $n \in N$, $m \in N$, or equivalently,

$$x = mn - 1. \tag{5}$$

Comparing (5) with (3) gives $mn - 1 = \frac{m^2 - n^2}{4}$, which simplifies to

$$m^2 - 4mn - (n^2 - 4) = 0. \tag{6}$$

Let's solve this equation as quadratic for m.

$$\frac{D}{4} = 4n^2 + n^2 - 4 = 5n^2 - 4.$$

It follows that $m = 2n \pm \sqrt{5n^2 - 4}$.

If $n = 1$, $\frac{D}{4} = 1$. The equation has two solutions, $m = 3$ and $m = 1$. The solution satisfying the conditions of the problem is $m = 3$. The second root $m = 1$ should be rejected, because substituting $n = 1$ and $m = 1$ into (3) and (4) leads to $x = 0$ and $y = 1$ (I hope readers can easily verify it), which is not a pair of natural solutions. Substituting $n = 1$ and $m = 3$ into (3) and (4) gives $x = 2$ and $y = 5$. Thus, $x = 2$, $y = 5$ is the first pair of natural solutions of the equation.

If $n = 2$, $\frac{D}{4} = 16$. The equation has two solutions, $m = 8$ and $m = 0$. The solution satisfying the conditions of the problem is $m = 8$. The second root $m = 0$ should be rejected, because substituting $n = 2$ and $m = 0$ into (3) and (4) leads to $x = -1$ and $y = 2$ (we leave it to readers to verify), which is not a pair of the natural solutions.

Substituting $n = 2$ and $m = 8$ into (3) and (4) gives $x = 15$ and $y = 34$. Hence, $x = 15$, $y = 34$ is the second pair of natural solutions of the equation. It was required to prove the existence of at least two natural solutions of the equation. We found two pairs $(2, 5)$ and $(15, 34)$ satisfying the equation. Thus, the proof of the second part of the problem is completed.

Proceeding to the third part, we are facing a more complicated problem. It has to be proved that the original equation has an infinite number of natural solutions. Solving the problem, we are going to do even more than asked; we will prove the statement and will show how to find the solutions.

To start, we may use the result found during the solution of the second part. We established that in order to get the natural solutions for x and y, we need to identify natural numbers m and n satisfying equation (6). Depending on values of n such that values of $\frac{D}{4}$ will become perfect squares, the quadratic equation (6) will have solutions for m. Substituting values of m and n into (3) and (4) will result in solutions for x and y. So, the goal is to analyze if it is possible to get the indefinitely many values of n such that $\frac{D}{4}$ will be perfect squares, allowing for indefinitely many solutions in natural m.

Let's find a few acceptable values of $\frac{D}{4} = 5n^2 - 4$ that are perfect squares and see if we can find a pattern:

$$n_1 = 1, \quad \frac{D}{4} = 1$$

$$n_2 = 2, \quad \frac{D}{4} = 16 = 4^2$$

$$n_3 = 5, \quad \frac{D}{4} = 121 = 11^2$$

$$n_4 = 13, \quad \frac{D}{4} = 841 = 29^2$$

$$n_5 = 34, \quad \frac{D}{4} = 5776 = 76^2.$$

Based on these few calculations, we can make the assumption that for the numbers of the form $n_{i+1} = 3n_i - n_{i-1}$ with $n_1 = 1$, $n_2 = 2$, the discriminants $\frac{D}{4}$ will be perfect squares. Letting $\left(\frac{D}{4}\right)_i = a_i$ for the sequence of numbers n_i, we can build the sequence of discriminants $a_i = 5n_i^2 - 4$, each of which is a perfect square of a natural number. If we manage to prove this, the third part of the problem will be solved. So, the third part of the problem is reduced to the following problem:

Given the sequence of natural numbers n_i such that $n_1 = 1$, $n_2 = 2$ and $n_{i+1} = 3n_i - n_{i-1}$, prove that each number in the sequence $a_i = 5n_i^2 - 4$ is a perfect square.

Clearly, the next step will be to translate the recursive formula for the sequence $\{n_i\}$ to an explicit formula. Consider the quadratic equation $r^2 - 3r + 1 = 0$.

$$D = 9 - 4 = 5.$$

$$r_1 = \frac{3 + \sqrt{5}}{2}, \quad r_2 = \frac{3 - \sqrt{5}}{2}.$$

Thus, the explicit formula should be

$$n_i = c \cdot \left(\frac{3 + \sqrt{5}}{2}\right)^i + d \cdot \left(\frac{3 - \sqrt{5}}{2}\right)^i.$$

We now need to select c and d to satisfy the conditions $n_1 = 1$, $n_2 = 2$. They are determined from the following system of the linear equations:

$$c \cdot \left(\frac{3 + \sqrt{5}}{2}\right)^1 + d \cdot \left(\frac{3 - \sqrt{5}}{2}\right)^1 = 1,$$

$$c \cdot \left(\frac{3 + \sqrt{5}}{2}\right)^2 + d \cdot \left(\frac{3 - \sqrt{5}}{2}\right)^2 = 2.$$

To make the calculations easier, introduce new variables

$$u = c \cdot \frac{3+\sqrt{5}}{2} \quad \text{and} \quad v = d \cdot \frac{3-\sqrt{5}}{2}.$$

The new system to solve is

$$u + v = 1,$$
$$\frac{3+\sqrt{5}}{2}u + \frac{3-\sqrt{5}}{2}v = 2.$$

Multiplying the first equation by $\frac{\sqrt{5}-3}{2}$ and adding to the second equation will eliminate v, and $\frac{\sqrt{5}-3}{2}u + \frac{3+\sqrt{5}}{2}u = \frac{\sqrt{5}-3}{2} + 2$. Solving this equation gives $u = \frac{\sqrt{5}+1}{2\sqrt{5}}$. Substituting this value for u in the first equation gives $v = \frac{\sqrt{5}-1}{2\sqrt{5}}$. Finally, the values for u and v are substituted back into $u = c \cdot \frac{3+\sqrt{5}}{2}$ and $v = d \cdot \frac{3-\sqrt{5}}{2}$ to get c and d. Multiplying the numerator and denominator of the fraction in each case by the conjugate number for $(3 + \sqrt{5})$ and $(3 - \sqrt{5})$ respectively for c and d and applying the formula for the difference of squares gives

$$c = \frac{\sqrt{5}+1}{2\sqrt{5}} \cdot \frac{2}{3+\sqrt{5}} = \frac{(\sqrt{5}+1)(3-\sqrt{5})}{\sqrt{5}(3+\sqrt{5})(3-\sqrt{5})}$$
$$= \frac{(\sqrt{5}+1)(3-\sqrt{5})}{\sqrt{5}(9-5)} = \frac{3\sqrt{5}+3-5-\sqrt{5}}{4\sqrt{5}}$$
$$= \frac{2\sqrt{5}-2}{4\sqrt{5}} = \frac{\sqrt{5}-1}{2\sqrt{5}},$$
$$d = \frac{\sqrt{5}-1}{2\sqrt{5}} \cdot \frac{2}{3-\sqrt{5}} = \frac{(\sqrt{5}-1)(3+\sqrt{5})}{\sqrt{5}(3-\sqrt{5})(3+\sqrt{5})}$$
$$= \frac{(\sqrt{5}-1)(3+\sqrt{5})}{\sqrt{5}(9-5)} = \frac{3\sqrt{5}-3+5-\sqrt{5}}{4\sqrt{5}}$$
$$= \frac{2\sqrt{5}+2}{4\sqrt{5}} = \frac{\sqrt{5}+1}{2\sqrt{5}}.$$

Thus, we arrived at the desired explicit formula for the sequence n_i as

$$n_i = \frac{\sqrt{5}-1}{2\sqrt{5}} \cdot \left(\frac{3+\sqrt{5}}{2}\right)^i + \frac{\sqrt{5}+1}{2\sqrt{5}} \cdot \left(\frac{3-\sqrt{5}}{2}\right)^i.$$

The goal is to investigate the sequence $a_i = 5n_i^2 - 4$. Substituting the expression for n_i into the formula for a_i yields the explicit formula for that sequence:

$$a_i = 5n_i^2 - 4 = 5\left(\frac{\sqrt{5}-1}{2\sqrt{5}}\cdot\left(\frac{3+\sqrt{5}}{2}\right)^i + \frac{\sqrt{5}+1}{2\sqrt{5}}\cdot\left(\frac{3-\sqrt{5}}{2}\right)^i\right)^2 - 4$$

$$= 5\cdot\left(\left(\frac{\sqrt{5}-1}{2\sqrt{5}}\right)^2\cdot\left(\frac{3+\sqrt{5}}{2}\right)^{2i} + 2\cdot\frac{\sqrt{5}-1}{2\sqrt{5}}\cdot\frac{\sqrt{5}+1}{2\sqrt{5}}\cdot\left(\frac{3+\sqrt{5}}{2}\right)^i\cdot\left(\frac{3-\sqrt{5}}{2}\right)^i + \left(\frac{\sqrt{5}+1}{2\sqrt{5}}\right)^2\cdot\left(\frac{3-\sqrt{5}}{2}\right)^{2i}\right) - 4$$

$$= 5\left(\frac{5-2\sqrt{5}+1}{20}\cdot\left(\frac{3+\sqrt{5}}{2}\right)^{2i} + 2\cdot\frac{(5-1)}{20}\cdot\left(\frac{9-5}{4}\right)^i + \frac{5+2\sqrt{5}+1}{20}\cdot\left(\frac{3-\sqrt{5}}{2}\right)^{2i}\right) - 4$$

$$= 5\left(\frac{3-\sqrt{5}}{10}\cdot\left(\frac{3+\sqrt{5}}{2}\right)^{2i} + \frac{2}{5}\cdot 1^i + \frac{3+\sqrt{5}}{10}\cdot\left(\frac{3-\sqrt{5}}{2}\right)^{2i}\right) - 4$$

$$= \frac{3-\sqrt{5}}{2}\cdot\left(\frac{3+\sqrt{5}}{2}\right)^{2i} + 2 + \frac{3+\sqrt{5}}{2}\cdot\left(\frac{3-\sqrt{5}}{2}\right)^{2i} - 4$$

$$= \frac{3-\sqrt{5}}{2}\cdot\frac{3+\sqrt{5}}{2}\cdot\left(\frac{3+\sqrt{5}}{2}\right)^{2i-1} + \frac{3+\sqrt{5}}{2}\cdot\frac{3-\sqrt{5}}{2}\cdot\left(\frac{3-\sqrt{5}}{2}\right)^{2i-1} - 2$$

$$= \frac{9-5}{4}\cdot\left(\frac{3+\sqrt{5}}{2}\right)^{2i-1} + \frac{9-5}{4}\cdot\left(\frac{3-\sqrt{5}}{2}\right)^{2i-1} - 2$$

$$= \left(\frac{3+\sqrt{5}}{2}\right)^{2i-1} + \left(\frac{3-\sqrt{5}}{2}\right)^{2i-1} - 2.$$

So,

$$a_i = \left(\frac{3+\sqrt{5}}{2}\right)^{2i-1} + \left(\frac{3-\sqrt{5}}{2}\right)^{2i-1} - 2.$$

We need to prove that every number in the sequence a_i is a perfect square of a natural number. This was already justified in problem 5 above. Indeed, the last expression for a_i is the explicit formula for the odd powers for terms a_i for the same sequence, as we considered in problem 5, and proved that every such term is a perfect square of some natural number. Therefore, we managed to prove that the sequence of the discriminants of the (6) is an infinite sequence of the perfect squares of some natural numbers. Thus, the (6) and the original equation have an infinite number of integer solutions. Having established that, we have to make a final step and justify that our solutions are not just integers, but natural numbers.

Going back to the quadratic equation (6), let's first evaluate the likely values of m for the values of $n \geq 3$. Considering the greater of the two roots for m (we may ignore the smaller root because it will have no effect in our analysis), we see that

$$m = 2n + \sqrt{5n^2 - 4} \geq 2 \cdot 3 + \sqrt{5 \cdot 3^2 - 4} = 6 + \sqrt{41} > 6 + 6 = 12,$$

which means that for all $n \geq 3$ there must exist respective natural numbers $m > 12$. Clearly, for natural $n \geq 3$ and $m > 12$, the value of x determined from (5), $x = mn - 1$, will be a natural number. Let's justify that y in this case is a natural number as well.

Indeed, if n is an even number, $n = 2p$, $p \in N$, then

$$\frac{D}{4} = 5n^2 - 4 = 5 \cdot 4p^2 - 4 = 4(5p^2 - 1)$$

is also an even number, and so is $\sqrt{\frac{D}{4}}$. Therefore, $m = 2n + \sqrt{\frac{D}{4}}$ is an even number because it is the sum of two even numbers. Hence, $m^2 + n^2$ is also an even number, as the sum of two even numbers. So, it has to be divisible by 2, proving that y, expressed in (4) as $y = \frac{m^2 + n^2}{2}$, has to be a natural number.

In case n is an odd number, $n = 2p + 1$, $p \in N$, $\frac{D}{4} = 5(2p+1)^2 - 4 = 20p^2 + 20p + 1$ is an odd number because it is the sum of two even numbers and 1. Then $\sqrt{\frac{D}{4}}$ has to be an odd number as well. So, $m = 2n + \sqrt{\frac{D}{4}}$ is an odd number, as the sum of an even and odd number. It follows that m^2 and n^2 are odd numbers and therefore their sum, $m^2 + n^2$, has to be an even number. Thus, it is divisible by 2 and so $y = \frac{m^2 + n^2}{2}$ has to be a natural number in this case as well.

So, to summarize, we managed to find indefinitely many natural values of n such that the discriminants $\frac{D}{4} = 5n^2 - 4$ become perfect squares and so the values of m, calculated as $m = 2n + \sqrt{5n^2 - 4}$, become natural as well. It follows that the values of x and y determined from (3) and (4), are natural numbers and the original equation has indefinitely many natural solutions, as was to be proved.

I am sure the readers realized that even though we separated it from the Main Problem, problem 5, as a matter of fact, was the essential part of the solution. It was extracted in a separate piece for convenience and for the better comprehension of the whole picture.

How challenging would it be to investigate what happens to the natural solutions of the equation for degree k greater than 2? We leave that exercise to the ambitious reader to pursue.

One may find the solution a bit cumbersome, yet it provides good demonstration of a useful and interesting technique, which is applicable to any constant-recursive sequences, among which are the well-known the *Lucas sequence*, the *Jacobsthal sequence*, the *Pell sequence* and probably the most intriguing of all, the *Fibonacci sequence.*

The Fibonacci sequence is a set of natural numbers where the first two terms are 1 and 1 and starting with the third term every number in the sequence equals the sum of the preceding two terms. In other words, the Fibonacci sequence is the infinite sequence consisting of natural numbers such that $F_1 = 1$, $F_2 = 1$, and $F_{n+1} = F_n + F_{n-1}$:

$$1, 1, 2, 3, 5, 8, 13, 21, 34, \ldots.$$

In some texts, it is customary to use the first two terms as 0 and 1, $F_0 = 0$, $F_1 = 1$. In this case the sequence becomes

$$0, 1, 1, 2, 3, 5, 8, 13, 21, 34, \ldots.$$

In the solution of the third part of the Main Problem we built the sequence of numbers n_i, $\{1, 2, 5, 13, 34, \ldots\}$, defined by the recursive formula $n_{i+1} = 3n_i - n_{i-1}$ with $n_1 = 1$ and $n_2 = 2$ to prove that the respective sequence of discriminants will be perfect squares. Comparing the numbers comprising each sequence, one can notice that every term in our sequence equals the corresponding odd term in the Fibonacci sequence:

$$1, 1, 2, 3, 5, 8, 13, 21, 34, \ldots$$
$$1, \quad 2, \quad 5, \quad 13, \quad 34, \ldots.$$

As we explore the properties of the Fibonacci numbers below and *Binet's Formula*, as the explicit formula for the Fibonacci sequence among them, it's not hard to justify the validity of that statement for any term of our sequence (a proof is offered in the "Solutions to Problems" section). Curiously, in the development of our solution, we arrived at the sequence of numbers, which turned out to be a subset of the Fibonacci numbers, namely, all its odd terms! Having established an interesting relationship between the odd Fibonacci numbers n_i and the numbers $a_i = 5n_i^2 - 4$, we concluded that for each number n_i, respective a_i is a perfect square. It's interesting that a similar relationship holds for the even Fibonacci numbers; in that case for each even Fibonacci number the respective number $5n^2 + 4$ is a perfect square (a proof is offered in the "Solutions to Problems" section). Amazingly, the converse statements are true as well. These results lead to an intriguing general question—what are the necessary and sufficient conditions for a natural number to be a Fibonacci number (a term in the Fibonacci sequence)? We suggest readers investigate the proof of the following quite surprising statement that enables one to recognize the Fibonacci numbers:

A positive integer F_n is a Fibonacci number if and only if $5F_n^2 + 4$ is a perfect square for even n or $5F_n^2 - 4$ is a perfect square for odd n.

There are several different proofs of this statement. I suggest comparing your own (if you manage to find it) to the one we offer in the "Solutions to Problems" section. It should be noted that during our proof we derived a few interesting formulas and properties of the Fibonacci numbers that are not mentioned below. You can add them to that list.

Another interesting observation is that we used in our calculations the number usually denoted as $\varphi = \frac{1+\sqrt{5}}{2}$, known under a few different names, among which are the golden ratio, the golden mean, the golden section, the divine proportion, etc. Two quantities are in the golden proportion if the ratio of the sum of two quantities to the larger of them equals the ratio of the larger to smaller quantities.

$$\frac{x+y}{y} = \frac{y}{x} = \frac{1+\sqrt{5}}{2} \approx 1.618034$$

A thoughtful reader should have noticed the impact of that number in the solution of problem 5. It was a key element in the introduction of an auxiliary sequence, which helped proving the problem's statement. This was not for the first time we saw the golden ratio. We came across the same number during the regular decagon construction. The length of the side of the regular decagon was derived as the reciprocal of the golden ratio. The golden ratio is one of the most ubiquitous irrational numbers known. It may not be as important and famous as π or e, but it does have its own unique and curious properties.

Dan Brown used the Fibonacci sequence as a plot element in his bestseller *The Da Vinci Code*. Thanks to his novel, many people, not familiar with math at all, became acquainted with its existence. It was brought into the mainstream of pop culture.

The Fibonacci sequence was known as early as the 6th century AD by Indian mathematicians. One of the most celebrated mathematicians of the Middle Ages, the Italian Leonardo Bonacci (Leonardo of Pisa, Leonardo Pisano Bigollo, Leonardo Fibonacci) (c. 1170–c. 1250), introduced the sequence of the Fibonacci numbers to the Western world while studying the reproduction of rabbits over the course of one year.

Fibonacci posed the following question:

If a pair of rabbits is placed in an enclosed area, how many rabbits will be born there if we assume that every month a pair of rabbits produces another pair, and that rabbits begin to bear young two months after their birth?

Fibonacci noticed that rabbit breeding pattern in the Fibonacci sequence. The "rabbit problem" is the first problem associated with the now-famous numbers.

Who could imagine that his observations would originate such stunning results! Since then his numbers and their properties captivated humanity for centuries. It is truly remarkable in how many different ways the golden ratio and the Fibonacci numbers are encountered in nature, art, engineering, architecture, and even in the human body.

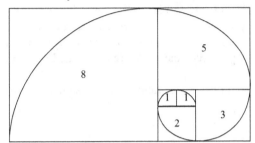 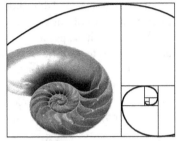

Just a few examples (some are disputed, but are still alluring): The number of petals in a flower follows the Fibonacci sequence; tree branches form or split in the Fibonacci sequence; the so called logarithmic spiral formed from a rectangle with sides in proportion equal to the golden ratio (nautilus shells follow the logarithmic spiral); reproductive dynamics of honey bees follow the Fibonacci sequence; animal flight patterns follow the Fibonacci sequence; amazingly, Fibonacci numbers appear as sums of oblique diagonals in Pascal's triangle; Fibonacci numbers are found in technical analysis of stock market transactions (the method is known as "Fibonacci retracement"); the golden ratio can be seen in the architecture of many ancient constructions, among which are the Egyptian Pyramids and the Parthenon; in geometry it has the reflection in the amazing golden rectangle and golden triangle, pentagon and pentagram; and even the human body follows the golden ratio rule (recall the famous Leonardo da Vinci's "Vitruvian Man" as an illustration of the golden ratio approximation). There exists a theory that when we judge random faces of unknown people, then the most attractive are deemed to be those that contain the golden ratio. One of the recent examples is presented in the *Daily Mail* July 27, 2017 issue that has an article about the methodology used by British plastic surgeon Dr. Julian DeSilva to determine the most handsome famous man. He concluded that it is George Clooney. Dr. DeSilva based his analysis on the golden ratio and digital facial mapping. In his theory Clooney's face has the lowest difference among all his other selections of 8.14% off a perfect golden ratio. Even though being beauty is subjective, the doctor's analysis gives an interesting explanation for the famous actor's face perfection. While many scientists are skeptical about the precision of applications of Fibonacci numbers and the golden ratio in nature, it is validly established in mathematics that the golden ratio is closely connected to the Fibonacci numbers. They have intriguing and unexpected applications in many branches of mathematics.

The equation considered in this chapter is a Diophantine equation. While studying the multivariable equations in the previous chapter, we mentioned David Hilbert's fundamental problem 10 regarding the solvability of Diophantine equations and whether there is a general algorithm providing the answer if an equation has integer solutions. In the development of his proof that such a general algorithm does not exist, Yuri Matiyasevich utilized a method involving properties of the Fibonacci numbers. He demonstrated how the Fibonacci numbers could be used in constructing a Diophantine equation whose solutions were such that one of its components grew exponentially with another of its components. Since the Fibonacci numbers and the golden ratio played a crucial role in our solution as well, in the remainder of this chapter we present a brief overview of some of their properties (mostly those pertaining to the analysis of Diophantine equation solutions) and demonstrate a special relationship between the golden ratio and the Fibonacci numbers.

To start, observe that any power of the golden ratio equals the sum of the two preceding smaller powers: $\varphi^{n+1} = \varphi^n + \varphi^{n-1}$.

Indeed, recalling that

$$\left(\frac{1+\sqrt{5}}{2}\right)^2 = \frac{1+2\sqrt{5}+5}{4} = \frac{6+2\sqrt{5}}{4} = \frac{3+\sqrt{5}}{2}$$

and factoring out the greatest common factor $\left(\frac{1+\sqrt{5}}{2}\right)^{n-1}$ yields

$$\varphi^n + \varphi^{n-1} = \left(\frac{1+\sqrt{5}}{2}\right)^n + \left(\frac{1+\sqrt{5}}{2}\right)^{n-1}$$
$$= \left(\frac{1+\sqrt{5}}{2}\right)^{n-1} \cdot \left(\frac{1+\sqrt{5}}{2} + 1\right)$$
$$= \left(\frac{1+\sqrt{5}}{2}\right)^{n-1} \cdot \frac{3+\sqrt{5}}{2}$$
$$= \left(\frac{1+\sqrt{5}}{2}\right)^{n-1} \cdot \left(\frac{1+\sqrt{5}}{2}\right)^2$$
$$= \left(\frac{1+\sqrt{5}}{2}\right)^{n+1} = \varphi^{n+1}.$$

This property leads to another surprising expression for the golden ratio as a continued fraction:

$$\varphi = \frac{\varphi^{n+1}}{\varphi^n} = \frac{\varphi^n + \varphi^{n-1}}{\varphi^n} = 1 + \frac{\varphi^{n-1}}{\varphi^n}$$
$$= 1 + \frac{1}{\frac{\varphi^n}{\varphi^{n-1}}} = 1 + \frac{1}{\frac{\varphi^{n-1}+\varphi^{n-2}}{\varphi^{n-1}}} = 1 + \frac{1}{1 + \frac{\varphi^{n-2}}{\varphi^{n-1}}}$$
$$= \cdots = 1 + \cfrac{1}{1 + \cfrac{1}{1 + \cfrac{1}{1 + \cdots}}}.$$

Furthermore, this implies that any power of the golden ratio can be reduced to the sum of an integer and an integer multiple of the golden ratio. For example, look at the first five calculations:

$$\varphi = \varphi^0 + \varphi^{-1} = 1 + \varphi^{-1},$$
$$\varphi^2 = \varphi^1 + \varphi^0 = \varphi + 1,$$
$$\varphi^3 = \varphi^2 + \varphi = \varphi + 1 + \varphi = 2\varphi + 1,$$
$$\varphi^4 = \varphi^3 + \varphi^2 = 2\varphi + 1 + \varphi + 1 = 3\varphi + 2,$$
$$\varphi^5 = \varphi^4 + \varphi^3 = (\varphi^3 + \varphi^2) + \varphi^3 = 2\varphi^3 + \varphi^2$$
$$= 2(\varphi^2 + \varphi^1) + \varphi^2 = 3\varphi^2 + 2\varphi^1 = 3(\varphi + 1) + 2\varphi = 5\varphi + 3.$$

It is amazing that the integer coefficients in each successive power of φ and the constants are consecutive numbers from the Fibonacci sequence, and the resulting expression can be written as $\varphi^n = F_n \cdot \varphi + F_{n-1}$, where F_n and F_{n-1} represent the nth and $(n-1)$st terms in the Fibonacci sequence. This can be proved by induction on n. Assume the statement holds for $n = k$, that is, $\varphi^k = F_k \cdot \varphi + F_{k-1}$. With that assumption we will prove that $\varphi^{k+1} = F_{k+1} \cdot \varphi + F_k$. Indeed,

$$\varphi^{k+1} = \varphi^k \cdot \varphi = (F_k \cdot \varphi + F_{k-1}) \cdot \varphi = F_k \cdot \varphi^2 + F_{k-1} \cdot \varphi$$
$$= F_k \cdot (\varphi + 1) + F_{k-1} \cdot \varphi = F_k \cdot \varphi + F_k + F_{k-1} \cdot \varphi$$
$$= \varphi(F_k + F_{k-1}) + F_k.$$

By definition the sum of two successive terms in the Fibonacci sequence equals the next term, so $F_k + F_{k-1} = F_{k+1}$, and replacing this into the last expression, we get that $\varphi^{k+1} = F_{k+1} \cdot \varphi + F_k$, which is what had to be proved. By mathematical induction the statement holds for any natural n.

Since the Fibonacci sequence is a classic example of a second order linear homogeneous recurrence, in order to get its explicit formula for the nth term, we can apply the usual technique and solve the quadratic equation $r^2 - r - 1 = 0$. It has roots $\frac{1+\sqrt{5}}{2}$ and $\frac{1-\sqrt{5}}{2}$. The explicit formula, therefore, will be

$$F_n = c \left(\frac{1 + \sqrt{5}}{2} \right)^n + d \left(\frac{1 - \sqrt{5}}{2} \right)^n,$$

where c and d are to be determined from the conditions $F_0 = 0$ and $F_1 = 1$:

$$c \left(\frac{1 + \sqrt{5}}{2} \right)^0 + d \left(\frac{1 - \sqrt{5}}{2} \right)^0 = 0,$$

$$c \left(\frac{1 + \sqrt{5}}{2} \right)^1 + d \left(\frac{1 - \sqrt{5}}{2} \right)^1 = 1.$$

From the first equation $c = -d$. Substituting this expression for c into the second equation gives

$$-d\left(\frac{1+\sqrt{5}}{2}\right)^1 + d\left(\frac{1-\sqrt{5}}{2}\right)^1 = 1,$$

from which $d = -\frac{1}{\sqrt{5}}$. Then, $c = \frac{1}{\sqrt{5}}$. It follows that the explicit formula for the Fibonacci sequence is

$$F_n = \frac{\left(\frac{1+\sqrt{5}}{2}\right)^n - \left(\frac{1-\sqrt{5}}{2}\right)^n}{\sqrt{5}}.$$

This is known as *Binet's Formula* after the French mathematician Jacques Binet (1786–1856), who first proved it in 1843; it's interesting to note that it was already known to Leonhard Euler (1707–1783), Abraham de Moivre (1667–1754), and Daniel Bernoulli (1700–1782) over a century earlier.

Using Binet's Formula, it's not hard to prove that the ratio between two successive terms in the Fibonacci sequence approaches a limit that equals the golden ratio. Indeed, simplifying and dividing the numerator and denominator of the ratio by $\left(\frac{1+\sqrt{5}}{2}\right)^{n+1}$ gives

$$\frac{F_{n+1}}{F_n} = \frac{\frac{\left(\frac{1+\sqrt{5}}{2}\right)^{n+1} - \left(\frac{1-\sqrt{5}}{2}\right)^{n+1}}{\sqrt{5}}}{\frac{\left(\frac{1+\sqrt{5}}{2}\right)^n - \left(\frac{1-\sqrt{5}}{2}\right)^n}{\sqrt{5}}} = \frac{\left(\frac{1+\sqrt{5}}{2}\right)^{n+1} - \left(\frac{1-\sqrt{5}}{2}\right)^{n+1}}{\left(\frac{1+\sqrt{5}}{2}\right)^n - \left(\frac{1-\sqrt{5}}{2}\right)^n}$$

$$= \frac{1 - \frac{\left(\frac{1-\sqrt{5}}{2}\right)^{n+1}}{\left(\frac{1+\sqrt{5}}{2}\right)^{n+1}}}{\frac{\left(\frac{1+\sqrt{5}}{2}\right)^n}{\left(\frac{1+\sqrt{5}}{2}\right)^{n+1}} - \frac{\left(\frac{1-\sqrt{5}}{2}\right)^n}{\left(\frac{1+\sqrt{5}}{2}\right)^{n+1}}} = \frac{1 - \left(\frac{1-\sqrt{5}}{1+\sqrt{5}}\right)^{n+1}}{\frac{1}{\frac{1+\sqrt{5}}{2}} - \left(\frac{1-\sqrt{5}}{1+\sqrt{5}}\right)^n \cdot \frac{1}{\frac{1+\sqrt{5}}{2}}}.$$

The numerator of the last fraction approaches 1 as $n \to \infty$ because $\left(\frac{1-\sqrt{5}}{1+\sqrt{5}}\right)^{n+1} \to 0$ as $n \to \infty$. The denominator of the fraction approaches $\frac{1}{\frac{1+\sqrt{5}}{2}}$ as $n \to \infty$, because $\left(\frac{1-\sqrt{5}}{1+\sqrt{5}}\right)^n \cdot \frac{1}{\frac{1+\sqrt{5}}{2}} \to 0$ as $n \to \infty$. Therefore, the whole fraction will be approaching

$$\frac{1}{\frac{1}{\frac{1+\sqrt{5}}{2}}} = \frac{1+\sqrt{5}}{2} = \varphi,$$

or, in other words, the ratios of successive Fibonacci numbers approximate the golden ratio:

$$\lim_{n \to \infty} \frac{F_{n+1}}{F_n} = \varphi.$$

A few more interesting relationships, which we leave for readers to prove:
The sum of the first n Fibonacci numbers is

$$F_1 + F_2 + \cdots + F_n = F_{n+2} - 1.$$

The sum of the first n odd terms of the Fibonacci numbers is

$$F_1 + F_3 + \cdots + F_{2n-1} = F_{2n}.$$

The sum of the first n even terms of the Fibonacci numbers is

$$F_2 + F_4 + \cdots + F_{2n} = F_{2n+1} - 1.$$

The sum of the squares of the first n Fibonacci numbers is

$$F_1^2 + F_2^2 + \cdots + F_n^2 = F_n \cdot F_{n+1}.$$

The difference of squares of the Fibonacci numbers is

$$F_{n+1}^2 - F_{n-1}^2 = F_{2n}.$$

The next property played an important role in getting the final solution to Hilbert's fundamental problem 10 and helped Yuri Matiyasevich to construct "equations in words and length." It is described by Zeckendorf's theorem (after the Belgian doctor and mathematician Édouard Zeckendorf (1901–1983)), a theorem about the representation of positive integers as sums of the Fibonacci numbers: *every natural number can be represented uniquely as the sum of one or more distinct Fibonacci numbers in such a way that the sum does not include any two consecutive Fibonacci numbers.* For example, we can express 30 as $30 = 1 + 8 + 21 = 1 + 3 + 5 + 21 = 1 + 1 + 2 + 5 + 8 + 13$. Out of these representations, Zeckendorf's representation is $30 = 1 + 8 + 21$, in which there are no consecutive Fibonacci numbers.

In the conclusion of our discussion of the Fibonacci numbers and their properties, let's recall the previous chapter, in which we saw using the Euclidean algorithm in the solution of Diophantine equations. One of the first practical applications of the Fibonacci numbers was in finding when the Euclidean algorithm requires n steps for a pair of natural numbers a and b. The French mathematician Gabriel Lamé (1795–1870) proved that the worst case input for the Euclidean algorithm is a pair of consecutive Fibonacci numbers, that is, the smallest values of a and b ($a > b$) for which the Euclidean algorithm requires n steps are neighboring Fibonacci numbers $a = F_{n+2}$ and $b = F_{n+1}$.

It will now be interesting to look back and revisit the generalization of the problem about cutting squares from a rectangle solved in the previous chapter. Let's recall the question: find natural numbers a and b such that it is possible to cut a rectangle of size $a \times b$ (a—length, b—width) into squares of n different sizes for the given natural number n.

This problem has an elegant and beautiful solution if you select the lengths of the sides of the rectangle to be neighboring Fibonacci numbers $a = F_{n+2}$ and $b = F_{n+1}$. Indeed, for each rectangle with sides F_k and F_{k-1} every time we will cut only one square with side length F_{k-1}. The remaining rectangle will have the sides F_{k-1} and F_{k-2}. According to Lamé's theorem, the suggested rectangle with sides $a = F_{n+2}$ and $b = F_{n+1}$ has the minimum possible length and width for the selected number n. In other words, if each of the numbers a and b does not exceed F_{n+2}, the $(n+2)$nd term in the Fibonacci sequence, then Euclidean algorithm produces the greatest common divisor of a and b, $\mathrm{GCD}(a, b)$, in no more than n steps. You may verify these observations, for example, for a rectangle with sides $a = F_9 = 34$ and $b = F_8 = 21$. It can be cut into squares of seven different sizes ($n = 7$) with the sides 21, 13, 8, 5, 3, 2, 1.

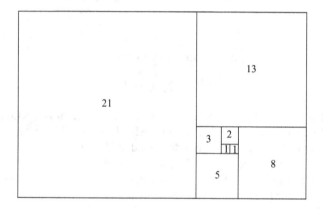

Before we leave the topic of the Fibonacci numbers and the golden ratio, we should recall the unexpected amazing discovery evolved from the Main Problem solution analysis, formulated above as the necessary and sufficient conditions for recognizing the Fibonacci numbers. It is particularly noteworthy that our proof presented in the "Solutions to Problems" section uses several of examined above relationships.

We gave a very brief overview of some interesting properties of the Fibonacci numbers and the golden ratio. There is an entire journal dedicated to their study, *The Fibonacci Quarterly* (http://www.fq.math.ca/), to which we can refer an ambitious reader to find out more about the amazing properties and various applications of the Fibonacci numbers.

I think that multivariable equations provide one of the most fascinating examples of algebraic equations that allow us to admire the beauty of algebra and number theory. We scratched only the surface; the topic has a rich history of great discoveries through the centuries by the prominent mathematicians. The study of Diophantine equations generated stunning results in number theory, differential calculus, topology, and encryption techniques that are based on Euler's generalization of Fermat's prime number theories.

In conclusion we mention an interesting variation to the Main Problem's part three. If we slightly modify the conditions and consider in the third part instead of the equation $x^2 + 6x + 1 = y^2$ another equation with the coefficient by x equal to 0 and use some nonsquare coefficient n for y, it becomes $x^2 + 1 = ny^2$. This is the typical so-called *Pell's equation*, or to be more precise, in this case a negative Pell's equation. Pell's equation (after the English mathematician John Pell (1611–1685)) or, as it is also called, Pell-Fermat equation, is any Diophantine equation of the form $x^2 - ny^2 = 1$, where n is a positive nonsquare integer and integer solutions are sought for x and y. These equations have been first studied by the Indian mathematician Brahmagupta in the 6th century. Another Indian mathematician Bhaskara II found the general solutions to Pell's equations in the 12th century. In Europe Pell's equation solutions were rediscovered in the 17th century. The outstanding French mathematician Pierre de Fermat studied them. It is amazing, that like Francois Viéte, whose work had much influence on him, Fermat was an attorney by education and was doing his mathematical research on the side of his major occupation. He made invaluable contributions in the field of mathematics. By many historians he is considered along with René Descartes as the most influential mathematician of the 17th century. His discoveries tremendously contributed to the development of number theory. Another prominent French mathematician, Joseph-Louis Lagrange (1736–1813), established the general theory of Pell's equations based on the solutions in continued fractions. If you recall problem 23 in chapter 6 on computing the cosine and sine of an angle nx, you should realize now that both expressions

$$\cos nx = 2 \cdot \cos x \cdot \cos(n-1)x - \cos(n-2)x \quad \text{and}$$
$$\sin nx = 2 \cdot \cos x \cdot \sin(n-1)x - \sin(n-2)x$$

represent recursive formulas similar to those studied in this chapter. It's interesting to note that the first formula trigonometrically defines the Chebyshev polynomials (after the Russian mathematician Pafnuty Lvovich Chebyshev (1821–1894)), which are also the solutions of certain Pell equations.

As you can see, you may find a rich set of explorations for various types of multivariable equations and their solutions, including the further investigation of the ideas discussed in this chapter.

Chapter 12

Systems of Nonlinear Equations

A *nonlinear system of equations* is a set of simultaneous equations in which the unknowns appear as variables of a polynomial of degree higher than one. To solve a system of equations means to find the values of variables such that all the equations are simultaneously satisfied. While nonlinear systems of equations are usually difficult to solve, there are some commonly used methods and techniques employed for simplification that will be discussed in this chapter.

First we consider *systems of nonlinear equations with two variables*.

The methods of substitution and elimination are the most commonly used methods. The idea behind the substitution method lies in expressing one of the variables in terms of the other from one of the equations and then substituting it into the second equation. The elimination method manipulates (usually with arithmetic operations) the equations in order to get rid of one variable and get an equation with another one. When the value of the second variable is found, you substitute it in any of the equations and find the first variable. If there are no common solutions of all the equations in the system, then the system has no solutions.

In the preceding two chapters we discussed various types of Diophantine and other multivariable equations and methods of their solutions. Often, when analyzing a Diophantine equation, you arrive at the solution of the system(s) of nonlinear equations with two or more variables. The following problem is a good example of such an analysis and serves as a link between the methods discussed earlier and those in this chapter.

Problem 1. Find all the integer solutions of the equation $y^3 - x^3 = 91$.

Solution. Factoring both sides, we can rewrite the equation as

$$(y - x)(y^2 + xy + x^2) = 13 \cdot 7.$$

Clearly, $y^2 + xy + x^2 > 0$ (I hope it is not hard for readers to prove this fact) and we know that 13 and 7 are prime numbers. Thus, the equality $y^3 - x^3 = 91$ holds only in the following four cases:

$$y - x = 1,$$
$$y^2 + xy + x^2 = 91,$$

or

$$y - x = 91,$$
$$y^2 + xy + x^2 = 1,$$

or

$$y - x = 7,$$
$$y^2 + xy + x^2 = 13,$$

or

$$y - x = 13,$$
$$y^2 + xy + x^2 = 7.$$

Expressing y in terms of x from the first equation as $y = x + 1$ and substituting into the left-hand side of the second equation of the first system, we have

$$(x + 1)^2 + x(x + 1) + x^2 = x^2 + 2x + 1 + x^2 + x + x^2 = 3x^2 + 3x + 1 = 91.$$

Solving this equation, we get

$$3x^2 + 3x + 1 = 91,$$
$$3x^2 + 3x - 90 = 0,$$

or equivalently,

$$x^2 + x - 30 = 0.$$
$$D = 1 + 120 = 121, \text{ and}$$
$$x = \frac{-1 \pm 11}{2},$$

from which $x_1 = 5$, $x_2 = -6$. Hence, $y_1 = 5 + 1 = 6$, $y_2 = -6 + 1 = -5$.

In the same manner solving the second system, we get

$$y = x + 91,$$
$$(x + 91)^2 + x(x + 91) + x^2 = 1.$$

The second equation is simplified to

$$x^2 + 91x + 2760 = 0.$$
$$D = 8281 - 11040 = -2759 < 0.$$

There are no real solutions to this equation.
Solving the third system, we get

$$y = x + 7,$$
$$(x + 7)^2 + x(x + 7) + x^2 = 13.$$

The second equation is simplified to $x^2 + 7x + 12 = 0$, solutions of which by Viète's formulas are $x_1 = -3$, $x_2 = -4$. Hence, $y_1 = -3 + 7 = 4$, $y_2 = -4 + 7 = 3$.
Solving the last system, we get

$$y = x + 13,$$
$$(x + 13)^2 + x(x + 13) + x^2 = 7.$$

The second equation is simplified to $x^2 + 13x + 54 = 0$. $D = 169 - 216 = -47 < 0$. No real solutions exist in this case.

Answer: $(5, 6), (-6, -5), (-3, 4), (-4, 3)$.

Problem 2. Solve the system of equations

$$x^2 - xy + 7x + 4y = 6,$$
$$xy - x^2 + x + 4y = 10.$$

Solution. Working with the system of nonlinear equations, you need to assess the most efficient way for simplifying the equations. Doing arithmetic operations with the equations helps in achieving that goal. In this problem adding the equations and replacing the first equation in the system by the resulting sum gives $8x + 8y = 16$, from which $y = 2 - x$. Next, substituting $y = 2 - x$ into the second equation yields

$$x \cdot (2 - x) - x^2 + x + 4 \cdot (2 - x) = 10,$$
$$2x - x^2 - x^2 + x + 8 - 4x - 10 = 0,$$
$$2x^2 + x + 2 = 0.$$

The discriminant of the quadratic equation is $D = 1 - 16 = -15 < 0$, which means the equation has no real solutions. Therefore, the system of equations has no solutions.

In some problems the straightforward expression of one variable in terms of the other one is not easily achievable or is not very useful. If that is the case, then some preliminary work is needed. You may want to introduce new variable(s), which would help to get a more efficient and elegant solution of the system.

Problem 3. Solve the system of equations

$$\frac{2x}{y} - \frac{3y}{x} = 5,$$
$$x^2 - y^2 = 8.$$

Solution. The introduction of a new variable in the first equation will simplify the solution. Let $t = \frac{x}{y}$ (1) and substitute it in the first equation. After solving for t, then substituting the value of t back into (1) will express x in terms y in the most efficient way.

$$2t - \frac{3}{t} = 5, \ (t \neq 0)$$

or equivalently,

$$2t^2 - 5t - 3 = 0.$$
$$D = 25 + 24 = 49.$$
$$t = \frac{5 \pm 7}{4}.$$

The solutions of this quadratic equation are $t_1 = 3$ and $t_2 = -\frac{1}{2}$. It follows that $\frac{x}{y} = 3$ or $\frac{x}{y} = -\frac{1}{2}$ and we need to consider two systems of equations:

$$\begin{aligned} x &= 3y, \\ x^2 - y^2 &= 8. \end{aligned} \quad \text{and} \quad \begin{aligned} x &= -\frac{1}{2}y, \\ x^2 - y^2 &= 8. \end{aligned}$$

Solving the first system by substituting the value of x from the first equation into the second equation gives $9y^2 - y^2 = 8$, which leads to $y^2 = 1$. Then $y = \pm 1$ and so $x = 3y = \pm 3$.

For the second system of equations we get that $\frac{1}{4}y^2 - y^2 = 8$, or equivalently, $-\frac{3}{4}y^2 = 8$. This equation has no real solutions. Thus, the second system has no solutions.

Answer: $(3, 1)$, $(-3, -1)$.

Problem 4. Solve the system of equations

$$x^2 - 3xy + y^2 = 19,$$
$$x^2 + 2xy - 3y^2 = 0.$$

Solution. Note that both equations in the given system are homogeneous. The pair $(0, 0)$ is not a solution of the first equation. Therefore, it is not a solution of the system and we may divide both sides of the second equation by $y^2 \neq 0$ and not lose any solutions of the system.

This gives $\left(\frac{x}{y}\right)^2 + 2\frac{x}{y} - 3 = 0$. Substituting $\frac{x}{y} = t$ leads to

$$t^2 + 2t - 3 = 0.$$

The solutions of this equation by Viète's formulas are $t_1 = 1, t_2 = -3$. Then $x = y$ or $x = -3y$. The next step is to consider two systems of equations:

$$
\begin{array}{ccc}
\begin{aligned}
x &= y, \\
x^2 - 3xy + y^2 &= 19.
\end{aligned}
& \text{and} &
\begin{aligned}
x &= -3y, \\
x^2 - 3xy + y^2 &= 19.
\end{aligned}
\end{array}
$$

Solving the first system by substituting x for y from the first equation gives

$$y^2 - 3y^2 + y^2 = 19,$$
$$-y^2 = 19.$$

No real solutions exist. Thus, the first system does not have any real solutions.

In the second case, $9y^2 + 9y^2 + y^2 = 19$, $19y^2 = 19$, from which $y = \pm 1$. Then $x = -3y = \mp 3$.

Answer: $(3, -1), (-3, 1)$.

In solving some systems of equations the combination of a few methods considered previously may be very helpful, that is, simplifying one or both equations, doing arithmetic operations with them, and introducing the new variables. A few commonly used substitutions are the following:

$$
\begin{array}{ccccc}
\begin{aligned}
x + y &= u, \\
xy &= v,
\end{aligned}
& \quad \text{or} \quad &
\begin{aligned}
x - y &= u, \\
xy &= v,
\end{aligned}
& \quad \text{or} \quad &
\begin{aligned}
x^2 + y^2 &= u, \\
xy &= v.
\end{aligned}
\end{array}
$$

Problem 5. Solve the system of equations

$$x^2 - 3x + y^2 - 3y = -2,$$
$$2x^2 - xy + 2y^2 = 20.$$

Solution. In each of the equations completing the square for the sum$(x + y)$ gives

$$(x + y)^2 - 2xy - 3(x + y) = -2,$$
$$2(x + y)^2 - 5xy = 20.$$

Introducing two new variables,

$$x + y = u, \quad xy = v \tag{1}$$

gives the system of equations

$$u^2 - 2v - 3u = -2,$$
$$2u^2 - 5v = 20.$$

Multiplying each side of the first equation by 5 and each side of the second equation by 2 yields

$$5u^2 - 10v - 15u = -10,$$
$$4u^2 - 10v = 40.$$

Subtracting the second equation from the first equation and solving the resulting quadratic equation for u yields

$$u^2 - 15u + 50 = 0.$$

The roots by Viète's formulas are $u_1 = 10$, $u_2 = 5$. Expressing v in terms of u from the second equation as $v = \frac{2u^2 - 20}{5}$ and substituting the values of u gives

$$v_1 = \frac{2 \cdot 10^2 - 20}{5} = \frac{180}{5} = 36; \quad v_2 = \frac{2 \cdot 5^2 - 20}{5} = \frac{30}{5} = 6.$$

Putting these values back in (1), we have

$$\begin{array}{ccc} x + y = 10, & & x + y = 5, \\ & \text{or} & \\ xy = 36. & & xy = 6. \end{array}$$

Solving the first system gives

$$x = 10 - y,$$
$$y(10 - y) = 36.$$

Thus, the second equation becomes

$$y^2 - 10y + 36 = 0.$$
$$D = 100 - 144 = -44 < 0.$$

The equation and hence the system have no real solutions.
    Solving the second system gives

$$x = 5 - y,$$
$$y(5 - y) = 6.$$

The second equation becomes

$$y^2 - 5y + 6 = 0.$$

Solutions are $y_1 = 3$, $y_2 = 2$. Then $x_1 = 2$, $x_2 = 3$.

Answer: $(2, 3)$, $(3, 2)$.

In some complicated systems of equations you need to work with equations simultaneously to simplify them as much as possible in order to see what the best choice for substitution is.

Problem 6. Solve the system of equations

$$x^5 - y^5 = 3{,}093,$$
$$x - y = 3.$$

Solution. Factoring the left-hand side of the first equation gives

$$(x - y)(x^4 + x^3y + x^2y^2 + xy^3 + y^4) = 3{,}093.$$

Using the given value of $x - y = 3$ from the second equation and substituting it in the modified first equation results in

$$3(x^4 + x^3y + x^2y^2 + xy^3 + y^4) = 3{,}093,$$

or equivalently, after dividing by 3,

$$x^4 + x^3y + x^2y^2 + xy^3 + y^4 = 1{,}031,$$

which is a homogeneous equation. The pair of numbers $x = 0$ and $y = 0$ does not satisfy this equation. Therefore, we may divide both sides by x^2y^2 and not lose any solutions.

$$\frac{x^2}{y^2} + \frac{x}{y} + 1 + \frac{y}{x} + \frac{y^2}{x^2} = \frac{1{,}031}{x^2y^2}.$$

Regrouping the terms gives

$$\left(\frac{x^2}{y^2} + \frac{y^2}{x^2}\right) + \left(\frac{x}{y} + \frac{y}{x}\right) + 1 = \frac{1{,}031}{x^2y^2}. \tag{1}$$

Introducing the new variable $t = \frac{x}{y} + \frac{y}{x}$ leads to

$$\frac{x^2}{y^2} + \frac{y^2}{x^2} = t^2 - 2. \tag{2}$$

Indeed,

$$t^2 = \left(\frac{x}{y} + \frac{y}{x}\right)^2 = \frac{x^2}{y^2} + \frac{y^2}{x^2} + 2,$$

so

$$\frac{x^2}{y^2} + \frac{y^2}{x^2} = t^2 - 2.$$

The original equation can be rewritten as

$$t^2 - 2 + t + 1 = \frac{1{,}031}{x^2y^2}.$$

As you can see, now it is possible to express the left-hand side of the last equation in terms of t. However, there still is a term x^2y^2 in the denominator of the fraction on the right-hand side. If we manage to express it in terms of t as well, then the equation will become a single variable equation and will be solvable.

Let's do some extra work using again the second equation of the original system:

$$t = \frac{x}{y} + \frac{y}{x},$$
$$x - y = 3.$$

The first equation can be rewritten as

$$\frac{x^2 + y^2}{xy} = t.$$

Completing the square in the numerator, we get

$$\frac{(x-y)^2 + 2xy}{xy} = t.$$

Substituting $x - y = 3$ into this equation yields $\frac{9 + 2xy}{xy} = t$, from which

$$\frac{1}{xy} = \frac{t-2}{9}. \tag{3}$$

Substituting the expressions in t for each variable from (2) and (3) into (1) gives

$$t^2 - 2 + t + 1 = \frac{1,031}{81} \cdot (t-2)^2.$$

After simplifications we get the quadratic equation

$$190t^2 - 841t + 841 = 0,$$
$$D = 841^2 - 760 \cdot 841 = 841 \cdot (841 - 760) = 841 \cdot 81 = 29^2 \cdot 9^2.$$

$t = \frac{841 \pm 261}{380}$. So, $t_1 = \frac{29}{19}$, and $t_2 = \frac{551}{190}$.

Finally, getting back to the two systems of equations for x and y gives

$$x - y = 3, \qquad \text{or} \qquad x - y = 3,$$
$$\frac{1}{xy} = \frac{\frac{29}{19} - 2}{9} \qquad\qquad \frac{1}{xy} = \frac{\frac{551}{190} - 2}{9}.$$

The first system is simplified to

$$x = y + 3,$$
$$xy = -19.$$

Substituting the expression for x into the second equation yields

$$y(y+3) + 19 = 0, \quad \text{or} \quad y^2 + 3y + 19 = 0.$$
$$D = 9 - 76 = -67 < 0.$$

The equation and hence the system have no real solutions.

The Equations World

The second system is simplified to

$$x = y + 3,$$
$$xy = 10.$$

Substituting the expression for x into the second equation yields

$$y(y+3) - 10 = 0, \quad \text{or} \quad y^2 + 3y - 10 = 0.$$
$$D = 9 + 40 = 49.$$

Then $y_1 = -5$, $y_2 = 2$. Substituting -5 and 2 for y into the first equation gives $x_1 = -2$, $x_2 = 5$.

Answer: $(-2, -5)$, $(5, 2)$.

Problem 7. Solve the system of equations

$$x^4 - y^4 = 15,$$
$$x^3 y - xy^3 = 6.$$

Solution. Usually as you assess each equation in the system, you should be able to see some hints. The expression on the left side of each equation is an open invitation to apply factoring with a further introduction of new variables:

$$(x^2 - y^2)(x^2 + y^2) = 15,$$
$$xy(x^2 - y^2) = 6.$$

The substitutions for new variables are

$$x^2 - y^2 = u, \qquad (1)$$
$$x^2 + y^2 = v. \qquad (2)$$

The first equation simplifies right away, $uv = 15$. Note that since $v > 0$ (it is the sum of two positive numbers), then $u > 0$ as well (since the product uv equals a positive number 15). We need to express the value of xy in terms of the new variables u and v in the second equation. In order to do that, let's first express each variable x and y in terms of u and v.

Adding (1) and (2) gives $x^2 = \frac{1}{2}(u + v)$, or equivalently, $x = \sqrt{\frac{1}{2}(u + v)}$.

Subtracting (1) from (2) gives $y^2 = \frac{1}{2}(v - u)$, or equivalently, $y = \sqrt{\frac{1}{2}(v - u)}$.
Thus, $xy = \frac{1}{2}\sqrt{v^2 - u^2}$.
    Now the system can be rewritten as

$$uv = 15,$$
$$\frac{1}{2}u\sqrt{v^2 - u^2} = 6.$$

$$v = \frac{15}{u},$$

$$u\sqrt{v^2 - u^2} = 12.$$

Squaring both sides of the second equation and substituting $v = \frac{15}{u}$ for v in the second equation, we get

$$v = \frac{15}{u},$$

$$u^2\left(\left(\frac{15}{u}\right)^2 - u^2\right) = 144.$$

$$v = \frac{15}{u},$$

$$225 - u^4 = 144.$$

$$v = \frac{15}{u},$$

$$u^4 = 81.$$

Solving the last system gives $u = \pm 3$, $v = \pm 5$. Since negative solutions do not satisfy the conditions of the problem, only the positive values are counted, $u = 3$, $v = 5$. Substituting $u = 3$ and $v = 5$ back into (1) and (2) gives

$$x^2 - y^2 = 3,$$

$$x^2 + y^2 = 5,$$

Adding the equations yields $2x^2 = 8$, or $x^2 = 4$. It follows $x = \pm 2$ and so $y = \pm 1$.

Answer: $(2, 1), (-2, -1), (2, -1), (-2, 1)$.

Sometimes the system of equations is written in an unconventional manner, as in the following problem offered in the November/December 1999 issue of the currently defunct magazine *Quantum*.

Problem 8. Solve the system of equations

$$\frac{x-1}{xy-3} = \frac{y-2}{xy-4} = \frac{3-x-y}{7-x^2-y^2}.$$

Solution. Curiously, the key to the solution of the system is hidden in the problem's presentation. There are three equal ratios, so we introduce a new variable t to be each of the given fractions. It follows that

$$x - 1 = t(xy - 3),$$
$$y - 2 = t(xy - 4),$$
$$3 - x - y = t(7 - x^2 - y^2).$$

Adding these equations gives the following equation

$$x - 1 + y - 2 + 3 - x - y = t(xy - 3 + xy - 4 + 7 - x^2 - y^2)$$

or equivalently, $t(x^2 + y^2 - 2xy) = 0$, which gives $t(x - y)^2 = 0$. Hence, either $t = 0$ or $x - y = 0$. Consider each case separately.
    If $t = 0$, then

$$x = 1,$$
$$y = 2.$$

If $x - y = 0$, or equivalently, $x = y$, then substituting y for x into the first equation $\frac{x-1}{xy-3} = \frac{y-2}{xy-4}$ gives

$$(x - 1)(x^2 - 4) = (x^2 - 3)(x - 2),$$
$$x^3 - x^2 - 4x + 4 = x^3 - 3x - 2x^2 + 6,$$
$$x^2 - x - 2 = 0.$$

Using Viète's formulas gives $x_1 = -1$, $x_2 = 2$. Then $y_1 = -1$, $y_2 = 2$. The solutions should be verified to satisfy the conditions $xy \neq 3$, $xy \neq 4$ and $x^2 + y^2 \neq 7$ (the restrictions on the domains of each expression in the system). The pair $(2, 2)$ does not belong to the domain $(xy \neq 4)$ of the original second expression in the system, $\frac{y-2}{xy-4}$, and has to be rejected. Therefore, the system has only two solutions $(1, 2)$; $(-1, -1)$.

Answer: $(1, 2)$, $(-1, -1)$.

    A useful application of Viéte's formulas appears in the solution of the systems of the equations (encountered a few times in the previous problems) of the type

$$ax \pm by = c,$$
$$xy = d.$$

Problem 9. Solve the system of equations

$$x + y = 7,$$
$$xy = 12.$$

Solution. Instead of solving this system by traditional methods, notice that according to Viéte's theorem, x and y have to be the roots of the quadratic equation

$z^2 - 7z + 12 = 0$. Thus, the solution of the system is reduced to the solution of the quadratic equation. The roots are $z_1 = 3$, $z_2 = 4$. Therefore, the solutions of the system are $x_1 = 3$, $y_1 = 4$ and $x_2 = 4$, $y_2 = 3$.

Answer: $(3, 4)$, $(4, 3)$.

Problem 10. Solve the system of equations

$$xy + x - y = 3,$$
$$x^2y - xy^2 = 2.$$

Solution. Rewrite the system as

$$(xy) + (x - y) = 3,$$
$$xy(x - y) = 2.$$

Consider the quadratic equation $z^2 - 3z + 2 = 0$, roots of which (xy) and $(x - y)$ satisfy Viète's formulas in the system. Solutions of the quadratic equation are $z_1 = 1$, $z_2 = 2$. Hence, the original system is equivalent to the following combination of two systems:

$$
\begin{array}{ccc}
x - y = 1, & & x - y = 2, \\
& \text{and} & \\
xy = 2, & & xy = 1.
\end{array}
$$

Rewrite the systems as

$$
\begin{array}{ccc}
x + (-y) = 1, & & x + (-y) = 2, \\
& \text{and} & \\
x(-y) = -2, & & x(-y) = -1.
\end{array}
$$

Each system in turn can be solved by applying Viète's formulas. For the first system we consider the quadratic equation $t^2 - t - 2 = 0$, solutions of which are $t_1 = -1$, $t_2 = 2$. It follows that $x_1 = -1$, $y_1 = -2$ and $x_2 = 2$, $y_2 = 1$. For the second system consider the quadratic equation $t^2 - 2t - 1 = 0$. $D = 4 + 4 = 8$, $t = \frac{2 \pm \sqrt{8}}{2}$, from which $t_1 = 1 + \sqrt{2}$, $t_2 = 1 - \sqrt{2}$. It follows that $x_1 = 1 + \sqrt{2}$, $y_1 = -1 + \sqrt{2}$ and $x_2 = 1 - \sqrt{2}$, $y_2 = -1 - \sqrt{2}$.

Answer: $(-1, -2)$, $(2, 1)$, $(1 + \sqrt{2}, -1 + \sqrt{2})$, $(1 - \sqrt{2}, -1 - \sqrt{2})$.

Completing the square method proves to be as effective in solving systems of nonlinear equations as it is in any of the problems studied in previous chapters.

Problem 11. Solve the system of equations

$$x + xy + y = 2 + 3\sqrt{2},$$
$$x^2 + y^2 = 6.$$

Solution. Multiplying the first equation by 2 and adding to the second equation gives

$$x^2 + y^2 + 2xy + 2(x + y) = 10 + 6\sqrt{2}.$$

It can be rewritten as

$$(x + y)^2 + 2(x + y) + 1 = 11 + 6\sqrt{2}.$$

Observing that the left and right sides each can be expressed as a perfect square, we get

$$(x + y + 1)^2 = (3 + \sqrt{2})^2.$$

It follows that $|x + y + 1| = 3 + \sqrt{2}$ and we need to consider two cases:

$$x + y + 1 = -3 - \sqrt{2} \quad \text{or} \quad x + y + 1 = 3 + \sqrt{2}.$$

Consider first $x + y + 1 = -3 - \sqrt{2}$ or equivalently, $x + y = -4 - \sqrt{2}$. Substituting this value into the first equation of the given system yields

$$xy - 4 - \sqrt{2} = 2 + 3\sqrt{2},$$

from which $xy = 6 + 4\sqrt{2}$. So, in the first case we get $x + y = -4 - \sqrt{2}$ and $xy = 6 + 4\sqrt{2}$. This implies that by Viète's formulas x and y have to be the solutions of the quadratic equation

$$t^2 + (4 + \sqrt{2})t + (6 + 4\sqrt{2}) = 0.$$

Its discriminant turns out to be a negative number. Indeed,

$$D = (4 + \sqrt{2})^2 - 4(6 + 4\sqrt{2}) = 16 + 8\sqrt{2} + 2 - 24 - 16\sqrt{2} = -6 - 8\sqrt{2} < 0.$$

Therefore, the quadratic equation has no real roots and the first case leads to no solutions in terms of x and y.

In the second case, $x + y = 2 + \sqrt{2}$. Therefore, $xy + 2 + \sqrt{2} = 2 + 3\sqrt{2}$, from which $xy = 2\sqrt{2}$. Consider now the quadratic equation

$$t^2 - (2 + \sqrt{2})t + 2\sqrt{2} = 0.$$
$$D = (2 + \sqrt{2})^2 - 8\sqrt{2} = 4 + 4\sqrt{2} + 2 - 8\sqrt{2} = 4 - 4\sqrt{2} + 2 = (2 - \sqrt{2})^2.$$
$$t = \frac{2 + \sqrt{2} \pm (2 - \sqrt{2})}{2}.$$

Hence, $t_1 = 2, t_2 = \sqrt{2}$. We conclude that the pairs of numbers $(2, \sqrt{2})$ and $(\sqrt{2}, 2)$ for x and y are the solutions of the original system of equations.

Answer: $(2, \sqrt{2})$ and $(\sqrt{2}, 2)$.

In chapter 1 we mentioned symmetrical systems of linear equations with two variables. Generally, a symmetrical system of algebraic equations with two variables is defined as the system

$$f_1(x, y) = 0,$$
$$f_2(x, y) = 0$$

such that $f_1(x, y)$ and $f_2(x, y)$ are symmetrical polynomials in x and y, i.e., $f_1(x, y)$ and $f_2(x, y)$ do not change if x is substituted for y and vice versa. The simple example of such system is the system solvable by Viète's formulas

$$x + y = a,$$
$$xy = b.$$

Symmetrical systems of algebraic equations provide an interesting technique for solving irrational equations of the type $\sqrt{a - f(x)} + \sqrt{b + f(x)} = c$, which we will cover here as an extension to the methods considered in chapter 5.

Introducing two new variables $u = \sqrt{a - f(x)}$ and $v = \sqrt{b + f(x)}$ and observing that $u^2 + v^2 = a - f(x) + b + f(x) = a + b$, transforms the original equation to a system of two equations

$$u + v = c,$$
$$u^2 + v^2 = a + b.$$

Since $(u + v)^2 = u^2 + v^2 + 2uv$, from which $uv = \frac{1}{2}((u + v)^2 - (u^2 + v^2)) = \frac{1}{2}(c^2 - a - b)$, it follows that the solution of the original equation is reduced to the solution of the system of symmetrical equations

$$u + v = c,$$
$$uv = \frac{1}{2}(c^2 - a - b).$$

Problem 12. Solve the equation

$$\sqrt{x + 5} + \sqrt{20 - x} = 7.$$

Solution. First, determine the domain of the equation by solving the system of inequalities

$$x + 5 \geq 0,$$
$$20 - x \geq 0$$

or equivalently,

$$x \geq -5,$$
$$x \leq 20.$$

The domain of the equation is the set of numbers $x \in [-5, 20]$.

Substituting $u = \sqrt{x+5}$ and $v = \sqrt{20-x}$ and observing that $u^2 + v^2 = x + 5 + 20 - x = 25$ gives

$$u + v = 7,$$
$$u^2 + v^2 = 25.$$

From this it is easy to get $uv = \frac{1}{2}(c^2 - a - b) = \frac{1}{2}(49 - 5 - 20) = 12$. Thus, the system can be rewritten as the symmetrical system

$$u + v = 7,$$
$$uv = 12.$$

The solutions of the last system are two pairs $(3, 4)$ and $(4, 3)$. Substituting $u = 3$ into $u = \sqrt{x+5}$ (or $v = 4$ into $v = \sqrt{20-x}$, which gives the same results) yields $\sqrt{x+5} = 3$, from which $x = 4$. Analogously, for $u = 4$, $\sqrt{x+5} = 4$, from which $x = 11$. Clearly, both solutions 4 and 11 are in the domain of the equation.

Answer: 4, 11.

Problem 13. Solve the system of equations

$$(x+y)(1-xy) = 2,$$
$$x^2 + y^2 = 4.$$

Solution. First, observe that the attempt to express one of the variables in terms of the other from one of the equations with further substitution into the second equation will result in the equation of the sixth degree. So, probably that is not a good idea. The hint to the solution is hidden in the second equation, $x^2 + y^2 = 4$. Recalling Viète's method for trigonometric substitution discussed in chapter 7 and letting $x = 2\sin\alpha$, $y = 2\cos\alpha$, $0 \leq \alpha < 2\pi$, if we substitute in the second equation, we get $4\sin^2\alpha + 4\cos^2\alpha = 4$ or, after dividing both sides by 4, $\sin^2\alpha + \cos^2\alpha = 1$.

The Pythagorean Identity justifies our selection of the trigonometric substitutions for x and y. Indeed, given the condition $x^2 + y^2 = 4$, we can state that there must exist an angle α, $0 \leq \alpha < 2\pi$, such that $x = 2\sin\alpha$, $y = 2\cos\alpha$. This leads to a rather unusual substitution. To simplify the solution we will use the trigonometric substitutions $x = 2\sin\alpha$, $y = 2\cos\alpha$, $0 \leq \alpha < 2\pi$. The first equation can be rewritten as $(2\sin\alpha + 2\cos\alpha)(1 - 2\sin\alpha \cdot 2\cos\alpha) = 2$. Dividing both sides by 2 and recalling that $2\sin\alpha \cdot \cos\alpha = \sin 2\alpha$, we get

$$(\sin\alpha + \cos\alpha)(1 - 2\sin 2\alpha) = 1.$$

Making simple modifications on the left-hand side and recalling from chapter 6 the identities for $\sin 3x$ and $\cos 3x$ as $\sin 3x = 3\sin x - 4\sin^3 x$

and $\cos 3x = 4\cos^3 x - 3\cos x$, we obtain

$$\sin \alpha + \cos \alpha - \sin \alpha \cdot 2 \sin 2\alpha - \cos \alpha \cdot 2 \sin 2\alpha$$
$$= \sin \alpha + \cos \alpha - \sin \alpha \cdot 4 \sin \alpha \cdot \cos \alpha - 4 \cos \alpha \cdot \sin \alpha \cdot \cos \alpha$$
$$= \sin \alpha + \cos \alpha - 4 \sin^2 \alpha \cdot \cos \alpha - 4 \cos^2 \alpha \cdot \sin \alpha$$
$$= \sin \alpha + \cos \alpha - 4(1 - \cos^2 \alpha) \cdot \cos \alpha - 4(1 - \sin^2 \alpha) \cdot \sin \alpha$$
$$= \sin \alpha + \cos \alpha - 4 \cos \alpha + 4 \cos^3 \alpha - 4 \sin \alpha + 4 \sin^3 \alpha$$
$$= (4 \cos^3 \alpha - 3 \cos \alpha) - (3 \sin \alpha - 4 \sin^3 \alpha) = \cos 3\alpha - \sin 3\alpha.$$

Hence, the first equation of the system can be rewritten as

$$\cos 3\alpha - \sin 3\alpha = 1.$$

To solve this equation we will apply the auxiliary angle technique covered in chapter 7. Dividing both sides of the equation by $\sqrt{1^2 + (-1)^2} = \sqrt{2}$ gives

$$\frac{1}{\sqrt{2}} \cos 3\alpha - \frac{1}{\sqrt{2}} \sin 3\alpha = \frac{1}{\sqrt{2}}.$$

Recalling that $\cos \frac{\pi}{4} = \sin \frac{\pi}{4} = \frac{1}{\sqrt{2}}$ and substituting $\cos \frac{\pi}{4}$ and $\sin \frac{\pi}{4}$ for $\frac{1}{\sqrt{2}}$ into the obtained equation gives

$$\cos \frac{\pi}{4} \cos 3\alpha - \sin \frac{\pi}{4} \sin 3\alpha = \frac{1}{\sqrt{2}};$$

using the identity for the cosine of a sum yields $\cos \left(3\alpha + \frac{\pi}{4}\right) = \frac{1}{\sqrt{2}}$. Therefore, we get $3\alpha + \frac{\pi}{4} = \pm \arccos \frac{1}{\sqrt{2}} + 2\pi n$, $n \in Z$. Noticing that $\arccos \frac{1}{\sqrt{2}} = \frac{\pi}{4}$ and dividing both sides by 3 gives $a = -\frac{\pi}{12} \pm \frac{\pi}{12} + \frac{2\pi n}{3}$, $n \in Z$.

Since $\alpha \in [0, 2\pi[$, let's review all the permissible values of α to consider:

$n = 0$. Then $\alpha = -\frac{\pi}{12} + \frac{\pi}{12} = 0$ or $\alpha = -\frac{\pi}{12} - \frac{\pi}{12} = -\frac{\pi}{6}$ (not satisfactory).

$n = 1$. Then $\alpha = -\frac{\pi}{12} + \frac{\pi}{12} + \frac{2\pi}{3} = \frac{2\pi}{3}$ or $\alpha = -\frac{\pi}{12} - \frac{\pi}{12} + \frac{2\pi}{3} = \frac{\pi}{2}$.

$n = 2$. Then $\alpha = -\frac{\pi}{12} + \frac{\pi}{12} + \frac{4\pi}{3} = \frac{4\pi}{3}$ or $\alpha = -\frac{\pi}{12} - \frac{\pi}{12} + \frac{4\pi}{3} = \frac{7\pi}{6}$.

$n = 3$. Then $\alpha = -\frac{\pi}{12} + \frac{\pi}{12} + \frac{6\pi}{3} = 2\pi$ (not satisfactory) or $\alpha = -\frac{\pi}{12} - \frac{\pi}{12} + \frac{6\pi}{3} = \frac{11\pi}{6}$.

So, the admissible values of α are $0, \frac{\pi}{2}, \frac{2\pi}{3}, \frac{7\pi}{6}, \frac{4\pi}{3}, \frac{11\pi}{6}$. Going back to the substitutions for x and y as $x = 2 \sin \alpha$, $y = 2 \cos \alpha$, we will be able to calculate the values for x and y for each of the values of α. We leave it for readers to verify that the solutions of the system are $(-\sqrt{3}, -1)$, $(-1, -\sqrt{3})$, $(-1, \sqrt{3})$, $(0, 2)$, $(\sqrt{3}, -1)$, $(2, 0)$.

Answer: $(-\sqrt{3}, -1)$, $(-1, -\sqrt{3})$, $(-1, \sqrt{3})$, $(0, 2)$, $(\sqrt{3}, -1)$, $(2, 0)$.

Systems of nonlinear equations with more than two variables

Problem 14. Solve the system of equations

$$x(y-1) = 3,$$
$$(3-y)z = 1,$$
$$(x-2)(2-z) = 1.$$

Solution. For the first two equations, let us write x and z in terms of y. Substituting $\frac{3}{y-1}$ for x and $\frac{1}{3-y}$ for z into the third equation gives

$$x = \frac{3}{y-1},$$

$$z = \frac{1}{3-y},$$

$$\left(\frac{3}{y-1} - 2\right)\left(2 - \frac{1}{3-y}\right) = 1.$$

Solve the last equation for y, $\frac{3-2y+2}{y-1} \cdot \frac{6-2y-1}{3-y} = 1$. It follows that

$$\frac{5-2y}{y-1} \cdot \frac{5-2y}{3-y} = 1, (y \neq 1, y \neq 3).$$
$$(5-2y)^2 = (y-1)(3-y),$$
$$25 - 20y + 4y^2 = -y^2 + 4y - 3,$$
$$5y^2 - 24y + 28 = 0.$$
$$D = 576 - 560 = 16.$$
$$y = \frac{24 \pm 4}{10}, \quad y_1 = 2, \quad y_2 = 2\frac{4}{5}.$$

Now we can back-solve for the corresponding values of x and z. It is important to keep track of the corresponding values of all the variables.

$$x_1 = \frac{3}{2-1} = 3, \quad x_2 = \frac{3}{2\frac{4}{5}-1} = \frac{5}{3} = 1\frac{2}{3}.$$

$$z_1 = \frac{1}{3-2} = 1, \quad z_2 = \frac{1}{3-2\frac{4}{5}} = 5.$$

Answer: $(3, 2, 1), (1\frac{2}{3}, 2\frac{4}{5}, 5)$.

Dealing with the systems of nonlinear equations with more than two variables usually involves various artificial tricks. A straightforward substitution rarely gives good results immediately. You may need to perform some manipulations with equations and introduce auxiliary variables establishing new relationships among the variables that simplify the system and make it solvable. A good example is provided by symmetrical systems of equations with three variables. By a symmetrical system of equations with three variables we will understand the system

$$P_1(x, y, z) = 0,$$
$$P_2(x, y, z) = 0,$$
$$P_3(x, y, z) = 0,$$

in which $P_1(x, y, z)$, $P_2(x, y, z)$, and $P_3(x, y, z)$ are polynomials symmetrical in variables x, y, and z, meaning each of the polynomials will not change by any permutations of the variables x, y, and z. The following substitutions are useful in simplifying the original system of equations:

$$x + y + z = u, \quad xy + xz + yz = v, \quad xyz = w.$$

The simplest symmetrical system of equations can be written as

$$\begin{aligned} x + y + z &= a, \\ xy + yz + zx &= b, \qquad\qquad (1) \\ xyz &= c. \end{aligned}$$

This system and the equation $t^3 - at^2 + bt - c = 0$ are linked with the following relationship: if t_1, t_2, t_3 are the roots of the equation $t^3 - at^2 + bt - c = 0$, then system (1) has solutions obtained as all possible permutations of the numbers t_1, t_2, and t_3 and has no other solutions. Conversely, if (x_0, y_0, z_0) is a solution of the system (1), then x_0, y_0, and z_0 are the solutions of the equation $t^3 - at^2 + bt - c = 0$.

Let's prove the direct statement. Assume that t_1, t_2, t_3 are the roots of the equation $t^3 - at^2 + bt - c = 0$. Applying Viète's formulas for a cubic equation gives

$$\begin{aligned} t_1 + t_2 + t_3 &= a, \\ t_1 t_2 + t_2 t_3 + t_3 t_1 &= b, \\ t_1 \cdot t_2 \cdot t_3 &= c. \end{aligned}$$

Therefore, (t_1, t_2, t_3) is the solution of the given system of equations. The other five solutions are obtained as various permutations of the numbers t_1, t_2, and t_3. We see that the direct statement is valid.

Consider now (x_0, y_0, z_0), a solution of system (1). It follows that x_0, y_0, and z_0 satisfy the equalities

$$\begin{aligned} x_0 + y_0 + z_0 &= a, \\ x_0 y_0 + y_0 z_0 + z_0 x_0 &= b, \\ x_0 y_0 z_0 &= c. \end{aligned}$$

Therefore, the polynomial $t^3 - at^2 + bt - c$ can be rewritten as

$$t^3 - at^2 + bt - c = t^3 - (x_0 + y_0 + z_0)t^2 + (x_0y_0 + y_0z_0 + z_0x_0)t - x_0y_0z_0$$
$$= (t - x_0)(t - y_0)(t - z_0),$$

which concludes the proof that x_0, y_0, and z_0 are the solutions of the equation $t^3 - at^2 + bt - c = 0$.

Problem 15. Solve the system of equations

$$x + y + z = 1,$$
$$x^2 + y^2 + z^2 = 1,$$
$$x^3 + y^3 + z^3 = 1.$$

Solution. Using the discussed technique, let's introduce the following substitutions:

$$x + y + z = u,$$
$$xy + xz + yz = v,$$
$$xyz = w.$$

By well-known identities, which have already been used before,

$$x^2 + y^2 + z^2 = (x + y + z)^2 - 2(xy + xz + yz) = u^2 - 2v.$$
$$x^3 + y^3 + z^3 = (z + y + z)^3 - 3(x + y + z)(xy + xz + yz) + 3xyz$$
$$= u^3 - 3uv + 3w.$$

It follows that the new system to solve is

$$u = 1,$$
$$u^2 - 2v = 1,$$
$$u^3 - 3uv + 3w = 1.$$

Solutions are $u = 1$, $v = 0$, $w = 0$. Therefore,

$$x + y + z = 1,$$
$$xy + xz + yz = 0,$$
$$xyz = 0.$$

From the last equation obviously one of the variables has to equal 0. Making the respective substitutions into the first and second equations leads to the conclusion that the solutions will be all permutations of the numbers 1, 0, 0.

Answer: : $(1, 0, 0)$, $(0, 1, 0)$, $(0, 0, 1)$.

Another type of symmetric system of nonlinear equations is the system

$$xy = a,$$
$$xz = b, \tag{2}$$
$$yz = c,$$

where a, b, and c are real numbers not equal to 0.

Recall that in chapter 1 we solved symmetric systems of linear equations by adding the equations in the system and then step by step subtracting each equation from the resulting sum. A similar approach (in this case by multiplying the equations) is efficient in the solution of the above system of nonlinear equations (2) as well. Multiplying all the equations gives $x^2y^2z^2 = abc$, from which $xyz = \pm\sqrt{abc}$. Dividing the last equation by each of the equations in the system yields the solutions. We can do this without losing any possible solutions because $a \neq 0, b \neq 0, c \neq 0$.

Problem 16. Solve the system of equations

$$x\left(\frac{y}{z} + \frac{z}{y}\right) = \frac{13}{6},$$
$$y\left(\frac{z}{x} + \frac{x}{z}\right) = \frac{20}{3},$$
$$z\left(\frac{x}{y} + \frac{y}{x}\right) = \frac{15}{2}.$$

Solution. Let's rewrite the system as

$$\frac{xy}{z} + \frac{xz}{y} = \frac{13}{6},$$
$$\frac{yz}{x} + \frac{xy}{z} = \frac{20}{3},$$
$$\frac{xz}{y} + \frac{zy}{x} = \frac{15}{2}.$$

Adding the equations and dividing by 2 gives $\frac{xy}{z} + \frac{xz}{y} + \frac{yz}{x} = \frac{49}{6}$. Subtracting from this equation in turn the first, the second, and the third equation of the system yields

$$\frac{yz}{x} = 6,$$
$$\frac{xz}{y} = \frac{3}{2},$$
$$\frac{xy}{z} = \frac{2}{3}.$$

Multiplying the three equations gives $xyz = 6$. Dividing it by each equation in the last system, we obtain

$$x^2 = 1,$$
$$y^2 = 4,$$
$$z^2 = 9.$$

It follows the solutions of the system are $(1, 2, 3)$; $(1, -2, -3)$; $(-1, 2, -3)$; $(-1, -2, 3)$.

Answer: $(1, 2, 3)$, $(1, -2, -3)$, $(-1, 2, -3)$, $(-1, -2, 3)$.

Problem 17. Solve the system of equations

$$y^3 - 9x^2 + 27x - 27 = 0,$$
$$z^3 - 9y^2 + 27y - 27 = 0,$$
$$x^3 - 9z^2 + 27z - 27 = 0.$$

Solution. This is a very interesting problem, whose solution needs the introduction of an auxiliary function $f(t) = t^2 - 3t + 3$. The system then can be rewritten as

$$y^3 = 9f(x),$$
$$z^3 = 9f(y),$$
$$x^3 = 9f(z).$$

Why did we do that? How would it help? What should be done next? It's not hard to notice that all three equations in the system look identical with the different variables' permutations. Somehow this fact has to be utilized in the solution of the problem. Analysis of the auxiliary function will clarify all the next steps to take. We see that $f(t)$ is a quadratic function. The first coefficient is $1 > 0$. The discriminant $D = 9 - 12 = -3 < 0$. Therefore, the range of the function is the set of all positive real numbers. It follows that $y^3 = 9f(x) > 0$, $z^3 = 9f(y) > 0$, and $x^3 = 9f(z) > 0$, from which $y > 0$, $z > 0$, and $x > 0$. Adding all three equations and regrouping terms, we have

$$(y^3 - 9y^2 + 27y - 27) + (x^3 - 9x^2 + 27x - 27) + (z^3 - 9z^2 + 27z - 27) = 0.$$

Notice that we have the sum of three cubes on the left-hand side:

$$(x - 3)^3 + (y - 3)^3 + (z - 3)^3 = 0. \tag{1}$$

Since elsewhere $x > 0$, there are only two choices for value of x:

$$x \geq 3 \quad \text{or} \quad 0 < x < 3.$$

Let's examine each case.

If $x \geq 3$, then $x^3 \geq 27$ and from the last equation of the original system we get that

$$9z^2 - 27z + 27 \geq 27,$$
$$9z^2 - 27z \geq 0,$$
$$9z(z - 3) \geq 0.$$

Since we know that $z > 0$ (from the above analysis of the function $f(y)$), then from the last inequality it follows that $z \geq 3$. Similarly, we conclude that the same restrictions should be valid for y, $y \geq 3$. So, for all three variables the following inequalities hold at the same time:

$$x \geq 3,$$
$$y \geq 3,$$
$$z \geq 3.$$

This implies that the equality (1) is possible only when all three variables equal 3,

$$x = y = z = 3.$$

In the second case, $0 < x < 3$. Then $x^3 < 27$ or $9z^2 - 27z + 27 < 27$,

$$9z^2 - 27z < 0,$$
$$9z(z - 3) < 0.$$

From the last inequality it follows that for the positive z,

$$z - 3 < 0, \quad \text{or} \quad z < 3.$$

In the same manner, we get that the same holds for the positive values of y, $y < 3$.

To summarize: if we assume that $0 < x < 3$, we come to the conclusion that at the same time $0 < y < 3$ and $0 < z < 3$, which contradicts the equality (1). It is impossible to get the sum of three negative numbers equal to 0. Hence, we conclude that $(3, 3, 3)$ is the only solution to the original system of equations.

Answer: $(3, 3, 3)$.

Problem 18. Solve the system of equations

$$x + y + z = 2,$$
$$2xy - z^2 = 4.$$

Solution. The system we are dealing with has two equations and three variables. Armed with the methods for solving multivariable equations, we should try one of those techniques. In this case it has to be completing the square. Expressing y in terms of x and z from the first equation as $y = 2 - x - z$ and substituting it into the second equation gives

$$2x(2 - x - z) - z^2 = 4.$$

Simplifying this equation yields $4x - 2x^2 - 2xz - z^2 - 4 = 0$ or equivalently,

$$2x^2 - 4x + 2xz + z^2 + 4 = 0.$$

After regrouping, we obtain $(x^2 - 4x + 4) + (x^2 + 2xz + z^2) = 0$. Completing the square results in $(x - 2)^2 + (x + z)^2 = 0$. The left-hand side is a nonnegative number as it is the sum of nonnegative numbers. It implies that equality to 0 is possible only for $x = 2$ and $z = -x = -2$. Substituting these values back into the first equation gives $2 + y - 2 = 2$, from which $y = 2$.

Answer: $(2, 2, -2)$.

Problem 19. Evaluate the expression $xy + zy$, knowing that x, y, and z are positive solutions of the system of equations

$$x^2 + y^2 = 16,$$
$$y^2 + z^2 = 48,$$
$$y^2 - xz = 0.$$

Solution. The conventional strategy to solve the problem is to find the numbers x, y, and z as the algebraic solutions to the system and then substituting those values in to evaluate $xy + zy$. Will it work? Absolutely, it will. We invite readers to work out this suggestion on their own.

There is another approach, which serves as a great demonstration of applying a geometrical interpretation in simplifying the solution of a purely algebraic problem.

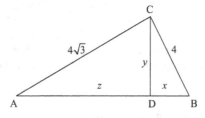

How about considering two right triangles BDC and ADC? The first has legs x and y and hypotenuse 4 (because it is given that $x^2 + y^2 = 16$ in the first equation of the system; by the converse of the Pythagorean Theorem, such a right triangle has to exist) and the second right triangle has legs z and y and hypotenuse $4\sqrt{3}$ (it has to exist since $y^2 + z^2 = 48$ from the second equation of the system). Let's put them together, so they have the leg CD as the common side. It follows that CD is the altitude of the newly formed triangle ACB. Rewrite the third equation in the system as $y^2 = xz$. Such a relationship between the altitude and the segments in which it splits the opposite side in a triangle is possible only in a right triangle. If readers are not familiar with this property, we hope they can easily prove it from the similarity of the triangles ADC and CDB. Thus, ACB has to be the right angle ($\angle C = 90°$). The area of the triangle ACB equals the sum of the areas of the right triangles BDC and ADC (since the area of a triangle equals half of its base times its height, each area is determined here as half the product of the legs):

$$S_{ACB} = S_{BDC} + S_{ADC} = \frac{1}{2}xy + \frac{1}{2}zy = \frac{1}{2}(xy + zy)$$

from which $xy + zy = 2S_{ACB}$. On the other hand, the area of the right triangle ACB equals

$$S_{ACB} = \frac{1}{2}AC \cdot CB = \frac{1}{2} \cdot 4\sqrt{3} \cdot 4 = 8\sqrt{3}.$$

Therefore, $xy + zy = 2S_{ACB} = 2 \cdot 8\sqrt{3} = 16\sqrt{3}$ and the problem is solved.

It is worth emphasizing a few important conclusions from this problem. As we saw, even though facing a system of equations in the problem, it does not necessarily imply the need to solve it in order to evaluate $xy + zy$. The geometric interpretation was enlightening. We were able to directly obtain the desired result by analyzing each equation and applying a geometrical interpretation to the whole system. When working with a system of equations, one should remember that the goal is to find the solutions that satisfy all the equations simultaneously. That's why we were able to put together two right triangles with a common leg, form the third right triangle, and take advantage of its properties. By doing this, we skipped the process of finding the separate values of x, y, and z and instead immediately evaluated $xy + zy$ in terms of the area of the right triangle. In solving a problem, it is critical to always keep in mind the question asked and, if possible, get to the answer by skipping the intermediate steps.

This non-standard technique might be useful in solving many purely geometric problems as well. The idea is to set up the system of equations from the given conditions. Analyze the equations and, if possible, skip the solution of the system and go directly to the desired element.

Problem 20. Medians drawn to the legs of a right triangle have lengths $\sqrt{52}$ and $\sqrt{73}$. Find the length of the hypotenuse of this triangle.

Solution.

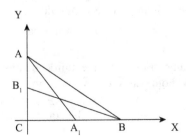

Consider the given right triangle ABC ($\angle C = 90°$) in a Cartesian coordinate system such that vertex C coincides with the origin, vertex A is located on the Y-axis, and vertex B is located on the X-axis. AA_1 and BB_1 are the given medians, $AA_1 = \sqrt{52}$, $BB_1 = \sqrt{73}$. We need to find AB.

Let's define each point by its coordinates: $A(0, y)$, $B(x, 0)$, $C(0, 0)$, $A_1(\frac{x}{2}, 0)$, $B_1(0, \frac{y}{2})$. By the formula for a distance between two points, the desired hypotenuse's length is $AB = \sqrt{x^2 + y^2}$. From the right triangles ACA_1 and BCB_1 we can set up the following equations representing the given medians' lengths in terms of the coordinates of the endpoints:

$$AA_1^2 = \left(\frac{x}{2}\right)^2 + y^2 = 52,$$
$$BB_1^2 = x^2 + \left(\frac{y}{2}\right)^2 = 73.$$

Adding these two equations yields $\frac{5}{4}x^2 + \frac{5}{4}y^2 = 125$, from which $x^2 + y^2 = 100$. Therefore, the hypotenuse $AB = \sqrt{100} = 10$, and we arrived at the desired result. There is no need to find x and y as soon as we get the hypotenuse's length directly by adding the equations from the system.

Systems of nonlinear equations with **more than three variables** are usually considered unconventional and often appear as tricky challenges on math Olympiads and contests. We will close this chapter with two such examples.

Problem 21. Given the system of equations

$$x_1 x_2 x_3 \cdot \ldots \cdot x_{2018} = 1,$$
$$x_1 - x_2 x_3 \cdot \ldots \cdot x_{2018} = 1,$$
$$x_1 x_2 - x_3 x_4 \cdot \ldots \cdot x_{2018} = 1,$$
$$\ldots$$
$$x_1 x_2 x_3 \cdot \ldots \cdot x_{2017} - x_{2018} = 1,$$

find the values of x_{25} satisfying the system.

Solution. Clearly, none of the variables can equal 0; otherwise their product will be 0, not 1, as it is given in the first equation of the system. Therefore, multiplying the second equation of the system by $x_1 \neq 0$ transforms it into the equivalent equation $x_1^2 - x_1 x_2 x_3 \cdot \ldots \cdot x_{2018} = x_1$. Substituting the value $x_1 x_2 x_3 \cdot \ldots \cdot x_{2018} = 1$ from the first equation of the system gives a quadratic equation for x_1, $x_1^2 - 1 = x_1$ or equivalently, $x_1^2 - x_1 - 1 = 0$. The roots of this equation are $\frac{1-\sqrt{5}}{2}$ and $\frac{1+\sqrt{5}}{2}$, $x_1 = \frac{1\pm\sqrt{5}}{2}$.

We repeat the same trick with the third equation of the system. This time we will multiply both sides of the equation by $(x_1 x_2) \neq 0$. We then arrive at the equation $(x_1 x_2)^2 - x_1 x_2 x_3 x_4 \cdot \ldots \cdot x_{2018} = x_1 x_2$. Again, noticing that $x_1 x_2 x_3 x_4 \cdot \ldots \cdot x_{2018} = 1$, we will rewrite the equation as $(x_1 x_2)^2 - 1 = x_1 x_2$ or equivalently, $(x_1 x_2)^2 - x_1 x_2 - 1 = 0$. Letting $x_1 x_2 = y$ and solving the quadratic equation $y^2 - y - 1 = 0$, we get $y = \frac{1 \pm \sqrt{5}}{2}$. So, we see that working with the third equation of the system results in $x_1 x_2 = \frac{1 \pm \sqrt{5}}{2}$. Recalling that $x_1 = \frac{1 \pm \sqrt{5}}{2}$, it's not hard to obtain the solutions for x_2:

If $x_1 = \frac{1+\sqrt{5}}{2}$ and $x_1 x_2 = \frac{1+\sqrt{5}}{2}$, then $x_2 = 1$.

If $x_1 = \frac{1-\sqrt{5}}{2}$ and $x_1 x_2 = \frac{1-\sqrt{5}}{2}$, then $x_2 = 1$.

If $x_1 = \frac{1+\sqrt{5}}{2}$ and $x_1 x_2 = \frac{1-\sqrt{5}}{2}$, then $x_2 = \frac{1-\sqrt{5}}{1+\sqrt{5}}$.

If $x_1 = \frac{1-\sqrt{5}}{2}$ and $x_1 x_2 = \frac{1+\sqrt{5}}{2}$, then $x_2 = \frac{1+\sqrt{5}}{1-\sqrt{5}}$.

Working the fourth and every following equation up to the 25th equation in the same manner, we will arrive at the values of x_{25} to be equal to the same results, $1, \frac{1-\sqrt{5}}{1+\sqrt{5}}, \frac{1+\sqrt{5}}{1-\sqrt{5}}$. In fact, every variable in this system of equations would have the same values.

Answer: $1, \frac{1-\sqrt{5}}{1+\sqrt{5}}, \frac{1+\sqrt{5}}{1-\sqrt{5}}$.

Problem 22. Find all natural numbers n such that there is a set of positive numbers $x_1, x_2, x_3, \ldots, x_n$ satisfying the system of equations

$$x_1 + x_2 + x_3 + \cdots + x_n = 9,$$

$$\frac{1}{x_1} + \frac{1}{x_2} + \frac{1}{x_3} + \cdots + \frac{1}{x_n} = 1.$$

Identify all such sets of numbers for each specific value of n.

Solution. Assume we managed to find the set of numbers $x_1, x_2, x_3, \ldots, x_n$, satisfying the given system of the equations for some specific natural number n. Let's denote

$$A = \sqrt[n]{x_1 \cdot x_2 \cdot \ldots \cdot x_n}.$$

Applying the AM–GM inequality to numbers $x_1, x_2, x_3, \ldots, x_n$ gives

$$x_1 + x_2 + x_3 + \cdots + x_n \geq n \cdot \sqrt[n]{x_1 \cdot x_2 \cdot \ldots \cdot x_n} = n \cdot A.$$

Since it is given in the first equation of the system that $x_1 + x_2 + x_3 + \cdots + x_n = 9$, we obtain,

$$n \cdot A \leq 9, \text{ from which } A \leq \frac{9}{n}. \tag{1}$$

Applying now the AM–GM inequality to numbers $\frac{1}{x_1}, \frac{1}{x_2}, \ldots, \frac{1}{x_n}$, we get

$$\frac{1}{x_1} + \frac{1}{x_2} + \frac{1}{x_3} + \cdots + \frac{1}{x_n} \geq n \cdot \sqrt[n]{\frac{1}{x_1} \cdot \frac{1}{x_2} \cdot \ldots \cdot \frac{1}{x_n}} = \frac{n}{\sqrt[n]{x_1 \cdot x_2 \cdot \ldots \cdot x_n}} = \frac{n}{A}.$$

It is given in the second equation of the system that

$$\frac{1}{x_1} + \frac{1}{x_2} + \frac{1}{x_3} + \cdots + \frac{1}{x_n} = 1. \text{ Therefore, } \frac{n}{A} \le 1, \text{ from which } A \ge n. \quad (2)$$

Comparing (1) and (2) yields $n \le A \le \frac{9}{n}$. Multiplying by a natural number n, $n > 0$, we obtain $n^2 \le A \cdot n \le 9$, which implies the inequality $n^2 \le 9$. Clearly, for natural n to satisfy the last inequality, it has to be either 1, 2, or 3. The value $n = 1$ should be rejected because the system has no solutions ($x_1 = 9$ and $\frac{1}{x_1} = 1$, which is impossible). If $n = 2$, we get the system of equations

$$x_1 + x_2 = 9,$$
$$\frac{1}{x_1} + \frac{1}{x_2} = 1.$$

Simplifying the second equation and substituting the value of $x_1 + x_2 = 9$ gives $\frac{x_2 + x_1}{x_1 \cdot x_2} = 1$ or equivalently, $x_1 \cdot x_2 = 9$. Hence, our system converts to

$$x_1 + x_2 = 9,$$
$$x_1 \cdot x_2 = 9.$$

By Viète's theorem, x_1 and x_2 have to be the roots of the quadratic equation

$$z^2 - 9z + 9 = 0.$$
$$D = 81 - 36 = 45.$$
$$z = \frac{9 \pm \sqrt{45}}{2} = \frac{9 \pm 3\sqrt{5}}{2}.$$

Both roots are positive numbers. Therefore, there are two pairs of positive solutions of the original system for $n = 2$, $\left(\frac{9+3\sqrt{5}}{2}, \frac{9-3\sqrt{5}}{2}\right)$ and $\left(\frac{9-3\sqrt{5}}{2}, \frac{9+3\sqrt{5}}{2}\right)$.

Finally, if $n = 3$ then the inequality $n^2 \le A \cdot n \le 9$ becomes $9 \le 3 \cdot \sqrt[3]{x_1 \cdot x_2 \cdot x_3} \le 9$, which leads to $\sqrt[3]{x_1 \cdot x_2 \cdot x_3} = 3$. The last equation has the only positive solutions $x_1 = x_2 = x_3 = 3$ satisfying the given system of equations. In other words, there is one solution of the given system, $(3, 3, 3)$, when $n = 3$.

Answer: When $n = 2$, there are two solutions $\left(\frac{9+3\sqrt{5}}{2}, \frac{9-3\sqrt{5}}{2}\right)$ and $\left(\frac{9-3\sqrt{5}}{2}, \frac{9+3\sqrt{5}}{2}\right)$. When $n = 3$, there is one solution $(3, 3, 3)$.

This chapter was devoted to methods and techniques for solving systems of nonlinear equations. We also exhibited in several problems that it is not less important to be able to properly set up the system of equations from the given conditions and manipulate them in a specific way to get to the desired result even without actually solving the system. As another important development, we learned that in some specific cases the solution of a single-variable equation might be simplified by introducing multiple variables and converting the equation to a system of equations.

The solutions presented are not unique. We encourage readers to always try a variety of approaches. Especially enriching is the strategy of using combinations of various methods and techniques investigated through the course of the book, such as highlighted in the solutions above: completion of the square, quadratic and cubic functions properties, application of homogeneous equations, degree reductions by various substitutions, introduction of auxiliary variables and auxiliary functions, exploiting classic inequalities, and many others. Since many mathematical ideas have both an algebraic and geometrical aspect, always keep your eyes open for possible links between algebraic and geometric interpretations of the items involved. Be creative!

Exercises

Problem 23. Solve the system of equations

$$x^2 - xy = 24,$$
$$y^2 - xy = -8.$$

Problem 24. Solve the system of equations

$$\sqrt[3]{x} \cdot \sqrt{y} + \sqrt{x} \cdot \sqrt[3]{y} = 12,$$
$$xy = 64.$$

Problem 25. Solve the system of equations

$$x^3 + y^3 = 65,$$
$$x^2 y + xy^2 = 20.$$

Problem 26. Solve the system of equations

$$x^{-1} + y^{-1} = 5,$$
$$x^{-2} + y^{-2} = 13.$$

Problem 27. Solve the system of equations

$$x^2 + 2y + \sqrt{x^2 + 2y + 1} = 1,$$
$$2x + y = 2.$$

Problem 28. Evaluate the expression $xy + 2yz + 3xz$, if x, y, and z are the positive solutions of the following system of equations

$$x^2 + xy + \frac{y^2}{3} = 25,$$

$$\frac{y^2}{3} + z^2 = 16,$$

$$z^2 + zx + x^2 = 9.$$

Problem 29. Solve the system of equations

$$x + y + z = 2,$$
$$xy + xz + yz = -1,$$
$$xyz = -2.$$

Chapter 13

Brainstorming

The highest form of pure thought is in mathematics
Plato.

This chapter is going to be the last one in the book. In each of the previous chapters we studied various types of equations and showed the methods of their solutions. It is much easier to solve a problem when you are told beforehand to what type the problem belongs to and what methods are to be applied. Then, basically, you just practice the applications. It is much more difficult if you face a problem and nobody told you what methods should be used in its solution. During the course of the book a lot of difficult problems have been discussed and solved. Hopefully, you should be able to recognize by now the relatives of specific equation families and find the key for their solutions. Here is an assortment of 12 more challenges, as many as there are previous chapters in the book, for your enjoyment. The problems deal with topics encountered earlier. It is strongly recommended, as you read each of the problems below, to try to solve the problem first. It's totally up to you how to approach the problems, what techniques or their combinations to apply. *Do not read* the suggested solutions until you manage to find your own. Have fun!

Problem 1. Solve the equation

$$\left((\sqrt{5}+\sqrt{3}) \cdot \sqrt{\frac{4-\sqrt{15}}{2}} \right)^{x} + 3^{0.5x} = \left(-\cos\frac{2\pi}{5} - \cos\frac{4\pi}{5} \right)^{-x}.$$

Solution. First, let's simplify the left-hand side.

Step 1. Multiplying the numerator and denominator of $\frac{4-\sqrt{15}}{2}$ by 2 gives

$$\sqrt{\frac{4-\sqrt{15}}{2}} = \sqrt{\frac{8-2\sqrt{15}}{4}} = \sqrt{\frac{5-2\sqrt{5}\cdot\sqrt{3}+3}{4}}$$

$$= \sqrt{\frac{(\sqrt{5}-\sqrt{3})^2}{4}} = \frac{\sqrt{5}-\sqrt{3}}{2}.$$

Step 2. Using the formula for the difference of squares,

$$(\sqrt{5}+\sqrt{3})(\sqrt{5}-\sqrt{3}) = (\sqrt{5})^2 - (\sqrt{3})^2 = 5-3 = 2,$$

so the left-hand side can be rewritten as

$$\left((\sqrt{5}+\sqrt{3})\cdot\frac{\sqrt{5}-\sqrt{3}}{2}\right)^x + 3^{0.5x} = \left(\frac{5-3}{2}\right)^x + 3^{0.5x} = 1+3^{0.5x}.$$

To modify the right-hand side, we will use the result derived in chapter 6, $\sin\frac{\pi}{10} = \frac{\sqrt{5}-1}{4}$. Thus,

$$\cos\frac{2\pi}{5} = \cos\left(\frac{\pi}{2}-\frac{\pi}{10}\right) = \sin\frac{\pi}{10} = \frac{\sqrt{5}-1}{4}.$$

Using the formula for cosine of a double angle gives

$$\cos\frac{4\pi}{5} = \cos^2\frac{2\pi}{5} - \sin^2\frac{2\pi}{5} = \cos^2\frac{2\pi}{5} - \left(1-\cos^2\frac{2\pi}{5}\right)$$

$$= 2\cos^2\frac{2\pi}{5} - 1 = 2\cdot\left(\frac{\sqrt{5}-1}{4}\right)^2 - 1 = 2\cdot\frac{6-2\sqrt{5}}{16} - 1$$

$$= \frac{3-\sqrt{5}}{4} - 1 = \frac{3-\sqrt{5}-4}{4} = -\frac{1+\sqrt{5}}{4}.$$

So, the right-hand side of the equation can be rewritten as

$$\left(-\cos\frac{2\pi}{5} - \cos\frac{4\pi}{5}\right)^{-x} = \left(-\frac{\sqrt{5}-1}{4} + \frac{1+\sqrt{5}}{4}\right)^{-x}$$

$$= \left(\frac{-\sqrt{5}+1+1+\sqrt{5}}{4}\right)^{-x} = \left(\frac{1}{2}\right)^{-x} = 2^x.$$

Thus, the given equation simplifies to an exponential equation

$$1+3^{0.5x} = 2^x,$$
$$1+3^{0.5x} = (2^2)^{0.5x},$$
$$1+3^{0.5x} = 4^{0.5x}.$$

Dividing both sides by $4^{0.5x}$ ($4^{0.5x} \neq 0$) yields

$$\left(\frac{1}{4}\right)^{0.5x} + \left(\frac{3}{4}\right)^{0.5x} = 1. \tag{*}$$

It's easy to see that $x = 2$ satisfies this equation:

$$\left(\frac{1}{4}\right)^{1} + \left(\frac{3}{4}\right)^{1} = 1 - \text{a true statement.}$$

Let's prove that there are no other solutions.
Indeed, if $x > 2$, then $\left(\frac{1}{4}\right)^{0.5x} + \left(\frac{3}{4}\right)^{0.5x} < \frac{1}{4} + \frac{3}{4} = 1$, which contradicts (*),
if $x < 2$, then $\left(\frac{1}{4}\right)^{0.5x} + \left(\frac{3}{4}\right)^{0.5x} > \frac{1}{4} + \frac{3}{4} = 1$, which contradicts (*).
Therefore, $x = 2$ is the only root of the original equation.

Answer: 2.

Problem 2. Solve the equation

$$(x^2 + x + 4)^2 + 8x(x^2 + x + 4) + 15x^2 = 0.$$

Solution. Use the substitution

$$x^2 + x + 4 = y. \tag{1}$$

It gives

$$y^2 + 8yx + 15x^2 = 0.$$

Solving this equation as a quadratic for y gives

$$D = 64x^2 - 60x^2 = 4x^2,$$
$$y = \frac{-8x \pm \sqrt{4x^2}}{2},$$

so $y_1 = -5x$, $y_2 = -3x$.
Substituting the values for y into the equality (1) leads to two equations

$$x^2 + x + 4 = -5x \quad \text{and} \quad x^2 + x + 4 = -3x.$$

Solving the first equation gives

$$x^2 + 6x + 4 = 0,$$
$$x = \frac{-6 \pm \sqrt{20}}{2},$$
$$x_1 = -3 + \sqrt{5}, \quad x_2 = -3 - \sqrt{5}.$$

Solving the second equation gives

$$x^2 + 4x + 4 = 0,$$
$$(x+2)^2 = 0,$$
$$x = -2.$$

Answer: $-3 + \sqrt{5}, -3 - \sqrt{5}, -2$.

Another option is to solve the homogeneous equation $y^2 + 8yx + 15x^2 = 0$ by dividing both sides of the equation by x^2, (it's not hard to see that $x \neq 0$) and continue solving the equation $\left(\frac{y}{x}\right)^2 + 8 \cdot \frac{y}{x} + 15 = 0$ by introducing one more new variable $t = \frac{y}{x}$. The result will be the same.

Problem 3. Solve the equation:

$$3^x + 3^{-x} - 2 = \log_3 \cos x.$$

Solution. First, notice that the domain of this equation is determined from the solutions of the inequality $\cos x > 0$. Therefore, the domain is all real numbers such that

$$-\frac{\pi}{2} + 2\pi k < x < \frac{\pi}{2} + 2\pi k, \quad k \in Z.$$

Modify the left-hand side of the equation as

$$3^x + 3^{-x} - 2 = 3^x + \frac{1}{3^x} - 2 = \frac{3^{2x} - 2 \cdot 3^x + 1}{3^x} = \frac{(3^x - 1)^2}{3^x}.$$

Since $(3^x - 1)^2 \geq 0$ and $3^x > 0$ for any real x, then $\frac{(3^x-1)^2}{3^x} \geq 0$, and we conclude the range of the left-hand side is the set of all real nonnegative numbers. On the other hand, the range of the right-hand side of the equation is the set of all real nonpositive numbers. Indeed, for any real x, $|\cos x| \leq 1$, therefore, $\log_3 \cos x \leq 0$. The only one common number in the ranges of both sides is 0. Therefore, the solutions will exist only when each side of the equation equals 0.

$$\frac{(3^x - 1)^2}{3^x} = 0,$$
$$\log_3 \cos x = 0.$$

From the first equation $3^x - 1 = 0$, or $x = 0$.
From the second equation $\cos x = 1$, or $x = 2\pi n, n \in Z$.
There is the only one common solution $x = 0$.

Answer: 0.

Problem 4. Solve the equation

a) $x^2 + 6x \sin(3xy) + 9 = \log \tan 1° + \log \tan 2° + \log \tan 3° + \cdots + \log \tan 89°$;

b) $x^2 + 6x \sin(3xy) + 9 = \log \tan 1° \cdot \log \tan 2° \cdot \log \tan 3° \cdot \ldots \cdot \log \tan 89°$.

Solution. To simplify the right-hand side of the equation a)

$$x^2 + 6x \sin(3xy) + 9 = \log \tan 1° + \log \tan 2° + \log \tan 3° + \cdots + \log \tan 89°$$

we will recall the property of logarithms, $\log_a(m \cdot n) = \log_a m + \log_a n$ and the properties of trigonometric functions, $\cot \alpha = \tan(90° - \alpha)$ and $\tan \alpha \cdot \cot \alpha = 1$. It follows that

$$\log \tan 1° + \log \tan 2° + \log \tan 3° + \cdots + \log \tan 89°$$
$$= \log(\tan 1° \cdot \tan 2° \cdot \tan 3° \cdot \ldots \cdot \tan 89°).$$

By regrouping the factors, we can rewrite it as

$$\log((\tan 1° \cdot \tan 89°) \cdot (\tan 2° \cdot \tan 88°) \cdot \ldots \cdot (\tan 44° \cdot \tan 46°) \cdot \tan 45°)$$
$$= \log((\tan 1° \cdot \tan(90° - 1°)) \cdot (\tan 2° \cdot \tan(90° - 2°)) \cdot \ldots \cdot (\tan 44° \cdot \tan(90° - 44°)) \cdot \tan 45°)$$
$$= \log((\tan 1° \cdot \cot 1°) \cdot (\tan 2° \cdot \cot 2°) \cdot \ldots \cdot (\tan 44° \cdot \cot 44°) \cdot 1)$$
$$= \log(1 \cdot 1 \cdot 1 \cdot \ldots \cdot 1) = \log 1 = 0.$$

Therefore, the original equation can be rewritten as

$$x^2 + 6x \sin(3xy) + 9 = 0.$$

Consider it as a quadratic equation for x. It follows its discriminant is

$$D = 36 \sin^2(3xy) - 36 = 36(\sin^2(3xy) - 1) = -36 \cos^2(3xy) \leq 0.$$

This quadratic equation has solutions only when $D = 0$ or equivalently, when $\cos(3xy) = 0$. If $\cos(3xy) = 0$, then $\sin(3xy) = \pm 1$. Consider two cases:

1) $\sin(3xy) = 1$, from which

$$3xy = \frac{\pi}{2} + 2\pi k, \; k \in Z; \quad xy = \frac{\pi}{6} + \frac{2\pi k}{3}, \; k \in Z. \qquad (1)$$

Thus, the equation
$$x^2 + 6x \sin(3xy) + 9 = 0$$

is modified to

$$x^2 + 6x + 9 = 0,$$
$$(x + 3)^2 = 0,$$
$$x = -3.$$

Substituting this value for x into (1) and dividing by -3 gives

$$y = -\frac{\pi}{18} + \frac{2\pi k}{9}, \quad k \in Z$$

(we don't have to change the sign in front of $\frac{2\pi k}{9}$ because k can be any integer and by selecting the negative values for k, we get the same answers when the sign is not changed).

2) $\sin(3xy) = -1$, from which

$$3xy = -\frac{\pi}{2} + 2\pi n, \ n \in Z; \quad xy = -\frac{\pi}{6} + \frac{2\pi n}{3}, \ n \in Z. \qquad (2)$$

Thus,

$$x^2 + 6x \sin(3xy) + 9 = 0 \text{ is modified to}$$
$$x^2 - 6x + 9 = 0,$$
$$(x-3)^2 = 0,$$
$$x = 3.$$

Substituting this value for x into (2) and dividing by 3 gives

$$y = -\frac{\pi}{18} + \frac{2\pi n}{9}, \quad n \in Z.$$

Answer: $\left(-3, -\frac{\pi}{18} + \frac{2\pi k}{9}, k \in Z\right), \left(3, -\frac{\pi}{18} + \frac{2\pi n}{9}, n \in Z\right).$

Let's now consider equation b)

$$x^2 + 6x \sin(3xy) + 9 = \log\tan 1° \cdot \log\tan 2° \cdot \log\tan 3° \cdots \log\tan 89°.$$

The equation looks difficult at first glance. However, this is the exact same equation as the previous one we just solved. The only difference is the expression on the right-hand side. The trick is in realizing that the right-hand side, as in the first equation, also equals 0, because one of the factors in the product equals 0. The right-hand side is the product of logarithms of tangents of all the angles from $1°$ through $89°$. Therefore, one of the factors is $\log\tan 45°$. Since $\tan 45° = 1$, $\log\tan 45° = \log 1 = 0$. Therefore, the whole product on the right-hand side equals 0 as well, and we get to the equation already solved in the first part!

Problem 5. Solve the equation

$$(16x^8 + 1)(y^8 + 1) = 16(xy)^4.$$

Solution. It's easy to verify that $x = 0$ and $y = 0$ do not satisfy the equation. Divide then both sides by $4(xy)^4$ to get

$$\frac{(16x^8 + 1)(y^8 + 1)}{4(xy)^4} = 4,$$

$$\frac{(16x^8 + 1)}{4x^4} \cdot \frac{(y^8 + 1)}{y^4} = 4,$$

$$\left(4x^4 + \frac{1}{4x^4}\right) \cdot \left(y^4 + \frac{1}{y^4}\right) = 4.$$

By the AM–GM inequality (see chapter 9), it follows that

$$4x^4 + \frac{1}{4x^4} \geq 2\sqrt{4x^4 \cdot \frac{1}{4x^4}} = 2, \quad \text{and} \quad y^4 + \frac{1}{y^4} \geq 2\sqrt{y^4 \cdot \frac{1}{y^4}} = 2.$$

Therefore, the left-hand side is greater than or equal to $2 \cdot 2 = 4$, with equality attainable only when each factor equals 2:

$$4x^4 + \frac{1}{4x^4} = 2,$$

$$y^4 + \frac{1}{y^4} = 2.$$

$$4x^4 + \frac{1}{4x^4} - 2 = 0,$$

$$y^4 + \frac{1}{y^4} - 2 = 0.$$

$$\frac{16x^8 - 2 \cdot 4x^4 + 1}{4x^4} = 0,$$

$$\frac{y^8 - 2 \cdot y^4 + 1}{y^4} = 0.$$

$$\frac{(4x^4 - 1)^2}{4x^4} = 0,$$

$$\frac{(y^4 - 1)^2}{y^4} = 0.$$

From the first equation, $(4x^4 - 1)^2 = 0$, or equivalently,

$$4x^4 - 1 = 0,$$

$$x^4 = \frac{1}{4},$$

$$x = \pm\sqrt{\frac{1}{2}}.$$

From the second equation, $(y^4 - 1)^2 = 0$, or equivalently,

$$y^4 - 1 = 0,$$
$$y = \pm 1.$$

Answer: $x = \pm\sqrt{\frac{1}{2}}$, $y = \pm 1$.

Problem 6. Find all the values of a for which the roots x_1, x_2, x_3 of the polynomial $x^3 - 6x^2 + ax + a$ satisfy the equality $(x_1 - 3)^3 + (x_2 - 3)^3 + (x_3 - 3)^3 = 0$. This problem was offered in 1983 on the Austrian national math Olympiad.

Solution. Let's introduce another variable y as $y = x - 3$. It follows that the numbers $y_1 = x_1 - 3$, $y_2 = x_2 - 3$, and $y_3 = x_3 - 3$ have to be the roots of the polynomial $(y + 3)^3 - 6(y + 3)^2 + a(y + 3) + a$. Simplifying this expression gives

$$(y + 3)^3 - 6(y + 3)^2 + a(y + 3) + a$$
$$= y^3 + 9y^2 + 27y + 27 - 6y^2 - 36y - 54 + ay + 3a + a$$
$$= y^3 + 3y^2 + (a - 9)y + 4a - 27.$$

Applying Viète's theorem to the last cubic polynomial, we get that its roots satisfy:

$$y_1 + y_2 + y_3 = -3,$$
$$y_1 y_2 + y_1 y_3 + y_2 y_3 = a - 9,$$
$$y_1 y_2 y_3 = 27 - 4a.$$

Our goal is to find the values of a such that the roots x_1, x_2, x_3 of the given polynomial satisfy the equality $(x_1 - 3)^3 + (x_2 - 3)^3 + (x_3 - 3)^3 = 0$. In the substitution for y the last equality is equivalent to $y_1^3 + y_2^3 + y_3^3 = 0$.

It's not hard to justify the following equality (we leave this exercise to readers):

$$y_1^3 + y_2^3 + y_3^3 = (y_1 + y_2 + y_3)^3 - 3(y_1 y_2 + y_1 y_3 + y_2 y_3)(y_1 + y_2 + y_3) + 3y_1 y_2 y_3.$$

Hence, substituting $y_1^3 + y_2^3 + y_3^3 = 0$, we get

$$(y_1 + y_2 + y_3)^3 - 3(y_1 y_2 + y_1 y_3 + y_2 y_3)(y_1 + y_2 + y_3) + 3y_1 y_2 y_3 = 0.$$

Finally, substituting the values from Viète's formulas into the last equality, we get the necessary and sufficient conditions for a parameter a to satisfy the problem's conditions: $(-3)^3 - 3(a - 9) \cdot (-3) + 3 \cdot (27 - 4a) = 0$. Solving the last equation gives $-27 + 9a - 81 + 81 - 12a = 0$, from which $-3a - 27 = 0$, yielding $a = -9$.

Answer: $a = -9$.

Problem 7. Solve the equation

$$\sqrt[3]{\sin^2 x} + \sqrt[3]{\cos^2 x} = \sqrt[3]{4}.$$

Solution. Applying the methods discussed in chapters 5 and 7 simplifies the solution. Cube both sides of the equation and apply the formula for the cube of the sum of two numbers: $(a+b)^3 = a^3 + b^3 + 3ab(a+b)$. Then

$$\underbrace{\sin^2 x + \cos^2 x}_{=1} + 3\sqrt[3]{\sin^2 x \cos^2 x} \cdot \underbrace{\left(\sqrt[3]{\sin^2 x} + \sqrt[3]{\cos^2 x}\right)}_{=\sqrt[3]{4}} = 4,$$

$$1 + 3 \cdot \sqrt[3]{4} \cdot \sqrt[3]{\sin^2 x \cos^2 x} = 4,$$

$3 \cdot \sqrt[3]{4} \cdot \sqrt[3]{\sin^2 x \cos^2 x} = 3$; dividing both sides by 3 gives $\sqrt[3]{4 \sin^2 x \cos^2 x} = 1$. Since $2 \sin x \cos x = \sin 2x$, it follows that $\sin^2 2x = 1$, from which $\sin 2x = \pm 1$. Then $2x = \frac{\pi}{2} + \pi k$, $k \in Z$, and finally $x = \frac{\pi}{4} + \frac{\pi k}{2}$, $k \in Z$.

Answer: $\frac{\pi}{4} + \frac{\pi k}{2}$, $k \in Z$.

Problem 8. Find the solutions of the equation $8x(2x^2 - 1)(8x^4 - 8x^2 + 1) = 1$ for all x such that $0 < x < 1$.

Solution. All attempts to solve this problem by conventional methods are bound to fail. We will engage some trigonometry to help us make the solution manageable. The hint is hidden in the restriction on x, that $0 < x < 1$. The range of the trigonometric function $\cos \alpha$ ($|\cos \alpha| \leq 1$) allows us to state that there exists an angle α such that $\cos \alpha = x$, and $-\frac{\pi}{2} < \alpha < \frac{\pi}{2}$ (values for which $\cos \alpha > 0$ because of the restriction on x that $x > 0$). This trick for Viète's method for trigonometric substitution was discussed in chapter 7; we saw its other application in chapter 12. It will be helpful in this case as well. The equation can be rewritten as

$$8 \cos \alpha (2 \cos^2 \alpha - 1)(8 \cos^4 \alpha - 8 \cos^2 \alpha + 1) = 1.$$

Applying the identity for the cosine of a double angle, $2\cos^2 \alpha - 1 = \cos 2\alpha$, and factoring gives

$$8 \cos \alpha \cdot \cos 2\alpha \cdot (8 \cos^2 \alpha (\cos^2 \alpha - 1) + 1) = 1,$$

which yields

$$8 \cos \alpha \cdot \cos 2\alpha \cdot (8 \cos^2 \alpha (- \sin^2 \alpha) + 1) = 1,$$
$$8 \cos \alpha \cdot \cos 2\alpha \cdot (1 - 8 \cos^2 \alpha \cdot \sin^2 \alpha) = 1.$$

Using the identity for the sine of a double angle, $2 \sin \alpha \cos \alpha = \sin 2\alpha$, the equation can be rewritten as

$$8 \cos \alpha \cdot \cos 2\alpha \cdot (1 - 2 \sin^2 2\alpha) = 1.$$

Noticing that $1 - 2\sin^2 2\alpha = \cos 4\alpha$ leads to

$$8\cos\alpha \cdot \cos 2\alpha \cdot \cos 4\alpha = 1.$$

To simplify the equation further, multiply both sides by $\sin\alpha$. Before doing that, we need to verify that $\sin\alpha \neq 0$. Indeed, if $\sin\alpha = 0$, then $\cos\alpha = \pm 1$ and the original equation becomes $\pm 8 \cdot 1 \cdot 1 = 1$, or $\pm 8 = 1$, which is not true. Therefore, $\sin\alpha \neq 0$ and we can proceed with the suggested transformation of the equation:

$$8\cos\alpha \cdot \sin\alpha \cdot \cos 2\alpha \cdot \cos 4\alpha = \sin\alpha.$$

Applying the identity for the sine of a double angle three times gives

$$4\sin 2\alpha \cdot \cos 2\alpha \cdot \cos 4\alpha = \sin\alpha,$$
$$2\sin 4\alpha \cdot \cos 4\alpha = \sin\alpha,$$
$$\sin 8\alpha = \sin\alpha,$$
$$\sin 8\alpha - \sin\alpha = 0.$$

Applying the formula for the difference of sines (see chapter 6) gives $2\sin\frac{7\alpha}{2} \cdot \cos\frac{9\alpha}{2} = 0$, which leads to the solution of two equations

$$\sin\frac{7\alpha}{2} = 0, \quad \text{or} \quad \cos\frac{9\alpha}{2} = 0.$$

From the first equation $\frac{7\alpha}{2} = \pi k$, $k \in Z$, leading to $\alpha = \frac{2\pi k}{7}$, $k \in Z$. We made the substitution $\cos\alpha = x$ for x located between 0 and 1. Therefore, to find x we need to determine all the values of the integer k, such that $x = \cos\frac{2\pi k}{7}$ and $0 < x < 1$. It follows that k should be selected from the solutions of the inequality $-\frac{\pi}{2} < \frac{2\pi k}{7} < \frac{\pi}{2}$ (the range of values for which the cosine is positive). Dividing both sides of the last inequality by $\frac{2\pi}{7}$ gives $-\frac{7}{4} < k < \frac{7}{4}$. The only integers satisfying this inequality are $-1, 0, 1$.

If $k = -1$, then $x = \cos\left(-\frac{2\pi}{7}\right) = \cos\frac{2\pi}{7}$. This is the first solution of the equation.

If $k = 0$, then $x = \cos 0 = 1$, which does not satisfy the equation since $0 < x < 1$.

If $k = 1$, then $x = \cos\frac{2\pi}{7}$, which coincides with the first solution of the equation.

Now turn to the second equation, $\cos\frac{9\alpha}{2} = 0$, and perform a similar analysis. $\frac{9\alpha}{2} = \frac{\pi}{2} + \pi k$, $k \in Z$, from which $\alpha = \frac{\pi}{9} + \frac{2\pi k}{9}$, $k \in Z$. The integer solutions have to satisfy the inequality $-\frac{\pi}{2} < \frac{\pi}{9} + \frac{2\pi k}{9} < \frac{\pi}{2}$. Subtracting $\frac{\pi}{9}$ from both sides yields $-\frac{11\pi}{18} < \frac{2\pi k}{9} < \frac{7\pi}{18}$. Dividing both sides of this inequality by $\frac{2\pi}{9}$ gives $-\frac{11}{4} < k < \frac{7}{4}$. The integers located between $-\frac{11}{4}$ and $\frac{7}{4}$ are $-2, -1, 0, 1$.

If $k = -2$, then $x = \cos\left(\frac{\pi}{9} - \frac{4\pi}{9}\right) = \cos\left(-\frac{3\pi}{9}\right) = \cos\frac{\pi}{3} = \frac{1}{2}$. This is the second solution of the equation.

If $k = -1$, then $x = \cos\left(\frac{\pi}{9} - \frac{2\pi}{9}\right) = \cos\left(-\frac{\pi}{9}\right) = \cos\frac{\pi}{9}$. This is the third solution of the equation.

If $k = 0$, then $x = \cos\left(\frac{\pi}{9} - 0\right) = \cos\frac{\pi}{9}$, which coincides with the above solution.

If $k = 1$, then $x = \cos\left(\frac{\pi}{9} + \frac{2\pi}{9}\right) = \cos\frac{\pi}{3} = \frac{1}{2}$, which coincides with the second solution of the equation.

So, the equation has the following solutions: $\cos\frac{2\pi}{7}$, $\frac{1}{2}$, and $\cos\frac{\pi}{9}$. We leave these results as the final answer to the equation. However, we suggest the ambitious reader make an extra effort and seek the numerical values of the first and the third solutions, $\cos\frac{2\pi}{7}$, $\cos\frac{\pi}{9}$, without the use of a calculator. It is a good exercise for refreshing the applications of trigonometric identities studied in chapter 6.

Answer: $\cos\frac{2\pi}{7}$, $\frac{1}{2}$, $\cos\frac{\pi}{9}$.

Problem 9. Solve the equation

$$x^3 + 1 = 2\sqrt[3]{2x - 1}.$$

This problem was offered by B. Kordemsky in "What's the best answer?" *Quantum*, July/August 1996.

Solution. Rewrite the equation as $x = \sqrt[3]{2\sqrt[3]{2x-1}-1}$ and introduce the auxiliary function $f(x) = \sqrt[3]{2x - 1}$. It follows that $x = f(f(x))$. Since the function $f(x)$ is increasing, the equations $f(f(x)) = x$ and $f(x) = x$ are equivalent. You can refer to the proof given in problem 14 of chapter 3. So, the original equation is reduced to the equation $f(x) = x$ or equivalently, $x = \sqrt[3]{2x-1}$. Cubing both sides of the last equation yields the equation $x^3 - 2x + 1 = 0$, which is easily solvable by factoring:

$$x^3 - x - x + 1 = 0,$$
$$(x^3 - x) - (x - 1) = 0,$$
$$x(x - 1)(x + 1) - (x - 1) = 0,$$
$$(x - 1)(x^2 + x - 1) = 0,$$
$$x - 1 = 0 \text{ or } x^2 + x - 1 = 0.$$

From the first equation, $x = 1$.

Solutions of the second equation are $x_1 = \frac{-1+\sqrt{5}}{2}$, $x_2 = \frac{-1-\sqrt{5}}{2}$.

Answer: 1, $\frac{-1+\sqrt{5}}{2}$, $\frac{-1-\sqrt{5}}{2}$.

Problem 10. Demonstrate that the number $\left(\frac{9+\sqrt{77}}{2}\right)^{2017} + \left(\frac{9-\sqrt{77}}{2}\right)^{2017}$ is a whole number and is divisible by 9.

Solution. Considering the sequence $a_n = \left(\frac{9+\sqrt{77}}{2}\right)^n + \left(\frac{9-\sqrt{77}}{2}\right)^n$, let's find its recursive formula using the method outlined in chapter 11.

The sum of $\frac{9+\sqrt{77}}{2}$ and $\frac{9-\sqrt{77}}{2}$ equals

$$\frac{9+\sqrt{77}}{2}+\frac{9-\sqrt{77}}{2}=\frac{9+\sqrt{77}+9-\sqrt{77}}{2}=\frac{18}{2}=9.$$

Using the formula for the difference of squares gives

$$\frac{9+\sqrt{77}}{2}\cdot\frac{9-\sqrt{77}}{2}=\frac{81-77}{4}=1.$$

Therefore, by Viète's theorem, the numbers $\frac{9+\sqrt{77}}{2}$ and $\frac{9-\sqrt{77}}{2}$ are the roots of the quadratic equation $r^2-9r+1=0$. The recursive formula for the sequence a_n is $a_n=9a_{n-1}-a_{n-2}$. Calculating the values of a_1 and a_2, we see that

$$a_1=\frac{9+\sqrt{77}}{2}+\frac{9-\sqrt{77}}{2}=9,$$

$$a_2=\left(\frac{9+\sqrt{77}}{2}\right)^2+\left(\frac{9-\sqrt{77}}{2}\right)^2=\frac{81+18\sqrt{77}+77+81-18\sqrt{77}+77}{4}$$

$$=\frac{316}{4}=79.$$

Therefore, the first two terms in the sequence are the whole numbers, and all the other terms starting with third satisfy $a_n=9a_{n-1}-a_{n-2}$. So, each term is the result of some simple arithmetic operations with whole numbers. Thus, all the terms in this sequence are integers. The number $\left(\frac{9+\sqrt{77}}{2}\right)^{2017}+\left(\frac{9-\sqrt{77}}{2}\right)^{2017}$ is the 2017th term in this sequence, therefore it has to be an integer as well. To prove that it is divisible by 9, it suffices to apply property 3 of the division of a sum of two powers with the natural odd exponent by the sum of their bases studied in chapter 4. Since the sum of the bases equals 9, it is divisible by 9. Thus, the sum of the powers must be divisible by 9 as well.

Problem 11. Solve the equation

$$\sqrt[4]{x^8-4x^4+6-4x^{-4}+x^{-8}}=x^{-2}-x^2.$$

Solution. Normally, one has to start the solution with determining the domain of the equation. Obviously, since there are some addends with x to a negative power, x cannot equal 0, $x\neq0$. Next, it has to be noted that the expression under the fourth degree root on the left-hand side should be a nonnegative number. However, at this point it's not a good idea to solve the inequality $x^8-4x^4+6-4x^{-4}+x^{-8}\geq0$ to get the permissible values of x. We will rather keep the thought in mind to get back to it later on. Squaring both sides of the equation, regrouping the terms, and noticing that $x^{-4} \cdot x^4 = 1$ gives

$$\sqrt{x^8 - 4x^4 + 6 - 4x^{-4} + x^{-8}} = x^{-4} - 2x^{-4} \cdot x^4 + x^4,$$

$$\sqrt{(x^8 + x^{-8}) - 4(x^4 + x^{-4}) + 6} = x^{-4} + x^4 - 2.$$

Making the substitution $x^{-4} + x^4 = y$ and squaring both sides yields $x^{-8} + x^8 = y^2 - 2$. Notice that in the substitution, y has to be greater than or equal to 2, $y \geq 2$. Indeed, $x^{-4} - 2(x^{-4} \cdot x^4) + x^4 = (x^{-2} - x^2)^2 \geq 0$, thus $x^{-4} + x^4 - 2 \geq 0$, or $y - 2 \geq 0$, which gives $y \geq 2$. It follows that

$$\sqrt{y^2 - 2 - 4y + 6} = y - 2,$$

$$\sqrt{y^2 - 4y + 4} = y - 2,$$

$$\sqrt{(y - 2)^2} = y - 2, \text{ or equivalently, } |y - 2| = y - 2.$$

Recalling that $y \geq 2$, we can drop the absolute value sign on the left and simplify the equation to $y - 2 = y - 2$. The last equality holds for any permissible values of y. In other words, for values of y greater than or equal to 2, any number will be a solution of the last equation.

Going back to the variable x, it becomes evident the decision not to investigate the solutions of the inequality $x^8 - 4x^4 + 6 - 4x^{-4} + x^{-8} \geq 0$ in determining the domain of the original equation. For any number x ($x \neq 0$) the expression on the left-hand side under the root sign will be nonnegative. Should we finalize the solution, concluding that any real number $x \neq 0$ is the root of the equation? The answer is no, not at this point. Don't forget that being a fourth degree root, the left-hand side must be a nonnegative number, which means that the expression on the right has to be nonnegative as well. It follows all the solutions of the equation have to satisfy:

$$x^{-2} - x^2 \geq 0,$$

$$x \neq 0.$$

Solving the inequality $x^{-2} - x^2 \geq 0$ gives $\frac{1 - x^4}{x^2} \geq 0$, or equivalently, $(x^2 - 1)(x^2 + 1) \leq 0$, which leads to $(x - 1)(x + 1) \leq 0$. Solutions of the last inequality are the numbers in the segment from -1 to 1, including the endpoints. Recalling that $x \neq 0$ yields the solutions of the original equation to be all real numbers from either of two segments $[-1, 0[$ or $]0, 1]$.

Answer: $x \in [-1, 0[\cup]0, 1]$.

Problem 12. Given a right angle and a segment with endpoints located on the sides of the angle. The segment moves in such a way that its endpoints slide on the sides of the given angle. With every move a square is built on the segment as on its side. Find the locus of the centers of the squares.

Solution. Assume AB is the given segment. Place it on the Cartesian coordinate plane in such a way that point A is located on the Y-axis and point B is located on the X-axis. When points A and B move along the Y- and X-axes, the segment will satisfy the conditions of the problem.

Let's denote by M on the Y-axis the point into which A will be moved when B coincides with the origin and by N on the X-axis the point into which B will be moved when A coincides with the origin. Then MO = NO = AB. We can say that OM and ON are the extreme locations of the segment AB in its sliding activity. Having established these segments, we built two squares NLKO and MFEO with ON and OM as their sides. Denote the points of intersection of their diagonals by O_1 and O_2 respectively. OL and OF are the bisectors of the right angles at O. Therefore, points O_1, O, and O_2 are located on the same straight line.

Let's now build an arbitrary square ABTP on the segment AB as on its side. Draw its diagonals AT and BP. The point of their intersection C has to be located on the same straight line as points O_1, O, and O_2. Let's prove it. Before we proceed, please note that with every move made by AB, two squares are drawn on AB as on the side—one is above AB and one is below AB. So, basically, two different sets of squares are generated. We first consider the case when the squares are built below AB (Figure 1). For the proof we will apply vector algebra techniques (see chapter 10).

Consider points A, B, and C and assign their coordinates as A$(0, n)$, B$(m, 0)$, and C(x, y), where m and n are positive constants, $m > 0$, $n > 0$. It follows that the coordinates of the vectors $\overrightarrow{CA}$ and $\overrightarrow{CB}$ are:

$$\overrightarrow{CA} = (-x, n - y), \quad \overrightarrow{CB} = (m - x, -y).$$

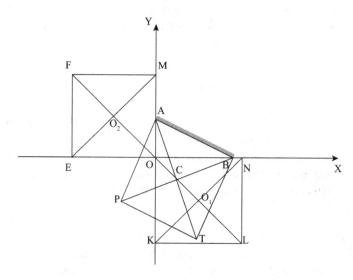

Figure 1

Vectors $\vec{CA}$ and $\vec{CB}$ are perpendicular, therefore their scalar product equals 0:

$$-x(m-x) - y(n-y) = 0, \quad \text{or} \quad x^2 - mx - ny + y^2 = 0. \tag{1}$$

The next step is to calculate the length of each vector:

$$\|\vec{CA}\| = \sqrt{x^2 + (n-y)^2},$$
$$\|\vec{CB}\| = \sqrt{(m-x)^2 + y^2}.$$

By the construction, the vectors have the same length. Therefore,

$$x^2 + (n-y)^2 = (m-x)^2 + y^2,$$

which is equivalent after simplifications to

$$2mx - 2ny = m^2 - n^2. \tag{2}$$

Combine (1) and (2) into a system of equations for two variables x and y:

$$x^2 - mx - ny + y^2 = 0,$$
$$2mx - 2ny = m^2 - n^2.$$

Recall now the techniques studied in chapter 12. Expressing x in terms of y from the second equation and then substituting it into the first equation gives

$$x = \frac{m^2 - n^2 + 2ny}{2m},$$
$$\left(\frac{m^2 - n^2 + 2ny}{2m}\right)^2 - m \cdot \frac{m^2 - n^2 + 2ny}{2m} - ny + y^2 = 0.$$

The second equation for y can be solved using the steps shown below.

$$\left(\frac{m^2 - n^2}{2m} + \frac{n}{m}y\right)^2 + y^2 - \frac{m^2 - n^2}{2} - ny - ny = 0,$$
$$\frac{(m^2 - n^2)^2}{4m^2} + \frac{n}{m} \cdot \frac{m^2 - n^2}{m} y + \frac{n^2}{m^2}y^2 + y^2 - \frac{m^2 - n^2}{2} - 2ny = 0.$$

Combining like terms, we get

$$y^2 \cdot \frac{n^2 + m^2}{m^2} - y \cdot \frac{n(m^2 + n^2)}{m^2} + \frac{(n^2 - m^2)(m^2 + n^2)}{4m^2} = 0.$$

Dividing both sides by $\frac{n^2 + m^2}{m^2} \neq 0$ and multiplying by 4 yields

$$4y^2 - 4yn + n^2 - m^2 = 0.$$

Solving the last equation as a quadratic in y gives

$$D = 16n^2 - 16(n^2 - m^2) = 16n^2 - 16n^2 + 16m^2 = 16m^2.$$

$y = \frac{4n \pm 4m}{8} = \frac{n \pm m}{2}$, from which $y_1 = \frac{n-m}{2}$, $y_2 = \frac{n+m}{2}$. Then $x_1 = \frac{m^2 - n^2 + 2n \cdot \frac{n-m}{2}}{2m} = -\frac{n-m}{2} = -y_1$; $x_2 = \frac{m^2 - n^2 + 2n \cdot \frac{n+m}{2}}{2m} = \frac{n+m}{2} = y_2.$

From the above results we see that when point C is located in the fourth quadrant, as it is in Figure 1, its coordinates satisfy the equation $y = -x$. The graph of the function $y = -x$ in the coordinate plane is a straight line, which is the bisector of the second and the fourth quadrants. Thus, C lies on the bisector. The locus of centers of squares built below the given segment as on the side is the segment O_1O_2, connecting the centers of the extreme squares NLKO and MFEO. To prove that O_1 and O_2 are the endpoints of the segment-locus of centers of squares built on AB, we need to show that C is located between O_1 and O_2. It suffices to prove that the abscissa of C, the center of any arbitrarily built square on AB, is located between the abscissa of O_1 and the abscissa of O_2. Recall that the endpoints of AB are defined by the coordinates as A$(0, n)$, B$(m, 0)$, where $m > 0$, $n > 0$. Then, the coordinates of M and N are M$(0, \sqrt{m^2 + n^2})$, N$(\sqrt{m^2 + n^2}, 0)$. Since O_1 is the center of the square NLKO, its coordinates are $x_{O_1} = \frac{x_N}{2} = \frac{\sqrt{m^2+n^2}}{2}$, $y_{O_1} = -\frac{y_M}{2} = -\frac{\sqrt{m^2+n^2}}{2}$; O_2 is the center of the square MFEO, and its coordinates are $x_{O_2} = -\frac{x_N}{2} = -\frac{\sqrt{m^2+n^2}}{2}$, $y_{O_2} = \frac{y_M}{2} = \frac{\sqrt{m^2+n^2}}{2}$. Consider the center of the arbitrarily built square C(x_C, y_C), where, as was found above, $x_C = -\frac{n-m}{2}$, $y_C = \frac{n-m}{2}$. Comparing the absolute values of the abscissas of C, O_1, and O_2, it is easy to see that $\left| -\frac{n-m}{2} \right| = \left| \frac{n-m}{2} \right| < \left| \frac{\sqrt{m^2+n^2}}{2} \right|$. To prove it, instead of comparing the positive numbers $\left| \frac{n-m}{2} \right|$ and $\left| \frac{\sqrt{m^2+n^2}}{2} \right|$, we can compare their squares and get $\left(\frac{\sqrt{m^2+n^2}}{2} \right)^2 - \left(\frac{n-m}{2} \right)^2 = \frac{m^2+n^2-m^2-n^2+2mn}{4} = \frac{mn}{2} > 0$ (since $m > 0$ and $n > 0$). Hence, $\left| \frac{n-m}{2} \right| < \left| \frac{\sqrt{m^2+n^2}}{2} \right|$, and noticing that $x_{O_2} < 0$, it indeed follows that $x_{O_2} < x_C < x_{O_1}$. As we proved, C, O_1, and O_2 are located on the straight line $y = -x$, and we got that the abscissa of C is between the abscissas of O_1 and O_2. Therefore, C has to lie between O_1 and O_2 on the segment O_1O_2, which concludes the proof that O_1O_2 is the desired locus of all centers of squares for the case when the squares are built below AB.

Consider now the second case when squares are built above AB. The extreme squares will coincide, and we get the square OMPN (Figure 2). From the above analysis it is clear that in this case point C is located in the first quadrant, its coordinates satisfy the equation of the straight line $y = x$ ($x_2 = y_2$ in our solutions of the system above), which means that C is located on the bisector of the first and the third quadrants.

Therefore, the locus of all the centers of the squares built on the given segment is the segment located on the bisector of the first and the third quadrants. While we easily identified the endpoints of the found segment as the centers O_1 and O_2 of the extreme squares during the investigation of the locus of centers of squares

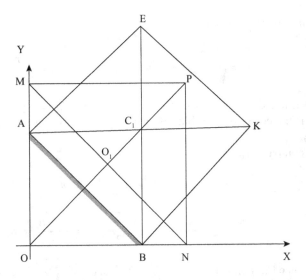

Figure 2

built on AB below AB, then for the second set of points, it is not an obvious choice at all. Our goal now is to specify the location of the endpoints of that segment.

As in the previous case, consider points $A(0, n_A)$ and $B(m_B, 0)$ ($m_A > 0$, $n_A > 0$) located on the coordinate lines. Out of all locations of AB in its sliding activity, we examine its position when AB $\parallel$ MN. As in the first case, select M and N on OY and OX such that $OM = ON = AB = \sqrt{m^2 + n^2}$. (We denoted the abscissa of B as m_B and the ordinate of A as n_A just to distinguish them from other m and n, the coordinates of an arbitrary location of AB in its sliding activity; still, $\sqrt{m_B^2 + n_A^2} = \sqrt{m^2 + n^2}$). In the right isosceles triangle AOB with the hypotenuse $AB = \sqrt{m^2 + n^2}$, we get

$$AO = n_A = BO = m_B = \frac{\sqrt{m^2 + n^2}}{\sqrt{2}} = \sqrt{\frac{m^2 + n^2}{2}}. \tag{1}$$

Let O_1 be the center of the square OMPN and C_1 be the center of the square AEKB. Let's prove that all the centers of all the other squares built above AB will be located between O_1 and C_1. Since these points are located on the straight line $y = x$, it suffices to prove that for any center C ($C \neq O_1$ and $C \neq C_1$) of the arbitrarily built square on AB, the abscissa of C satisfies the inequality $x_{O_1} < x_C < x_{C_1}$.

Since O_1 is the center of the square OMPN, its coordinates are

$$x_{O_1} = \frac{x_N}{2} = \frac{\sqrt{m^2 + n^2}}{2}, \quad y_{O_1} = \frac{y_M}{2} = \frac{\sqrt{m^2 + n^2}}{2}.$$

Using (1), the coordinates of C_1 are

$$x_{C_1} = x_B = m_B = \sqrt{\frac{m^2 + n^2}{2}} \quad \text{and} \quad y_{C_1} = y_A = n_A = \sqrt{\frac{m^2 + n^2}{2}}.$$

As we have already proved (see the solutions $x_2 = \frac{n+m}{2} = y_2$ of the system above), C as the center of an arbitrarily built square on $\overline{AB}$, has the coordinates $x_C = y_C = \frac{m+n}{2}$.

First compare x_{O_1} and x_C:

$$x_{O_1} = \frac{\sqrt{m^2 + n^2}}{2} = \frac{\sqrt{m^2 + 2mn + n^2 - 2mn}}{2}$$

$$= \frac{\sqrt{(m+n)^2 - 2mn}}{2} < \frac{\sqrt{(m+n)^2}}{2} = \frac{m+n}{2} = x_C.$$

Second, we need to compare x_C and x_{C_1}, $\frac{m+n}{2}$ and $\sqrt{\frac{m^2+n^2}{2}}$. Since each of the numbers x_C and x_{C_1} are positive, $x_{C_1} > 0$ and $x_C > 0$, we can apply the proof derived in chapter 9 for comparing the arithmetic mean and the quadratic mean, $\frac{m+n}{2} \leq \sqrt{\frac{m^2+n^2}{2}}$. We infer from the last inequality that $x_C < x_{C_1}$. Summarizing the results, we see that indeed $x_{O_1} < x_C < x_{C_1}$ and so, points O_1 and C_1 are the endpoints of the segment consisting of all the centers of the squares in question.

We can now conclude that in any case the locus of centers of the squares consists of two segments $O_1 C_1$ and $O_1 O_2$ located on the straight lines, one of which is the bisector of the right angle and the second one passes through the angle's vertex perpendicular to the first line.

The purely algebraic solution to this geometrical problem was obtained using Cartesian coordinates. Identifying any point in space by its coordinates and employing vector algebra techniques allowed setting up two equations, solutions of which clarified the location of the centers of squares in question. We translated the problem's conditions into algebra, solved a system of equations, and then translated back into geometry to describe the locus of points as two segments located on perpendicular lines. Furthermore, by using the applied techniques, we were able to identify the endpoints of each such segment. This is not the only way to solve the problem. We encourage readers to search for other solution(s) including the geometrical one. You then will be able to compare the effectiveness and efficiency of the methods and probably will have a better appreciation of the vector algebra technique.

The last equation in the book is solved. We truly hope that you enjoyed our journey into the equations world and its amazing relationships and characteristics that show the beauty of mathematics. The evolution of mathematics is closely connected with the development and investigation of various equations. The study of equations through the centuries contributed tremendously to the discovery and introduction of new kinds of numbers, starting from positive whole numbers and going to fractions, negative, irrational, transcendent, and complex or imaginary numbers. Historically, many of the most important mathematical problems arose from issues that are practical in origin, as geometrical problems of divvying up lots of land, calculating perimeters and areas of various figures. While the mathematicians in ancient Babylonia and China knew how to solve equations (even quadratics, as we mentioned in chapter 2), they considered only positive solutions. Negative numbers were rejected since unknowns usually represented lengths. Even two different positive solutions of a quadratic equation were perceived as illogical. The ancient Greeks concentrated on geometry and did not pay much attention to algebra. Euclid developed geometrical solutions to quadratic equations by formulating methods for finding lengths. Diophantus considered only positive integer answers to equations' solutions. He called the equation $4x + 20 = 0$ absurd, since it has a negative solution. The great Indian mathematician Brahmagupta was the first to introduce negative solutions to a quadratic equation, referring to them as to "debts," contrary to positive solutions, which he called "fortunes." The legend attributes the discovery of irrational numbers to Hippasus of Metapontum, a Pythagorean philosopher who lived in 5th century BC. It was so shocking at that time that he was sentenced to death and drowned at sea for revealing the new numbers non-consistent with the Pythagorean's doctrine that "all things are numbers." The discovery of the general formula for the solution of cubic equations brought to life imaginary numbers. Geralomo Cardano called them "sophistic" numbers and viewed them as useless. Another Italian mathematician, Rafael Bombelli (1526–1572), was the first to realize that one cannot avoid dealing with imaginary numbers. The concept of the one-dimensional number line was extended further to the two-dimensional complex plane. From these historical lessons we learn the importance of the study of equations in the evolution and development of various mathematical disciplines, and establishing the close links and connections among algebra, geometry, trigonometry, number theory, and calculus.

The best way to learn mathematics is to *do* mathematics. The more problems you solve, the more knowledge you gain, the more comfortable you feel, and the more efficient you become in your discoveries of solutions. One can continue to endlessly pursue other methods and algorithms for various equations' solutions; however, we feel that we have captured the essence of the major practical techniques considered. After all, we should leave some of these mathematical puzzles for the reader to discover.

> The enchanting charms of this sublime science reveal only to those who have the courage to go deeply into it.
>
> —Carl Friedrich Gauss

Solutions to Problems

Chapter 1

Problem 29. Diophantus's epitaph.

Solution. If we denote the length of his life by x, then the following equation encompasses the major events in his life described in the epitaph:

$$\frac{x}{6} + \frac{x}{12} + \frac{x}{7} + 5 + \frac{x}{2} + 4 = x,$$

Finding the common denominator gives

$$\frac{14x + 7x + 12x + 420 + 42x + 336}{84} = x,$$

$75x + 756 = 84x$, or equivalently, $9x = 756$, from which $x = 84$.

Answer: Diophantus was 84 years old when he died.

Problem 30. Assume the given numbers are x and y. Then $x - y = n$ and $\frac{x}{y} = n$, where n is a natural number. Expressing x in terms of y from each of the above equations, we get $x = y + n$, $x = y \cdot n$. Therefore $y + n = yn$, from which $y = \frac{n}{n-1}$. Rewriting the last expression for y as $y = \frac{n}{n-1} = 1 + \frac{1}{n-1}$ leads to the conclusion that for y and n natural numbers, n has to equal 2, $n = 2$. It follows $y = 1 + 1 = 2$ and so, $x = y + n = 4$.

Answer: The given numbers are 4 and 2.

Problem 31. Assume Bryan would need to drive with the speed of x m/h to catch Alex in 4 hours by 12 noon. Then he will cover the distance $4x$ miles. Alex was driving 5 hours and he covered the distance till their meeting of $50 \cdot 5 = 250$ miles. Therefore,

$$4x = 250,$$
$$x = 62.5.$$

Answer: Bryan would need to travel at a speed of 62.5 m/h.

Problem 32. Putting the merchant's original estate as x pounds, we get the annual increase for the first year expressed as $\frac{1}{3}(x - 100)$ pounds, making his balance at the end of year one equal to

$$x - 100 + \frac{1}{3}x - \frac{100}{3} = \frac{4}{3}x - \frac{400}{3}.$$

During the second year the increase was $\frac{1}{3} \cdot \left(\frac{4}{3}x - \frac{400}{3} - 100\right)$ pounds, and so his balance at the end of the second year became

$$\left(\frac{4}{3}x - \frac{400}{3} - 100\right) + \frac{1}{3} \cdot \left(\frac{4}{3}x - \frac{400}{3} - 100\right) = \frac{16}{9}x - \frac{2800}{9}.$$

The increase during the third year was $\frac{1}{3}\left(\frac{16}{9}x - \frac{2800}{9} - 100\right)$ pounds, which brought the total estate at the end of the third year to

$$\left(\frac{16}{9}x - \frac{2800}{9} - 100\right) + \frac{1}{3}\left(\frac{16}{9}x - \frac{2800}{9} - 100\right) = \frac{64}{27}x - \frac{14800}{27}.$$

Since his original estate doubled, we get the equation

$$\frac{64}{27}x - \frac{14800}{27} = 2x,$$
$$64x - 54x = 14800,$$

from which $x = 1480$.

Answer: The merchant had 1480 pounds at the beginning of year one.

Problem 33. Solve the system of linear equations

$$x + y - z = 0,$$
$$x - y + z = 2,$$
$$-x + y + z = 4.$$

Solution. Add all three equations to get $x + y + z = 6$. Subtract this equation in turn from the first, the second, and the third equation to find $z = 3, y = 2, x = 1$.

Answer: $(1, 2, 3)$.

Problem 34. Let x be the number of nickels he received; then $(14 - x)$ will be the number of dimes. We obtain the equation

$$5x + 10(14 - x) = 100,$$
$$5x + 140 - 10x = 100,$$
$$5x = 40,$$
$$x = 8.$$

Answer: He got 8 nickels in exchange.

Problem 35. Assume x new tubes were opened. They worked $9 - 4 - 2 = 3$ hours. The amount of water poured through the additional tubes in 3 hours is the same as would have poured through the saved 2 hours through 12 old tubes. Therefore, $3x = 12 \cdot 2$. Thus, $x = 8$.

Answer: 8 new tubes were opened.

Problem 36. The table below summarizes the conditions of the problem assuming the son is x years old now.

	Father	Son
Age now	$2x$	x
Age 10 years ago	$2x - 10$	$x - 10$

Since 10 years ago the father was 20 years older than the son, then their ages 10 years ago can be expressed as $2x - 10 = x - 10 + 20$, from which $x = 20$.

Answer: The son is 20 years old and the father is 40 years old now.

Chapter 2

Problem 15. The answer is yes. If the first player selects any three different integers such that their sum equals 0, then regardless of the order chosen by the second player, the equation will have two distinct roots, one of which will be 1 and the second root $\frac{c}{a}$. Indeed, assume our equation is written as $ax^2 + bx + c = 0$, and the coefficients are chosen such that $a + b + c = 0$. Then for $x = 1$, we get $a \cdot 1^2 + b \cdot 1 + c = 0$, which means one of the roots must be 1. The second root is calculated from Viète's formulas.

Problem 16. According to Viète's formulas, the following equalities must hold for the equation $x^2 - x - q = 0$:

$$x_1 + x_2 = 1,$$
$$x_1 \cdot x_2 = -q.$$

Let's find the sum of the cubes of the roots. First, we'll find the cube of the sum of $(x_1 + x_2)$ and then we will express it from there.

$$(x_1 + x_2)^3 = 1, \tag{1}$$

On the other hand,

$$(x_1 + x_2)^3 = x_1^3 + x_2^3 + 3\underbrace{x_1 \cdot x_2}_{-q}\underbrace{(x_1 + x_2)}_{1} = x_1^3 + x_2^3 - 3q. \tag{2}$$

Comparing (1) and (2) gives

$$x_1^3 + x_2^3 - 3q = 1, \quad x_1^3 + x_2^3 = 3q + 1.$$

Finally, in order for the sum of the cubes to equal 19, q has to satisfy the equation $3q + 1 = 19$, from which $q = 6$.

Problem 17. If $2b^2 - 9ac = 0$, the $ac = \frac{2b^2}{9}$. Let's now multiply both sides of the equation by a and rewrite it as:

$$a^2 x^2 + abx + ac = 0,$$

$$a^2 x^2 + abx + \frac{2b^2}{9} = 0,$$

$$9a^2 x^2 + 9abx + 2b^2 = 0.$$

Solving the last equation as a quadratic in x leads to

$$D = 81a^2 b^2 - 4 \cdot 9a^2 \cdot 2b^2 = 9a^2 b^2.$$

$$x = \frac{-9ab \pm \sqrt{9a^2 b^2}}{18a^2} = \frac{-9ab \pm 3ab}{18a^2}.$$

Thus $x_1 = -\frac{b}{3a}$, $x_2 = -\frac{2b}{3a}$. The ratio of the roots is $\frac{x_1}{x_2} = \frac{-\frac{b}{3a}}{-\frac{2b}{3a}} = \frac{1}{2}$.

Problem 18. Let's solve the given equation:

$$\frac{1}{x+p} + \frac{1}{x-p} = \frac{1}{a},$$

$$\frac{x-p+x+p}{x^2 - p^2} = \frac{1}{a},$$

$$x^2 - p^2 = 2ax,$$

$$x^2 - 2ax - p^2 = 0.$$

Finding the discriminant of the quadratic equation gives $D = 4a^2 + 4p^2 > 0$. D is a positive number as it is the sum of positive and nonnegative numbers. Therefore, the equation has two distinct real solutions.

Problem 19. Assume the legs' lengths are x_1 and x_2, which are the roots of the equation $ax^2 + bx + c = 0$. Then, by Viète's formulas,

$$x_1 + x_2 = -\frac{b}{a},$$

$$x_1 \cdot x_2 = \frac{c}{a}.$$

The hypotenuse of the triangle equals the square root of the sum of the squares of the legs. So, let's find $x_1^2 + x_2^2$.

$$\left(\underbrace{x_1 + x_2}_{-\frac{b}{a}}\right)^2 = x_1^2 + x_2^2 + 2\underbrace{x_1 \cdot x_2}_{\frac{c}{a}}.$$

Therefore,

$$x_1^2 + x_2^2 = \left(-\frac{b}{a}\right)^2 - 2\frac{c}{a} = \frac{b^2 - 2ac}{a^2}.$$

If we denote the hypotenuse by y, then

$$y = \sqrt{\frac{b^2 - 2ac}{a^2}} = \frac{\sqrt{b^2 - 2ac}}{|a|}.$$

Problem 20. First, notice that if $a = 0$, the equation becomes linear, and will have only one solution. If $a \neq 0$, then we are dealing with quadratic equation $ax^2 - 2x + 3 = 0$, which has only one solution when its discriminant equals 0.

$$D = 4 - 12a,$$
$$4 - 12a = 0,$$

so $a = \frac{1}{3}$.

Answer: $0, \frac{1}{3}$.

Problem 21.
$$x^2 + 2(p+1)x + 9p - 5 = 0.$$

Both roots have to be negative, therefore $x_1 \leq x_2 < 0$, and the following conditions have to be satisfied:

$D \geq 0,$

$a \cdot f(0) > 0,$

$0 > -\dfrac{2(p+1)}{2}.$

$(p+1)^2 - 9p + 5 \geq 0,$

$9p - 5 > 0,$

$p > -1.$

$p^2 - 7p + 6 \geq 0,$

$p > \dfrac{5}{9},$

$p > -1.$

$(p-1)(p-6) \geq 0,$

$p > \dfrac{5}{9},$

$p > -1.$

The segments with common solutions are $]\frac{5}{9}, 1] \cup [6, +\infty[$.

Chapter 3

Problem 16. Solve the equation

$$\frac{x^2 + x - 5}{x} + \frac{3x}{x^2 + x - 5} + 4 = 0.$$

Solution. First, determine the domain of the equation: $x \neq 0$ and $x^2 + x - 5 \neq 0$. Solving the quadratic equation $x^2 + x - 5 = 0$, we exclude $x = \frac{-1 \pm \sqrt{21}}{2}$. Let's now introduce a new variable $y = \frac{x^2 + x - 5}{x}$. It follows that $y + \frac{3}{y} + 4 = 0$, $y^2 + 4y + 3 = 0$. $y_1 = -1$, $y_2 = -3$. The next step is to solve each of the two equations:

$$\frac{x^2 + x - 5}{x} = -1 \quad \text{and} \quad \frac{x^2 + x - 5}{x} = -3.$$

Solving the first equation gives

$$x^2 + 2x - 5 = 0,$$

$$D = 4 + 20 = 24 \quad \text{and} \quad x = \frac{-2 \pm \sqrt{24}}{2} = -1 \pm \sqrt{6}.$$

Thus, $x_1 = -1 + \sqrt{6}$, $x_2 = -1 - \sqrt{6}$.

Solving the second equation gives $x^2 + 4x - 5 = 0$. Using Viète's formulas gives $x_3 = 1$, $x_4 = -5$.

Answer: $-1 + \sqrt{6}, -1 - \sqrt{6}, 1, -5.$

Problem 17. Solve the equation

$$(x+3)(x-1)(x+2)(x+6) = -20.$$

Solution. Rewrite the equation as

$$(x+3)(x+2)(x-1)(x+6) = -20,$$
$$(x^2+5x+6)(x^2+5x-6) = -20.$$

The substitution to make is $x^2 + 5x = y$. It follows that

$$(y+6)(y-6) = -20,$$
$$y^2 - 36 = -20,$$
$$y^2 = 16,$$

from which $y = 4$ or $y = -4$. Therefore, $x^2 + 5x = 4$ or $x^2 + 5x = -4$. We have to solve two quadratic equations:

$$x^2 + 5x - 4 = 0 \quad \text{and} \quad x^2 + 5x + 4 = 0.$$

Solving the first equation, we get $x_1 = \frac{-5+\sqrt{41}}{2}$, $x_2 = \frac{-5-\sqrt{41}}{2}$.
Solving the second equation, we get $x_3 = -1$, $x_4 = -4$.

Answer: $\frac{-5\pm\sqrt{41}}{2}, -1, -4$.

Problem 18. Solve the equation

$$\frac{4x}{4x^2 - 8x + 7} + \frac{3x}{4x^2 - 10x + 7} = 1.$$

Solution. First, we figure out the domain of the equation by excluding all values of x for which $4x^2 - 8x + 7 = 0$ or $4x^2 - 10x + 7 = 0$. In each case the discriminant of the quadratic polynomial is negative, $D < 0$; therefore, the domain is the set of all real numbers. Clearly, $x \neq 0$. By dividing numerator and denominator of each fraction by x ($x \neq 0$), we will obtain:

$$\frac{4}{4x - 8 + \frac{7}{x}} + \frac{3}{4x - 10 + \frac{7}{x}} = 1.$$

The substitution is $y = 4x + \frac{7}{x}$. It follows that

$$\frac{4}{y-8} + \frac{3}{y-10} = 1, \quad (y \neq 8, y \neq 10).$$

$y^2 - 25y + 144 = 0$; solving the equation, we find $y_1 = 16$, $y_2 = 9$.
To find x we have to solve two equations, $4x + \frac{7}{x} = 16$ and $4x + \frac{7}{x} = 9$.
The first equation becomes $4x^2 - 16x + 7 = 0$, from which $x_1 = 3.5$, $x_2 = 0.5$.

The second equation becomes $4x^2 - 9x + 7 = 0$. $D = 81 - 112 = -31 < 0$. No solutions exist.

Answer: $3.5, 0.5$.

Problem 19. Solve the equation

$$(x+3)^4 + (x+5)^4 = 16.$$

Solution. The substitution to make is $x = y - \frac{3+5}{2} = y - 4$,

$$(y-1)^4 + (y+1)^4 = 16,$$
$$y^4 - 4y^3 + 6y^2 - 4y + 1 + y^4 + 4y^3 + 6y^2 + 4y + 1 = 16,$$
$$2y^4 + 12y^2 - 14 = 0,$$
$$y^4 + 6y^2 - 7 = 0.$$

Another substitution is $z = y^2$, $z \geq 0$. $z^2 + 6z - 7 = 0$. The roots of this equation are $z_1 = 1$ and $z_2 = -7 < 0$ (has to be rejected). Therefore, $y^2 = 1$, from which $y = \pm 1$ and so, $x = 1 - 4 = -3$ or $x = -1 - 4 = -5$.

Answer: $-3, -5$.

Problem 20. Solve the equation

$$6x^4 - 13x^3 + 12x^2 - 13x + 6 = 0.$$

Solution. Obviously, $x \neq 0$. Dividing both sides of the equation by x^2 gives

$$6x^2 - 13x + 12 - \frac{13}{x} + \frac{6}{x^2} = 0,$$
$$6\left(x^2 + \frac{1}{x^2}\right) - 13\left(x + \frac{1}{x}\right) + 12 = 0.$$

Substitution: $x + \frac{1}{x} = y$,   so $x^2 + \frac{1}{x^2} = y^2 - 2$.   It   follows   that $6(y^2 - 2) - 13y + 12 = 0$, $6y^2 - 13y = 0$, from which $y = 0$ or $y = \frac{13}{6}$. Therefore,

$$x + \frac{1}{x} = 0 \quad \text{or} \quad x + \frac{1}{x} = \frac{13}{6}.$$

The first equation has no roots.
The second equation transforms to $6x^2 - 13x + 6 = 0$. $x_1 = \frac{3}{2}, x_2 = \frac{2}{3}$.

Answer: $\frac{3}{2}, \frac{2}{3}$.

Problem 21. Evaluate $\sqrt[3]{9+\sqrt{80}}+\sqrt[3]{9-\sqrt{80}}$.

Solution. Let $\sqrt[3]{9+\sqrt{80}}+\sqrt[3]{9-\sqrt{80}}=a$. Cubing both sides of the equality and applying the formula $(a+b)^3 = a^3+b^3+3ab(a+b)$ gives

$$a^3 = \left(\sqrt[3]{9+\sqrt{80}}+\sqrt[3]{9-\sqrt{80}}\right)^3$$

$$= (9+\sqrt{80})+(9-\sqrt{80})+3\sqrt[3]{9+\sqrt{80}}\cdot\sqrt[3]{9-\sqrt{80}}\cdot\left(\sqrt[3]{9+\sqrt{80}}+\sqrt[3]{9-\sqrt{80}}\right)$$

$$= 9+9+3\sqrt[3]{9^2-(\sqrt{80})^2}\cdot a = 18+3\sqrt[3]{81-80}\cdot a = 18+3a.$$

To find a, we need to solve the equation $a^3-3a-18=0$.

Noting that one root is 3, we factor the trinomial as

$$(a-3)(a^2+3a+6)=0.$$

The discriminant of the quadratic trinomial (a^2+3a+6) is negative, $D = 9-24 = -15 < 0$, therefore, cubic equation has no real roots other than 3. We conclude that $\sqrt[3]{9+\sqrt{80}}+\sqrt[3]{9-\sqrt{80}}=3$.

Problem 22. See if the number $x = \sqrt[3]{4+\sqrt{80}}-\sqrt[3]{\sqrt{80}-4}$ is one of the roots of the equation $x^3+12x-8=0$.

Solution. To solve the problem, we will use the formula for the cube of the difference of two numbers:

$$(a-b)^3 = a^3-b^3-3ab(a-b).$$

Let's find x^3:

$$x^3 = \left(\sqrt[3]{4+\sqrt{80}}-\sqrt[3]{\sqrt{80}-4}\right)^3$$

$$= 4+\sqrt{80}-(\sqrt{80}-4)-3\sqrt[3]{(4+\sqrt{80})\cdot(\sqrt{80}-4)}\cdot\underbrace{\left(\sqrt[3]{4+\sqrt{80}}-\sqrt[3]{\sqrt{80}-4}\right)}_{=x}$$

$$= 8-3\sqrt[3]{80-16}\cdot x = 8-3\cdot 4x = 8-12x.$$

Therefore, $x^3 = 8-12x$, or $x^3+12x-8=0$, and we conclude that the given value of x satisfies the given equation. Thus, $\sqrt[3]{4+\sqrt{80}}-\sqrt[3]{\sqrt{80}-4}$ is the root of the given equation.

Chapter 4

Problem 17. Given $P(x) = x^3+a_1x^2+a_2x+a_3$, find the coefficients a_1, a_2, and a_3 such that $P(x)$ divides evenly into $(x-1)$ and $(x+2)$ and has the remainder 10 after its division by $(x+1)$.

Solution. According to the conditions of the problem, $P(1) = 0$, $P(-2) = 0$, and $P(-1) = 10$. Therefore, we get a system of three linear equations:

$$1 + a_1 + a_2 + a_3 = 0,$$
$$-8 + 4a_1 - 2a_2 + a_3 = 0,$$
$$-1 + a_1 - a_2 + a_3 = 10.$$

$$a_1 + a_2 + a_3 = -1,$$
$$4a_1 - 2a_2 + a_3 = 8,$$
$$a_1 - a_2 + a_3 = 11.$$

By subtracting equation 3 from equation 1, we find a_2 right away, $a_2 = -6$. Substituting the value of a_2 into the first and the second equations and subtracting those two equations, we find that $a_1 = -3$, $a_3 = 8$.

Answer: $a_1 = -3$, $a_2 = -6$, $a_3 = 8$.

Problem 18. Find the coefficients p and q such that polynomial $P(x) = x^5 - 3x^4 + px^3 + qx^2 - 5x - 5$ divides evenly into $x^2 - 1$.

Solution. Since $x^2 - 1 = (x - 1)(x + 1)$, then for $P(x)$ to be evenly divisible by $(x^2 - 1)$, the following equalities must hold: $P(1) = 0$, $P(-1) = 0$.

$$P(1) = 1 - 3 + p + q - 5 - 5 = p + q - 12,$$
$$P(-1) = -1 - 3 - p + q + 5 - 5 = -p + q - 4,$$

Now we need to solve the system of two linear equations

$$p + q - 12 = 0,$$
$$-p + q - 4 = 0.$$

Adding the equations gives $2q = 16$, from which $q = 8$. Substituting this value of q into either of the equations gives $p = 4$.

Answer: $p = 4$, $q = 8$.

Problem 19. Find the coefficients a, b, and c such that polynomial $x^4 - x^3 + ax^2 + bx + c$ divides evenly into $x^3 - 2x^2 - 5x + 6$.

Solution. If the first polynomial is divisible evenly by the second given polynomial, then there must exist a linear binomial $(x + m)$, such that

$$x^4 - x^3 + ax^2 + bx + c = (x^3 - 2x^2 - 5x + 6)(x + m).$$

$$x^4 - x^3 + ax^2 + bx + c = x^4 + (m-2)x^3 + (-2m-5)x^2 + (6-5m)x + 6m.$$

Using the method of undefined coefficients, we conclude that

$$-1 = m - 2,$$
$$a = -2m - 5,$$
$$b = 6 - 5m,$$
$$c = 6m.$$

Solving the above system of equations, we get that $m = 1, a = -7, b = 1, c = 6$.

Problem 20. Prove that for any natural number n, $9^{n+1} + 2^{6n+1}$ is divisible by 11.

Solution.

$$9^{n+1} + 2^{6n+1} = 9 \cdot 9^n + 2 \cdot 64^n + 2 \cdot 9^n - 2 \cdot 9^n$$
$$= (9 \cdot 9^n + 2 \cdot 9^n) + (2 \cdot 64^n - 2 \cdot 9^n) = 11 \cdot 9^n + 2(64^n - 9^n).$$

The first addend is divisible by 11. The second addend is divisible by 11 as well because $64 - 9 = 55$, which is divisible by 11. Therefore, the sum is divisible by 11.

Problem 21. Prove that for any natural number n, $5^{2n+1} + 2^{n+4} + 2^{n+1}$ is divisible by 23.

Solution.

$$5^{2n+1} + 2^{n+4} + 2^{n+1} = 5^{2n} \cdot 5 + 2^n \cdot 16 + 2^n \cdot 2 = 25^n \cdot 5 + 2^n \cdot 18$$
$$= 25^n \cdot 5 + 2^n \cdot 18 + 25^n \cdot 18 - 25^n \cdot 18$$
$$= 25^n(5 + 18) - 18(25^n - 2^n) = 25^n \cdot 23 - 18(25^n - 2^n).$$

Each number in this difference is divisible by 23. Therefore, the difference will be divisible by 23 as well.

Problem 22. Given that α, β, and γ are the roots of the equation $x^3 + px^2 + qx + r = 0$, express in terms of $p, q,$ and r:

a) $\alpha^2 + \beta^2 + \gamma^2$,

b) $\alpha^3 + \beta^3 + \gamma^3$,

c) $\alpha^4 + \beta^4 + \gamma^4$.

Solution. We'll start with problem a). We know that

$$(\alpha + \beta + \gamma)^2 = \alpha^2 + \beta^2 + \gamma^2 + 2(\alpha\beta + \alpha\gamma + \beta\gamma). \qquad (*)$$

By Viète's formulas for a cubic equation,

$$\alpha + \beta + \gamma = -p,$$
$$\alpha\beta + \alpha\gamma + \beta\gamma = q.$$

Therefore, substituting the values into (*), we get that

$$\alpha^2 + \beta^2 + \gamma^2 = (-p)^2 - 2q = p^2 - 2q.$$

Problem b). It is given that numbers α, β, and γ are the roots of the equation $x^3 + px^2 + qx + r = 0$. Therefore, these numbers should satisfy the equation, which leads to

$$\alpha^3 + p\alpha^2 + q\alpha + r = 0,$$
$$\beta^3 + p\beta^2 + q\beta + r = 0,$$
$$\gamma^3 + p\gamma^2 + q\gamma + r = 0.$$

Adding the equalities and substituting the values from Viète's formulas and the expression for $\alpha^2 + \beta^2 + \gamma^2 = p^2 - 2q$ in problem a) above gives the expression for $\alpha^3 + \beta^3 + \gamma^3$:

$$\alpha^3 + \beta^3 + \gamma^3 = -p(\alpha^2 + \beta^2 + \gamma^2) - q(\alpha + \beta + \gamma) - 3r$$
$$= -p(p^2 - 2q) + pq - 3r = -p^3 + 3pq - 3r.$$

Problem c). It is given that numbers α, β, and γ are the roots of the equation $x^3 + px^2 + qx + r = 0$. Therefore, they satisfy the following equation as well:

$$x^4 + px^3 + qx^2 + rx = 0.$$

It follows that

$$\alpha^4 + p\alpha^3 + q\alpha^2 + r\alpha = 0,$$
$$\beta^4 + p\beta^3 + q\beta^2 + r\beta = 0,$$
$$\gamma^4 + p\gamma^3 + q\gamma^2 + r\gamma = 0.$$

Adding all the equalities and substituting values from Viète's formulas and the results obtained in a) and b) gives the expression for $\alpha^4 + \beta^4 + \gamma^4$:

$$\alpha^4 + \beta^4 + \gamma^4 = -p(\alpha^3 + \beta^3 + \gamma^3) - q(\alpha^2 + \beta^2 + \gamma^2) - r(\alpha + \beta + \gamma)$$
$$= -p(-p^3 + 3pq - 3r) - q(p^2 - 2q) + rp$$
$$= p^4 - 4p^2q + 2q^2 + 4rp.$$

Chapter 5

Problem 18. Solve the equation

$$\sqrt{x-5}+\sqrt{x+3} = \sqrt{2x+4}.$$

Solution. The domain of the equation is the set of all real numbers such that $x \geq 5$. Squaring both sides gives

$$x-5+2\sqrt{x-5}\sqrt{x+3}+x+3 = 2x+4,$$
$$2\sqrt{x-5}\sqrt{x+3} = 6,$$
$$\sqrt{(x-5)(x+3)} = 3.$$

Squaring both sides leads to

$$x^2-2x-15 = 9,$$
$$x^2-2x-24 = 0.$$

There are two solutions: $x = 6$, $x = -4$. The negative root -4 does not belong to the equation's domain, so we reject it. Verifying 6, we see that it satisfies the equation: $\sqrt{6-5}+\sqrt{6+3} = 4, \sqrt{12+4} = 4; 4 = 4$.

Answer: 6.

Problem 19. Solve the equation

$$x - \sqrt[3]{x^2-x-1} = 1.$$

Solution. Rewrite the equation as $\sqrt[3]{x^2-x-1} = x-1$, Cubing both sides of the equation gives

$$x^2-x-1 = (x-1)^3,$$
$$x^2-x-1 = x^3-3x^2+3x-1,$$
$$x^3-4x^2+4x = 0,$$
$$x(x^2-4x+4) = 0,$$
$$x(x-2)^2 = 0,$$
$$x = 0, x = 2.$$

Answer: 0, 2.

Problem 20. Solve the equation

$$\sqrt{x + \sqrt{x + 11}} + \sqrt{x - \sqrt{x + 11}} = 4.$$

Solution. Squaring both sides gives

$$x + \sqrt{x + 11} + \underbrace{2\sqrt{x + \sqrt{x + 11}} \cdot \sqrt{x - \sqrt{x + 11}}}_{\text{difference of squares}} + x - \sqrt{x + 11} = 16,$$

Noticing that

$$\sqrt{x + \sqrt{x + 11}} \cdot \sqrt{x - \sqrt{x + 11}} = \sqrt{x^2 - (\sqrt{x + 11})^2} = \sqrt{x^2 - x - 11},$$

we get

$$2x + 2\sqrt{x^2 - x - 11} = 16,$$
$$\sqrt{x^2 - x - 11} = 8 - x.$$

Squaring both sides leads to

$$x^2 - x - 11 = x^2 - 16x + 64,$$
$$15x = 75,$$
$$x = 5.$$

It must be verified by substituting into the original equation.

$$\sqrt{5 + \sqrt{5 + 11}} + \sqrt{5 - \sqrt{5 + 11}} = 3 + 1 = 4.$$

So, $4 = 4$. It does satisfy it. Hence, it is a valid solution.

Answer: 5.

Problem 21. Solve the equation

$$\sqrt{\frac{x + 5}{x}} + 4\sqrt{\frac{x}{x + 5}} = 4.$$

Solution. The domain of the equation is the set of real numbers such that $x < -5$ or $x > 0$, $x \in \ ]-\infty, -5[\cup]0, +\infty[$. The substitution to make is $y = \sqrt{\frac{x + 5}{x}} \geq 0$. Then, $y + \frac{4}{y} = 4$, or $y^2 - 4y + 4 = 0$, $(y - 2)^2 = 0$, from which $y = 2$.

Then $\sqrt{\frac{x+5}{x}} = 2$. Squaring both sides gives

$$x + 5 = 4x,$$
$$x = \frac{5}{3}.$$

Checking the solution, we confirm that it is valid.

$$\sqrt{\frac{\frac{5}{3}+5}{\frac{5}{3}} + 4 \cdot \sqrt{\frac{\frac{5}{3}}{\frac{5}{3}+5}}} = \sqrt{\frac{20}{5}} + 4 \cdot \sqrt{\frac{5}{20}} = 2 + 2 = 4; \, 4 = 4.$$

Answer: $\frac{5}{3}$.

Problem 22. Solve the equation

$$\sqrt[4]{x+8} - \sqrt[4]{x-8} = 2.$$

Solution is similar to the solution of problem 12.

Answer: $x = 8$.

Problem 23. Solve the equation

$$\sqrt{x^3+x^2-1} + \sqrt{x^3+x^2+2} = 3.$$

Solution. Multiplying both sides of the equation by the expression conjugate to the left-hand side gives

$$\left(\sqrt{x^3+x^2-1} + \sqrt{x^3+x^2+2}\right) \cdot \left(\sqrt{x^3+x^2-1} - \sqrt{x^3+x^2+2}\right)$$
$$= 3\left(\sqrt{x^3+x^2-1} - \sqrt{x^3+x^2+2}\right),$$
$$x^3+x^2-1-x^3-x^2-2 = 3\left(\sqrt{x^3+x^2-1} - \sqrt{x^3+x^2+2}\right),$$
$$\sqrt{x^3+x^2-1} - \sqrt{x^3+x^2+2} = -1.$$

Recalling the original equation, we get the following system

$$\sqrt{x^3+x^2-1} - \sqrt{x^3+x^2+2} = -1,$$
$$\sqrt{x^3+x^2-1} + \sqrt{x^3+x^2+2} = 3.$$

Adding the equations gives $2\sqrt{x^3 + x^2 - 1} = 2$, or after dividing by 2 and squaring both sides,

$$x^3 + x^2 - 1 = 1,$$
$$(x^3 - 1) + (x^2 - 1) = 0,$$
$$(x - 1)(x^2 + x + 1) + (x - 1)(x + 1) = 0,$$
$$(x - 1)(x^2 + x + 1 + x + 1) = 0,$$
$$(x - 1)(x^2 + 2x + 2) = 0,$$
$$x - 1 = 0, \text{ or } x^2 + 2x + 2 = 0.$$

From the first equation, $x = 1$. The second equation has no solutions. Verify the solution: $\sqrt{1^3 + 1^2 - 1} + \sqrt{1^3 + 1^2 + 2} = 1 + 2 = 3; 3 = 3$.

Answer: 1.

Problem 24. Solve the equation

$$\frac{(34 - x)\sqrt[3]{x + 1} - (x + 1)\sqrt[3]{34 - x}}{\sqrt[3]{34 - x} - \sqrt[3]{x + 1}} = 30.$$

Solution. First, find the domain of the equation. Solving the equation $\sqrt[3]{34 - x} - \sqrt[3]{x + 1} = 0$ and excluding its roots will define the domain of the given equation (the denominator cannot equal 0). $\sqrt[3]{34 - x} = \sqrt[3]{x + 1}$. Cubing both sides gives $34 - x = x + 1$, from which $2x = 33$ or $x = 16.5$. Therefore, the domain of the equation is the set of all real numbers except 16.5, $x \neq 16.5$.

In the numerator factoring out the common factor $\sqrt[3]{(34 - x)(x + 1)}$ yields

$$\frac{\sqrt[3]{(34 - x)(x + 1)}\left(\sqrt[3]{(34 - x)^2} - \sqrt[3]{(x + 1)^2}\right)}{\sqrt[3]{34 - x} - \sqrt[3]{x + 1}} = 30.$$

Factoring the second factor in the numerator as the difference of squares and then canceling out the common factor in numerator and denominator leads to

$$\frac{\sqrt[3]{(34 - x)(x + 1)}\left(\sqrt[3]{34 - x} - \sqrt[3]{x + 1}\right)\left(\sqrt[3]{34 - x} + \sqrt[3]{x + 1}\right)}{\sqrt[3]{34 - x} - \sqrt[3]{x + 1}} = 30,$$
$$\sqrt[3]{(34 - x)(x + 1)}\left(\sqrt[3]{34 - x} + \sqrt[3]{x + 1}\right) = 30.$$

Multiplying both sides by 3 and then adding $\underbrace{-x + x}_{=0}$ on the left-hand side and adding 35 to both sides gives

$$\underbrace{34 - x}_{a^3} + \underbrace{x + 1}_{b^3} + 3\underbrace{\sqrt[3]{(34 - x)(x + 1)}}_{ab}\left(\underbrace{\sqrt[3]{34 - x}}_{a} + \underbrace{\sqrt[3]{x + 1}}_{b}\right) = 90 + 35.$$

Using $(a+b)^3 = a^3 + b^3 + 3ab(a+b)$ allows to simplify the left-hand side to

$$(\sqrt[3]{34-x} + \sqrt[3]{x+1})^3 = 125,$$
$$\sqrt[3]{34-x} + \sqrt[3]{x+1} = 5.$$

Cubing both sides gives

$$34 - x + x + 1 + 3\sqrt[3]{(34-x)(x+1)}\underbrace{\left(\sqrt[3]{34-x} + \sqrt[3]{x+1}\right)}_{=5} = 125,$$

$$35 + 3 \cdot 5 \cdot \sqrt[3]{(34-x)(x+1)} = 125,$$
$$15\sqrt[3]{(34-x)(x+1)} = 90,$$
$$\sqrt[3]{(34-x)(x+1)} = 6,$$

cubing both sides gives

$$(34-x)(x+1) = 216,$$
$$-x^2 + 33x + 34 - 216 = 0,$$
$$x^2 - 33x + 182 = 0.$$
$$D = 1089 - 728 = 361.$$

Hence, $x = \frac{33 \pm 19}{2}$, which yields $x_1 = 26$, $x_2 = 7$.

Answer: 26, 7.

Chapter 6

Problem 16. Evaluate without a calculator the value of $\cos 18°$.

Solution.

$$\cos 18° = \sqrt{1 - \sin^2 18°} = \sqrt{1 - \left(\frac{-1+\sqrt{5}}{4}\right)^2}$$

$$= \sqrt{1 - \frac{3-\sqrt{5}}{8}} = \sqrt{\frac{5+\sqrt{5}}{8}}.$$

Hint—use the results from problem 2.

Problem 17. Evaluate without a calculator the value of

$$\sin 10° \cdot \sin 20° \cdot \sin 30° \cdot \sin 40° \cdot \sin 50° \cdot \sin 60° \cdot \sin 70° \cdot \sin 80°.$$

Solution. Applying the result of problem 6 twice gives

$$\sin 10° \cdot \sin 50° \cdot \sin 70° = \frac{1}{4} \sin 30° = \frac{1}{8},$$

$$\sin 20° \cdot \sin 40° \cdot \sin 80° = \frac{1}{4} \sin 60° = \frac{\sqrt{3}}{8}.$$

Substituting the above results into the original product, we will get

$$\frac{1}{8} \cdot \sin 30° \cdot \sin 60° \cdot \frac{\sqrt{3}}{8} = \frac{3}{256}.$$

Problem 18. Evaluate $\sin^3 x - \cos^3 x$, if $\sin x - \cos x = n$.

Solution. First, note that $(\sin x - \cos x)^2 = n^2$, from which

$$\sin^2 x - 2 \sin x \cos x + \cos^2 x = n^2,$$

or equivalently, $1 - 2 \sin x \cos x = n^2$ and so, $\sin x \cos x = \frac{1-n^2}{2}$. Applying the formula for the difference of cubes gives

$$\sin^3 x - \cos^3 x = (\sin x - \cos x)(\sin^2 x + \sin x \cos x + \cos^2 x)$$

$$= n(1 + \sin x \cos x) = n\left(1 + \frac{1-n^2}{2}\right)$$

$$= n \cdot \frac{3 - n^2}{2} = \frac{3n - n^3}{2}.$$

Problem 19. Does there exist an angle γ such that $\cos \gamma = a + \frac{1}{a}$ $(a \ne 0)$?

Solution. Obviously, $a + \frac{1}{a} > 2$. As we know, the range of the function $y = \cos x$ consists of all real numbers not greater in absolute value than 1. Therefore, such an angle γ does not exist.

Problem 20. Evaluate $\sin(x + y) \sin(x - y)$ if $\sin x = -\frac{1}{3}$ and $\sin y = -\frac{1}{2}$, where $\frac{3\pi}{2} < x < 2\pi$ and $\frac{3\pi}{2} < y < 2\pi$.

Solution. First, let's find the values of $\cos x$ and $\cos y$ and then substitute all the values into the modified expression for $\sin(x + y) \sin(x - y)$.

$$\cos x = \sqrt{1 - \frac{1}{9}} = \frac{\sqrt{8}}{3}, \quad \cos y = \sqrt{1 - \frac{1}{4}} = \frac{\sqrt{3}}{2}.$$

$$\sin(x + y) \sin(x - y) = (\sin x \cos y + \sin y \cos x)(\sin x \cos y - \sin y \cos x)$$

$$= -\frac{5}{36}.$$

Problem 21. Does there exist an angle α such that $\tan\alpha = 2+\sqrt{3}$ and $\cot\alpha = 2-\sqrt{3}$?

Solution. $(2+\sqrt{3})(2-\sqrt{3}) = 4-3 = 1$. Since $\tan\alpha \cdot \cot\alpha = 1$ for any angle α (for which each function is defined), then such an angle must exist.

Problem 22. Prove that for all x such that $0 < x < \frac{\pi}{4}$

$$1 - \tan x + \tan^2 x - \tan^3 x + \cdots = \frac{\sqrt{2}\cos x}{2\sin\left(\frac{\pi}{4}+x\right)}.$$

Solution. If $0 < x < \frac{\pi}{4}$, then $|\tan x| < 1$. The expression on the left-hand side of the equality is an infinite geometric sequence with the ratio $r = -\tan x$, such that $r < 1$. Therefore, the sum of this geometric series can be calculated by the formula

$$1 - \tan x + \tan^2 x - \tan^3 x + \cdots = \frac{1}{1-r} = \frac{1}{1+\tan x}.$$

Let's modify the right-hand side of the last expression to

$$\frac{1}{1+\tan x} = \frac{1}{1+\frac{\sin x}{\cos x}} = \frac{\cos x}{\cos x + \sin x}.$$

Multiplying the numerator and denominator of the last fraction by $\frac{\sqrt{2}}{2}$ and using the formula for the sine of the sum of two angles and the fact that $\sin\frac{\pi}{4} = \cos\frac{\pi}{4} = \frac{\sqrt{2}}{2}$ gives

$$\frac{\cos x}{\cos x + \sin x} = \frac{\frac{\sqrt{2}}{2}\cos x}{\frac{\sqrt{2}}{2}\cos x + \frac{\sqrt{2}}{2}\sin x} = \frac{\frac{\sqrt{2}}{2}\cos x}{\sin\frac{\pi}{4}\cos x + \cos\frac{\pi}{4}\sin x}$$

$$= \frac{\sqrt{2}\cos x}{2\sin\left(\frac{\pi}{4}+x\right)}.$$

Therefore, the statement

$$1 - \tan x + \tan^2 x - \tan^3 x + \cdots = \frac{\sqrt{2}\cos x}{2\sin\left(\frac{\pi}{4}+x\right)}$$

holds true for any x such that $0 < x < \frac{\pi}{4}$.

Problem 23. Prove that for any natural n the following identities are true:

$$\cos nx = 2\cdot\cos x\cdot\cos(n-1)x - \cos(n-2)x;$$
$$\sin nx = 2\cdot\cos x\cdot\sin(n-1)x - \sin(n-2)x.$$

Proof. First, consider the identity for cosine nx in terms of the cosines of $(n-1)x$ and $(n-2)x$. Expanding the right-hand side, using the compound angles formula for cosine and the formulas for cosine and sine of a double angle, we get

$$2 \cdot \cos x \cdot \cos(n-1)x - \cos(n-2)x$$
$$= 2 \cdot \cos x \cdot \cos(nx - x) - \cos(nx - 2x)$$
$$= 2 \cdot \cos x \cdot (\cos nx \cdot \cos x + \sin nx \cdot \sin x) - (\cos nx \cdot \cos 2x + \sin nx \cdot \sin 2x)$$
$$= 2 \cdot \cos^2 x \cdot \cos nx + 2 \cdot \cos x \cdot \sin x \cdot \sin nx$$
$$- \cos nx \cdot \underbrace{\left(2 \cdot \cos^2 x - 1\right)}_{\cos 2x} - \sin nx \cdot \underbrace{2 \sin x \cdot \cos x}_{\sin 2x}$$
$$= 2 \cdot \cos^2 x \cdot \cos nx + 2 \cdot \cos x \cdot \sin x \cdot \sin nx - 2 \cdot \cos^2 x \cdot \cos nx + \cos nx$$
$$- 2 \cdot \cos x \cdot \sin x \cdot \sin nx$$
$$= \cos nx,$$

as was to be proved.

We will work the second identity in the similar manner:

$$2 \cdot \cos x \cdot \sin(n-1)x - \sin(n-2)x$$
$$= 2 \cdot \cos x \cdot (\sin nx \cdot \cos x - \cos nx \cdot \sin x) - (\sin nx \cdot \cos 2x - \cos nx \cdot \sin 2x)$$
$$= 2 \cdot \cos^2 x \cdot \sin nx - 2 \cdot \cos x \cdot \sin x \cdot \cos nx - \sin nx \cdot \underbrace{\left(2 \cos^2 x - 1\right)}_{\cos 2x}$$
$$+ \underbrace{2 \sin x \cdot \cos x}_{\sin 2x} \cdot \cos nx$$
$$= 2 \cdot \cos^2 x \cdot \sin nx - 2 \cdot \cos x \cdot \sin x \cdot \cos nx - 2 \cdot \cos^2 x \cdot \sin nx + \sin nx$$
$$+ 2 \cdot \cos x \cdot \sin x \cdot \cos nx$$
$$= \sin nx,$$

as was to be proved. □

Chapter 7

Problem 17. Solve the equation

$$\frac{1}{2}(\cos 5x + \cos 7x) - \cos^2 2x + \sin^2 3x = 0.$$

Solution. Using formulas for the sum of cosines and power reducing formulas gives

$$\frac{1}{2} \cdot 2 \cos 6x \cos x - \frac{1}{2} - \frac{1}{2}\cos 4x + \frac{1}{2} - \frac{1}{2}\cos 6x = 0,$$
$$\cos 6x \cos x - \frac{1}{2}(\cos 4x + \cos 6x) = 0.$$

Using the formula for the sum of cosines gives

$$\cos 6x \cos x - \cos 5x \cos x = 0,$$
$$\cos x (\cos 6x - \cos 5x) = 0,$$
$$\cos x \cdot 2 \sin \frac{11x}{2} \cdot \sin \frac{-x}{2} = 0,$$
$$\cos x \cdot \sin \frac{11x}{2} \cdot \sin \frac{x}{2} = 0,$$
$$\cos x = 0, \text{ or } \sin \frac{11x}{2} = 0, \text{ or } \sin \frac{x}{2} = 0,$$
$$x = \frac{\pi}{2} + \pi k, k \in Z \text{ or } x = \frac{2\pi n}{11}, n \in Z \text{ or } x = 2\pi m, m \in Z.$$

Solutions of the last equation are included among the solutions of the second equation.

Answer: $x = \frac{\pi}{2} + \pi k, k \in Z, x = \frac{2\pi n}{11}, n \in Z.$

Problem 18. Solve the equation

$$2(\cos 4x - \sin x \cos 3x) = \sin 4x + \sin 2x.$$

Solution.

$$2\left(\cos 4x - \frac{1}{2}(\sin 4x - \sin 2x)\right) - \sin 4x - \sin 2x = 0,$$
$$2\cos 4x - \sin 4x + \sin 2x - \sin 4x - \sin 2x = 0,$$
$$2\cos 4x - 2\sin 4x = 0,$$
$$\sin 4x - \cos 4x = 0,$$

verifying that $\cos 4x \neq 0$ and dividing both sides by $\cos 4x$ gives

$$\tan 4x = 1,$$
$$4x = \frac{\pi}{4} + \pi n, n \in Z.$$
$$x = \frac{\pi}{16} + \frac{\pi n}{4}, n \in Z.$$

Answer: $\frac{\pi}{16} + \frac{\pi n}{4}, n \in Z.$

Problem 19. Solve the equation

$$\tan(70° + x) + \tan(20° - x) = 2.$$

Solution. We will use the formula for the sum of tangents of two angles:

$$\tan x + \tan y = \frac{\sin(x + y)}{\cos x \cos y}.$$

It follows that

$$\frac{\sin 90°}{\cos(70° + x)\cos(20° - x)} = 2,$$

$$\cos(70° + x)\cos(20° - x) = \frac{1}{2},$$

$$\frac{1}{2}(\cos(50° + 2x) + \cos 90°) = \frac{1}{2},$$

$$\cos(50° + 2x) = 1,$$

$$x = -25° + 180°n, n \in Z.$$

Answer: $-25° + 180°n, n \in Z.$

Problem 20. Evaluate $\cos(2\arctan 2) - \sin(4\arctan 3)$.

Solution. Let $\arctan 2 = \alpha$, so $\tan \alpha = 2, 0 \le \alpha < \frac{\pi}{2}$.
    We need to find $\cos 2\alpha$.

$$\cos^2 \alpha = \frac{1}{1 + \tan^2 \alpha} = \frac{1}{5}, \text{ so } \cos \alpha = \frac{1}{\sqrt{5}}.$$

$$\sin \alpha = \sqrt{1 - \cos^2 \alpha} = \frac{2}{\sqrt{5}},$$

$$\cos 2\alpha = \cos^2 \alpha - \sin^2 \alpha = -\frac{3}{5}.$$

Now let $\arctan 3 = \beta$, so $\tan \beta = 3, 0 \le \beta < \frac{\pi}{2}$.
    We need to find $\sin 4\beta$.

$$\sin 4\beta = 2\sin 2\beta \cdot \cos 2\beta = 4\sin \beta \cdot \cos \beta \cdot \cos 2\beta.$$

We need to calculate each factor.

$$\cos \beta = \frac{1}{\sqrt{1 + \tan^2 \beta}} = \frac{1}{\sqrt{10}}, \quad \sin \beta = \sqrt{1 - \frac{1}{10}} = \frac{3}{\sqrt{10}},$$

$$\cos 2\beta = \cos^2 \beta - \sin^2 \beta = -\frac{8}{10} = -\frac{4}{5},$$

$$\sin 4\beta = 4\sin \beta \cdot \cos \beta \cdot \cos 2\beta = 4 \cdot \frac{3}{\sqrt{10}} \cdot \frac{1}{\sqrt{10}} \cdot \left(-\frac{4}{5}\right) = -\frac{24}{25}.$$

Finally,

$$\cos(2\arctan 2) - \sin(4\arctan 3) = -\frac{3}{5} - \left(-\frac{24}{25}\right) = \frac{9}{25}.$$

Answer: $\frac{9}{25}$.

Problem 21. Solve the equation

$$\sin^2 4x + \cos^2 x = 2\sin 4x \cos^4 x.$$

Solution. Rewrite the equation as

$$\sin^2 4x - 2\sin 4x \cos^4 x = -\cos^2 x.$$

Now, add $\cos^8 x$ to both sides of the equation:

$$\sin^2 4x - 2\sin 4x \cos^4 x + \cos^8 x = \cos^8 x - \cos^2 x,$$
$$(\sin 4x - \cos^4 x)^2 = -\cos^2 x(1 - \cos^6 x).$$

The left-hand side of the last equation is a perfect square, so it is a nonnegative number. The right-hand side is a nonpositive number (since $\cos^2 x \geq 0$ and $1 - \cos^6 x \geq 0$). Therefore, the equality holds only when the left-hand side and the right-hand side equal 0 simultaneously:

$$\sin 4x - \cos^4 x = 0,$$
$$\cos^2 x(1 - \cos^6 x) = 0.$$

Let's first solve the second equation because it looks easier.

$$\cos^2 x = 0 \text{ or } 1 - \cos^6 x = 0.$$

From the first equation, $x = \frac{\pi}{2} + \pi n, n \in Z$.
From the second equation, $x = \pi m, m \in Z$.
Now, instead of solving the equation $\sin 4x - \cos^4 x = 0$, we will check if the solutions already obtained satisfy it.

$$\sin 4\left(\frac{\pi}{2} + \pi n\right) - 0 = \sin(2\pi + 4\pi n) = 0,$$

which means that the set of solutions $x = \frac{\pi}{2} + \pi n, n \in Z$ satisfies both equations.

$$\sin 4\pi m - \cos^4 \pi m = 0 - 1 = -1 \neq 0.$$

Thus, the set of solutions $x = \pi m, m \in Z$ does not satisfy both equations.

Answer: $x = \frac{\pi}{2} + \pi n, n \in Z$.

Problem 22. Solve the equation

$$\sin x + \sqrt{3}\cos x = 1.$$

Solution. Multiply both sides of the equation by $\frac{1}{2}$:

$$\frac{1}{2}\sin x + \frac{\sqrt{3}}{2}\cos x = \frac{1}{2},$$

$$\sin x \cos\frac{\pi}{3} + \cos x \sin\frac{\pi}{3} = \frac{1}{2},$$

$$\sin\left(x + \frac{\pi}{3}\right) = \frac{1}{2},$$

$$x + \frac{\pi}{3} = (-1)^n \cdot \frac{\pi}{6} + \pi n, n \in Z \quad \text{or} \quad x = (-1)^n \cdot \frac{\pi}{6} - \frac{\pi}{3} + \pi n, n \in Z.$$

Answer: $(-1)^n \cdot \frac{\pi}{6} - \frac{\pi}{3} + \pi n, n \in Z.$

Problem 23. Solve the equation

$$\cos x + 2\cos 2x + \cdots + n\cos nx = \frac{n^2 + n}{2},$$

where n is a natural number.

Solution. Note that since the range of the cosine function does not exceed 1 in absolute value, the following inequalities are true for the addends on the left-hand side of the equation:

$$\cos x \le 1, 2\cos 2x \le 2, \ldots, n\cos nx \le n.$$

Therefore,

$$\cos x + 2\cos 2x + \cdots + n\cos nx \le 1 + 2 + \cdots + n = \frac{n^2 + n}{2}.$$

The equality is possible if and only if

$$\cos x = 1,$$
$$\cos 2x = 1,$$
$$\cdots$$
$$\cos nx = 1.$$

Solutions of each equation are

$$x = 2\pi k, k \in Z; x = \pi k, k \in Z; \ldots; x = \frac{2\pi k}{n}, k \in Z.$$

Since we need to identify the common solution satisfying each equation, then the answer is $x = 2\pi k, k \in Z$.

Answer: $2\pi k, k \in Z.$

Chapter 8

Problem 19. Solve the equation

$$3^{2x+4} + 45 \cdot 6^x - 9 \cdot 2^{2x+2} = 0.$$

Solution. Rewrite the equation as

$$81 \cdot 3^{2x} + 45 \cdot 2^x \cdot 3^x - 36 \cdot 2^{2x} = 0.$$

Divide both sides by 2^{2x}:

$$81 \cdot \frac{3^{2x}}{2^{2x}} + 45 \cdot \frac{3^x}{2^x} - 36 = 0.$$

The substitution is $y = \frac{3^x}{2^x}$, $y > 0$. It follows that

$$81y^2 + 45y - 36 = 0,$$
$$9y^2 + 5y - 4 = 0,$$
$$y = -1 \text{ or } y = \frac{4}{9}.$$

The first solution has to be rejected because it does not satisfy the condition $y > 0$. Substitute the second solution for y to find x: $\frac{3^x}{2^x} = \frac{4}{9}$ or equivalently, $\left(\frac{3}{2}\right)^x = \left(\frac{3}{2}\right)^{-2}$. Thus, $x = -2$.

Answer: -2.

Problem 20. Solve the equation

$$2^{\sin^2 x} + 4 \cdot 2^{\cos^2 x} = 6.$$

Solution. Rewrite the equation:

$$2^{\sin^2 x} + 4 \cdot 2^{1-\sin^2 x} = 6,$$
$$2^{\sin^2 x} + 4 \cdot 2 \cdot 2^{-\sin^2 x} - 6 = 0.$$

The substitution is $y = 2^{\sin^2 x}$, $y > 0$.
　Solving the equation $y + \frac{8}{y} - 6 = 0$, or equivalently, $y^2 - 6y + 8 = 0$ gives $y = 4$ or $y = 2$. Therefore, $2^{\sin^2 x} = 4$ or $2^{\sin^2 x} = 2$.
　From the first equation, $\sin^2 x = 2$. There are no solutions, since a sine can't exceed 1 in absolute value.
　From the second equation $\sin^2 x = 1$, so $\sin x = \pm 1$ and $x = \frac{\pi}{2} + \pi k, k \in Z$.

Answer: $\frac{\pi}{2} + \pi k, k \in Z$.

Problem 21. Solve the equation

$$2 + 3\log_5 2 - x = \log_5(3^x - 5^{2-x}).$$

Solution.

$$2 - x = \log_5(3^x - 5^{2-x}) - \log_5 2^3,$$
$$2 - x = \log_5 \frac{(3^x - 5^{2-x})}{8},$$
$$5^{2-x} = \frac{(3^x - 5^{2-x})}{8},$$
$$8 \cdot 5^{2-x} = 3^x - 5^{2-x}.$$

Transferring 5^{2-x} to the left side gives

$$9 \cdot 5^{2-x} = 3^x,$$
$$9 \cdot 25 \cdot 5^{-x} = 3^x,$$
$$3^x \cdot 5^x = 9 \cdot 25,$$
$$15^x = 15^2,$$

from which $x = 2$.

    Now, verify the solutions. The left-hand side becomes $2 + 3\log_5 2 - 2 = 3\log_5 2 = \log_5 2^3 = \log_5 8$. The right-hand becomes $\log_5(3^2 - 5^{2-2}) = \log_5(9 - 1) = \log_5 8$. So, $\log_5 8 = \log_5 8$.

Answer: 2.

Problem 22. Solve the equation

$$\log(81 \cdot \sqrt[3]{3^{x^2-8x}}) = 0.$$

Solution. The domain of the equation is the set of all real numbers.

$$\log\left(3^4 \cdot 3^{\frac{x^2-8x}{3}}\right) = 0,$$
$$3^{4+\frac{x^2-8x}{3}} = 1,$$
$$3^{\frac{x^2-8x+12}{3}} = 3^0,$$
$$x^2 - 8x + 12 = 0,$$

from which $x = 2$ or $x = 6$.

Answer: 2, 6.

Problem 23. Solve the equation

$$3\log_x 4 + 2\log_{4x} 4 + 3\log_{16x} 4 = 0.$$

Solution. The domain of this equation is the set of real numbers such that $x > 0$, $x \neq 1$. Clearly, we can transform to a logarithm with base 4.

$$\frac{3}{\log_4 x} + \frac{2}{\log_4 4x} + \frac{3}{\log_4 16x} = 0,$$

$$\frac{3}{\log_4 x} + \frac{2}{1 + \log_4 x} + \frac{3}{2 + \log_4 x} = 0.$$

The substitution is $y = \log_4 x$, $y \neq 0$, $y \neq -1$, $y \neq -2$.

$$\frac{3}{y} + \frac{2}{1+y} + \frac{3}{2+y} = 0.$$

After a few simplifications we get to the equation $4y^2 + 8y + 3 = 0$. Solutions are $y = -\frac{1}{2}$ or $y = -\frac{3}{2}$. Then $\log_4 x = -\frac{1}{2}$ or $\log_4 x = -\frac{3}{2}$, from which $x = \frac{1}{2}$ or $x = \frac{1}{8}$.

Answer: $\frac{1}{2}, \frac{1}{8}$.

Problem 24. Solve the equation

$$(x+1)^{x^2+3x} = (x+1)^{10x-12}.$$

Solution. Step 1. Solve the equation for the equality of bases, $x + 1 = 1$. Thus, $x = 0$.

Step 2. Solve the equation for the equality of exponents, $x^2 + 3x = 10x - 12$,

$$x^2 - 7x + 12 = 0.$$
$$x = 3 \text{ or } x = 4.$$

Answer: 0, 3, 4.

Problem 25. Solve the equation

$$(x^2 - x - 1)^{x^2-1} = 1.$$

Solution. Step 1. Solve the equation for the equality of bases,

$$x^2 - x - 1 = 1,$$
$$x^2 - x - 2 = 0,$$
$$x = -1 \text{ or } x = 2.$$

Step 2. Solve the equation for the equality of the exponents,

$$x^2 - 1 = 0,$$
$$x = -1 \text{ or } x = 1.$$

Answer: $-1, 1, 2$.

Chapter 10

Problem 19. Solve the equation

$$2 + y^2 - 2y = 1 - \cos^2 \frac{x}{y}.$$

Solution. Rewrite the equation as

$$2 + y^2 - 2y - 1 = -\cos^2 \frac{x}{y},$$

$$y^2 - 2y + 1 = -\cos^2 \frac{x}{y},$$

$$(y - 1)^2 = -\cos^2 \frac{x}{y}.$$

The left-hand side of the equation is a nonnegative number, while the right-hand side is a nonpositive number. The equality is possible only when each side equals to 0.

$$y - 1 = 0,$$

$$\cos^2 \frac{x}{y} = 0.$$

$$y = 1,$$

$$\cos x = 0.$$

Then, $x = \frac{\pi}{2} + \pi k, k \in z.$

Answer: $x = \frac{\pi}{2} + \pi k, k \in Z, y = 1.$

Problem 20. Solve the equation

$$\left(2 + \frac{1}{\cos^2 x}\right)(4 - 2\cos^4 x) = 1 + 5\sin 3y.$$

Solution. Since $|\sin 3y| \leq 1$, $1 + 5\sin 3y \leq 1 + 5 = 6$. On the other hand, $2 + \frac{1}{\cos^2 x} \geq 2 + 1 = 3$ and $4 - 2\cos^4 x \geq 4 - 2 = 2$. Thus,

$$\left(2 + \frac{1}{\cos^2 x}\right)(4 - 2\cos^4 x) \geq 6.$$

Therefore, the equality will be possible only when

$$2 + \frac{1}{\cos^2 x} = 3,$$

$$4 - 2\cos^4 x = 2,$$

$$1 + 5\sin 3y = 6.$$

It follows

$$\cos^2 x = 1,$$
$$\cos^4 x = 1,$$
$$\sin 3y = 1.$$

Solutions of the system are

$$x = \pi n, n \in Z,$$
$$y = \frac{\pi}{6} + \frac{2\pi k}{3}, k \in Z.$$

Answer: $x = \pi n, n \in Z, y = \frac{\pi}{6} + \frac{2\pi k}{3}, k \in Z.$

Problem 21. Solve the equation

$$2^{|x|} - \cos y + \ln(1 + x^2 + |y|) = 0.$$

Solution. Rewrite the equation as

$$2^{|x|} + \ln(1 + x^2 + |y|) = \cos y.$$

The right-hand side of the equation does not exceed 1, $\cos y \leq 1$.

Let's consider the left-hand side of the equation and analyze it. Since $|x| \geq 0$, $2^{|x|} \geq 1$. Also $\ln \underbrace{\left(1 + x^2 + |y|\right)}_{\geq 0} \geq \ln 1 = 0$. It follows that $2^{|x|} + \ln(1 + x^2 + |y|) \geq 1$. Thus, the equality is possible only when the following system of equations has solutions:

$$2^{|x|} = 1,$$
$$\ln(1 + x^2 + |y|) = 0,$$
$$\cos y = 1.$$

The solution of the first equation is $x = 0$.
The solution of the second equation is $y = 0$.
The solution of the third equation is $y = 2\pi k, k \in Z$.
Therefore, the common solutions are $x = 0, y = 0$.

Answer: $x = 0, y = 0$.

Problem 22. Solve the equation

$$\sqrt{2x - y} + \sqrt{2z - x} + \sqrt{2y - z} = \frac{\sqrt{81(2x - y)(2z - x)(2y - z)}}{\sqrt{(2x - y)(2z - x)} + \sqrt{(2x - y)(2y - z)} + \sqrt{(2z - x)(2y - z)}}.$$

The solution is similar to the solution of problem 13.

Answer: (k, k, k), where $k > 0$.

Problem 23. Solve the equation

$$\sqrt{(1+u)^2+(v-3)^2}+\sqrt{(u+2)^2+9v^2}=|2v+3|.$$

Solution. Let's consider three points in a plane Cartesian coordinate system $A(u,3)$, $B(-1,v)$, and $C(u+1,-2v)$. $AB=\sqrt{(1+u)^2+(v-3)^2}$, $BC=\sqrt{(u+2)^2+9v^2}$, and $AC=\sqrt{1+(2v+3)^2}$. Due to the *triangle inequality*, for any three points A, B, and C, the sum of the lengths of any two sides must be greater than the length of the remaining side. The equality $AB+BC=AC$ holds when three points are collinear. Therefore, in any case, the following inequality holds for A, B, and C: $AB+BC\geq AC$. It implies that

$$\sqrt{(1+u)^2+(v-3)^2}+\sqrt{(u+2)^2+9v^2}\geq\sqrt{1+(2v+3)^2}>\sqrt{(2v+3)^2}=|2v+3|.$$

We conclude that the original equation has no solutions.

Problem 24. Solve the equation

$$x^2+xy-\sqrt{3}x+y^2-\sqrt{3}y+1=0.$$

Solution. Rewrite the equation as

$$x^2+x(y-\sqrt{3})+(y^2-\sqrt{3}y+1)=0$$

and solve it as a quadratic in x.

$$D=(y-\sqrt{3})^2-4(y^2-\sqrt{3}y+1)=y^2-2\sqrt{3}y+3-4y^2+4\sqrt{3}y-4$$
$$=-3y^2+2\sqrt{3}y-1=-(\sqrt{3}y-1)^2.$$

The original equation has solutions when $D\geq 0$, or $-(\sqrt{3}y-1)^2\geq 0$. Since the left-hand side of the last inequality is a nonpositive number, it follows that the condition $D\geq 0$ is satisfied only for $y=\frac{1}{\sqrt{3}}$. Substituting this value into the original equation gives

$$x^2+\frac{1}{\sqrt{3}}x-\sqrt{3}x+\left(\frac{1}{\sqrt{3}}\right)^2-\sqrt{3}\cdot\frac{1}{\sqrt{3}}+1=0,$$

which simplifies to $x^2-\frac{2}{\sqrt{3}}x+\frac{1}{3}=0$, or equivalently, $\left(x-\frac{1}{\sqrt{3}}\right)^2=0$. The solution of the last equation is $x=\frac{1}{\sqrt{3}}$.

Answer: $x=\frac{1}{\sqrt{3}}$, $y=\frac{1}{\sqrt{3}}$.

Problem 25. Solve the equation

$$(x^2-4x+5)(y^2-6y+10)=1.$$

Solution. Each of the factors on the left-hand side is a quadratic trinomial with the first coefficient 1, which is a positive number. Recalling the properties of a quadratic function, the minimum value of each factor will be achieved at the vertex of the parabola for each quadratic function (the parabola opens upward). Let's consider the function $f(x) = x^2 - 4x + 5$. It has a minimum when $x = \frac{-b}{2a} = \frac{4}{2} = 2$. For the second function, $f(y) = y^2 - 6y + 10$, its minimum value is attained when $y = \frac{-b}{2a} = \frac{6}{2} = 3$. Therefore, the left-hand side of the equation is greater than or equal to $(2^2 - 4 \cdot 2 + 5)(3^2 - 6 \cdot 3 + 10) = 1 \cdot 1 = 1$ with equality possible only when $x = 2$, $y = 3$. Hence, the only solutions of the given equation are $x = 2$, $y = 3$.

Answer: $x = 2$, $y = 3$.

Chapter 11

Let's consider the proof that the sequence of numbers n_i, $\{1, 2, 5, 13, 34, \ldots\}$, defined by the recursive formula $n_{i+1} = 3n_i - n_{i-1}$ with $n_1 = 1$ and $n_2 = 2$, represents the sequence of odd terms in the Fibonacci sequence.

The explicit formula for the sequence n_i was obtained during the solution of the Main Problem as

$$n_i = \frac{\sqrt{5}-1}{2\sqrt{5}} \cdot \left(\frac{3+\sqrt{5}}{2}\right)^i + \frac{\sqrt{5}+1}{2\sqrt{5}} \cdot \left(\frac{3-\sqrt{5}}{2}\right)^i.$$

Multiplying the numerator and denominator of the first addend by $\sqrt{5}+1$ and of the second addend by $\sqrt{5}-1$ and recalling that $\frac{3+\sqrt{5}}{2} = \left(\frac{\sqrt{5}+1}{2}\right)^2$ and $\frac{3-\sqrt{5}}{2} = \left(\frac{1-\sqrt{5}}{2}\right)^2$ gives

$$\frac{(\sqrt{5}-1)\cdot(\sqrt{5}+1)}{2\sqrt{5}\cdot(\sqrt{5}+1)} \cdot \left(\frac{3+\sqrt{5}}{2}\right)^i + \frac{(\sqrt{5}+1)\cdot(\sqrt{5}-1)}{2\sqrt{5}\cdot(\sqrt{5}-1)} \cdot \left(\frac{3-\sqrt{5}}{2}\right)^i$$

$$= \frac{5-1}{2\sqrt{5}\cdot(\sqrt{5}+1)} \cdot \left(\frac{\sqrt{5}+1}{2}\right)^{2i} + \frac{5-1}{2\sqrt{5}\cdot(\sqrt{5}-1)} \cdot \left(\frac{1-\sqrt{5}}{2}\right)^{2i}$$

$$= \frac{4}{2\sqrt{5}\cdot(\sqrt{5}+1)} \cdot \left(\frac{\sqrt{5}+1}{2}\right)^{2i} + \frac{4}{2\sqrt{5}\cdot(\sqrt{5}-1)} \cdot \left(\frac{1-\sqrt{5}}{2}\right)^{2i}$$

$$= \frac{2\cdot(\frac{\sqrt{5}+1}{2})^{2i}}{\sqrt{5}\cdot(\sqrt{5}+1)} + \frac{2\cdot(\frac{1-\sqrt{5}}{2})^{2i}}{\sqrt{5}\cdot(\sqrt{5}-1)}$$

$$= \frac{2\cdot(\sqrt{5}-1)\cdot(\frac{\sqrt{5}+1}{2})^{2i} + 2\cdot(\sqrt{5}+1)\cdot(\frac{1-\sqrt{5}}{2})^{2i}}{\sqrt{5}\cdot((\sqrt{5})^2 - 1^2)}$$

$$= \frac{2\cdot(\sqrt{5}-1)\cdot(\frac{\sqrt{5}+1}{2})\cdot(\frac{\sqrt{5}+1}{2})^{2i-1} + 2\cdot(\sqrt{5}+1)\cdot(\frac{1-\sqrt{5}}{2})\cdot(\frac{1-\sqrt{5}}{2})^{2i-1}}{\sqrt{5}\cdot 4}$$

$$= \frac{2 \cdot \frac{5-1}{2} \cdot \left(\frac{\sqrt{5}+1}{2}\right)^{2i-1} - 2 \cdot \frac{5-1}{2} \cdot \left(\frac{1-\sqrt{5}}{2}\right)^{2i-1}}{\sqrt{5} \cdot 4}$$

$$= \frac{4 \cdot \left(\frac{\sqrt{5}+1}{2}\right)^{2i-1} - 4 \cdot \left(\frac{1-\sqrt{5}}{2}\right)^{2i-1}}{\sqrt{5} \cdot 4}$$

$$= \frac{\left(\frac{\sqrt{5}+1}{2}\right)^{2i-1} - \left(\frac{1-\sqrt{5}}{2}\right)^{2i-1}}{\sqrt{5}}.$$

The last expression is *Binet's Formula* for the odd numbers $n = 2i - 1$ of the Fibonacci sequence. The proof is completed.

During the solution of the third part of the Main Problem we determined that for odd Fibonacci numbers, as proved above,

$$n_i = \frac{\sqrt{5}-1}{2\sqrt{5}} \cdot \left(\frac{3+\sqrt{5}}{2}\right)^i + \frac{\sqrt{5}+1}{2\sqrt{5}} \cdot \left(\frac{3-\sqrt{5}}{2}\right)^i$$

$$= \frac{\left(\frac{\sqrt{5}+1}{2}\right)^{2i-1} - \left(\frac{1-\sqrt{5}}{2}\right)^{2i-1}}{\sqrt{5}},$$

the number $5n_i^2 - 4$ is a perfect square.

Let's show now that for even Fibonacci numbers, $n_i = \frac{\left(\frac{\sqrt{5}+1}{2}\right)^{2i} - \left(\frac{1-\sqrt{5}}{2}\right)^{2i}}{\sqrt{5}}$, the respective number $5n_i^2 + 4$ has to be a perfect square. In fact, by doing this, we will complete the proof of the direct statement of the necessary and sufficient conditions for recognizing a positive integer as a Fibonacci number formulated in chapter 11. Indeed,

$$5n_i^2 + 4 = 5 \cdot \left(\frac{\left(\frac{\sqrt{5}+1}{2}\right)^{2i} - \left(\frac{1-\sqrt{5}}{2}\right)^{2i}}{\sqrt{5}}\right)^2 + 4$$

$$= 5 \cdot \frac{\left(\left(\frac{\sqrt{5}+1}{2}\right)^2\right)^{2i} - 2 \cdot \left(\frac{\sqrt{5}+1}{2}\right)^{2i} \cdot \left(\frac{1-\sqrt{5}}{2}\right)^{2i} + \left(\left(\frac{\sqrt{5}-1}{2}\right)^2\right)^{2i}}{5} + 4$$

$$= \left(\frac{5+2\sqrt{5}+1}{4}\right)^{2i} - 2 \cdot \left(\frac{5-1}{4}\right)^{2i} + \left(\frac{5-2\sqrt{5}+1}{4}\right)^{2i} + 4$$

$$= \left(\frac{3+\sqrt{5}}{2}\right)^{2i} - 2 \cdot 1 + \left(\frac{3-\sqrt{5}}{2}\right)^{2i} + 4$$

$$= \left(\frac{3+\sqrt{5}}{2}\right)^{2i} + 2 + \left(\frac{3-\sqrt{5}}{2}\right)^{2i}.$$

Note that $\frac{3+\sqrt{5}}{2} \cdot \frac{3-\sqrt{5}}{2} = \frac{9-5}{4} = 1$. Therefore, we can rewrite the above expression as a perfect square:

$$\left(\frac{3+\sqrt{5}}{2}\right)^{2i} + 2 + \left(\frac{3-\sqrt{5}}{2}\right)^{2i}$$

$$= \left(\frac{3+\sqrt{5}}{2}\right)^{2i} + 2 \cdot \underbrace{\left(\frac{3+\sqrt{5}}{2}\right)^{i} \cdot \left(\frac{3-\sqrt{5}}{2}\right)^{i}}_{=1} + \left(\frac{3-\sqrt{5}}{2}\right)^{2i}$$

$$= \left(\left(\frac{3+\sqrt{5}}{2}\right)^{i} + \left(\frac{3-\sqrt{5}}{2}\right)^{i}\right)^{2}.$$

In problem 5 it was proved that $\left(\frac{3+\sqrt{5}}{2}\right)^{i} + \left(\frac{3-\sqrt{5}}{2}\right)^{i} - 2$ is a natural number. Then $\left(\frac{3+\sqrt{5}}{2}\right)^{i} + \left(\frac{3-\sqrt{5}}{2}\right)^{i}$ is a natural number as well. Hence, we conclude that for any even Fibonacci number n_i the respective number $5n_i^2 + 4$ has to be the perfect square of some integer.

Let's now revisit the statement about the necessary and sufficient conditions for recognizing a natural number as a Fibonacci number:

A positive integer F_n is a Fibonacci number if and only if $5F_n^2 + 4$ is a perfect square for even n or $5F_n^2 - 4$ is a perfect square for odd n.

As we just completed the proof that if F_n is a Fibonacci number, then $5F_n^2 + 4$ is a perfect square for even n or $5F_n^2 - 4$ is a perfect square for odd n, we now have to prove the converse statement, that if one or both numbers $5F_n^2 \pm 4$ are perfect squares, then the number F_n has to be an nth term in the Fibonacci sequence, an even term when $5F_n^2 + 4$ is a perfect square or an odd term when $5F_n^2 - 4$ is a perfect square. Instead of taking a straightforward approach to the proof (given $5F_n^2 \pm 4$ is a perfect square, prove that F_n is the nth term in the Fibonacci sequence), we will do it a little differently. Working with Binet's Formula, we will modify it to evaluating the golden ratio φ in terms of the Fibonacci numbers and transforming to the equivalent expression, which, as it will be shown, holds only under the conditions that $5F_n^2 \pm 4$ is a perfect square. This will justify our converse statement.

Binet's Formula, as it was derived in chapter 11, establishes a relationship between the Fibonacci number F_n, the nth term in the Fibonacci sequence, and the golden ratio. Let's now rearrange it, so we can express a power of a golden ratio in terms of a Fibonacci number. Recalling that $\frac{1+\sqrt{5}}{2} = \varphi$ and observing that $\frac{1+\sqrt{5}}{2} \cdot \frac{1-\sqrt{5}}{2} = \frac{1-5}{4} = -1$, we see that $\frac{1-\sqrt{5}}{2} = -\frac{1}{\varphi}$. Hence, we can rewrite Binet's Formula for the Fibonacci numbers $F_n = \dfrac{\left(\frac{1+\sqrt{5}}{2}\right)^{n} - \left(\frac{1-\sqrt{5}}{2}\right)^{n}}{\sqrt{5}}$ as $F_n = \dfrac{\varphi^{n} - \left(-\frac{1}{\varphi}\right)^{n}}{\sqrt{5}}$.

For even powers n this formula becomes

$$F_n = \frac{\varphi^n - \left(\frac{1}{\varphi}\right)^n}{\sqrt{5}} \qquad (1)$$

and for odd powers n it can be written as

$$F_n = \frac{\varphi^n + \left(\frac{1}{\varphi}\right)^n}{\sqrt{5}}. \qquad (2)$$

Consider first the case for even powers and transform (1) as

$$\varphi^{2n} - \sqrt{5}F_n \cdot \varphi^n - 1 = 0.$$

Substituting $\varphi^n = y$ into the last expression, it can be rewritten as a quadratic equation for y, $y^2 - \sqrt{5}F_n \cdot y - 1 = 0$, which we will use to express φ^n in terms of F_n. Its solutions are $y = \frac{\sqrt{5}F_n \pm \sqrt{5F_n^2 + 4}}{2}$. Going back to our substitution for y, we see that $\varphi^n = \frac{\sqrt{5}F_n \pm \sqrt{5F_n^2 + 4}}{2}$. Note that

$$\sqrt{5F_n^2 + 4} > \sqrt{5F_n^2} = \sqrt{5}F_n,$$

so $\sqrt{5}F_n - \sqrt{5F_n^2 + 4} < 0$. Clearly, $\varphi^n > 0$, hence the negative value $\varphi^n = \frac{\sqrt{5}F_n - \sqrt{5F_n^2 + 4}}{2}$ has to be rejected, and we are left with $\varphi^n = \frac{\sqrt{5}F_n + \sqrt{5F_n^2 + 4}}{2}$. Analogous results will be obtained by examining the case for odd powers in (2). In that case, after solving the quadratic equation $\varphi^{2n} - \sqrt{5}F_n \cdot \varphi^n + 1 = 0$, φ^n will be expressed in terms of F_n as $\varphi^n = \frac{\sqrt{5}F_n \pm \sqrt{5F_n^2 - 4}}{2}$. It is easy to verify that $\varphi^n = \frac{\sqrt{5}F_n - \sqrt{5F_n^2 - 4}}{2}$ does not hold and has to be rejected. Indeed, for example, when $n = 1$, $\varphi^1 = \frac{1+\sqrt{5}}{2}$, while $\frac{\sqrt{5}F_1 - \sqrt{5F_1^2 - 4}}{2} = \frac{\sqrt{5} \cdot 1 - \sqrt{5 \cdot 1 - 4}}{2} = \frac{\sqrt{5} - 1}{2}$. We see that $\frac{1+\sqrt{5}}{2} \neq \frac{\sqrt{5}-1}{2}$. Hence, for odd powers, the expression for φ^n in terms of F_n is $\varphi^n = \frac{\sqrt{5}F_n + \sqrt{5F_n^2 - 4}}{2}$. Therefore, combining the cases, we arrive at

$$\varphi^n = \frac{\sqrt{5}F_n + \sqrt{5F_n^2 \pm 4}}{2}. \qquad (3)$$

It was proved in chapter 11 that $\varphi^n = F_n \cdot \varphi + F_{n-1}$, so substituting this expression for φ^n into (3) gives $F_n \cdot \varphi + F_{n-1} = \frac{\sqrt{5}F_n + \sqrt{5F_n^2 \pm 4}}{2}$. Recalling that $\varphi = \frac{1+\sqrt{5}}{2}$, this can be rewritten as

$$F_n \cdot \frac{1+\sqrt{5}}{2} + F_{n-1} = \frac{\sqrt{5}F_n + \sqrt{5F_n^2 \pm 4}}{2},$$

$$\frac{1}{2}F_n + \frac{\sqrt{5}}{2}F_n + F_{n-1} = \frac{\sqrt{5}}{2}F_n + \frac{\sqrt{5F_n^2 \pm 4}}{2}.$$

Cancelling $\frac{\sqrt{5}}{2}F_n$ on both sides, we arrive at $\frac{1}{2}F_n + F_{n-1} = \frac{\sqrt{5F_n^2 \pm 4}}{2}$ or equivalently, $F_n + 2F_{n-1} = \sqrt{5F_n^2 \pm 4}$. At this point we can stop and draw important conclusions to finish the proof. However, I suggest we go one step further and modify it to a more presentable and interesting form, utilizing the fact that for Fibonacci numbers, $F_n + F_{n-1} = F_{n+1}$. Note that

$$F_n + 2F_{n-1} = F_n + F_{n-1} + F_{n-1} = (F_n + F_{n-1}) + F_{n-1} = F_{n+1} + F_{n-1}.$$

Hence, we obtain

$$F_{n+1} + F_{n-1} = \sqrt{5F_n^2 \pm 4}. \tag{4}$$

The left side of the last expression is a natural number as it is the sum of two natural numbers. Therefore, equality holds only when the right side is a natural number as well, which is possible only when the number under the square root sign is a perfect square. So, analyzing the last expression (4), we see that in order for F_n to be a Fibonacci number and for Binet's Formula to hold (since we utilized it to get to the last equality), $5F_n^2 \pm 4$ has to be a perfect square.

In other words, since Binet's Formula holds for all natural n, a positive integer F_n is a Fibonacci number only if one of the numbers $5F_n^2 + 4$ (for even n) or $5F_n^2 - 4$ (for odd n) is a perfect square, as was to be proved.

To conclude our analysis, we rewrite (3) as

$$n = \log_\varphi \frac{\sqrt{5}F_n + \sqrt{5F_n^2 \pm 4}}{2}.$$

This allows us to identify the position in the sequence of a specific Fibonacci number F_n. Furthermore, we can see from (4) that the square root of $5F_n^2 + 4$ or $5F_n^2 - 4$ equals the sum of the preceding and the following terms for the considered Fibonacci number F_n; an interesting result on its own, which we can add to the list of the properties of the Fibonacci numbers.

To demonstrate the above property, let's consider, for example, the ninth term in the Fibonacci sequence, $F_9 = 34$. Since it is an odd term, $5F_9^2 - 4$ has to be a perfect square. We get $5F_9^2 - 4 = 5 \cdot 34^2 - 4 = 5776 = 76^2$. Hence, $\sqrt{5F_9^2 - 4} = 76$. Observing that the eighth and tenth terms in the Fibonacci sequence are $F_8 = 21$ and $F_{10} = 55$, it's easy to verify that $F_{10} + F_8 = \sqrt{5F_9^2 - 4}$. Indeed, $21 + 55 = 76$.

Back in my first year as an undergraduate, my favorite calculus professor and scientific adviser told freshmen that while becoming a mathematician, one usually goes through several development stages. First, gaining an extensive knowledge in mathematics, you learn how to apply it in solving problems starting from simple ones to more and more complicated tasks. Passing this stage, you take one step further into deriving conclusions from solved problems and posing new questions, while formulating your own problems based on the results. Finally, at the last stage, you look for answers to the inquiries by solving new problems, and if lucky, even make generalizations. As a matter of fact, when working on the Main Problem from chapter 11, we went step by step through all these stages. We solved three variations of the problem, starting with the relatively simple problem for $k = 1$, progressing to a more difficult one of finding two solutions when $k = 2$, and finally solving the complicated case 3 by proving that the equation has indefinitely many natural solutions when $k = 2$. All these steps represented just the first stage, after passing which we came across a few amazing outcomes from our solutions. By selecting the sequence of numbers n_i, $\{1, 2, 5, 13, 34, \ldots\}$, which clarified the path to our solution, we realized that unexpectedly we worked with a subset of the Fibonacci numbers, all its odd terms, and we concluded that for each n_i the respective $a_i = 5n_i^2 - 4$ is a perfect square. Isn't it natural to inquire if a similar property would hold for even Fibonacci numbers, i.e., should $a_i = 5n_i^2 + 4$ be a perfect square for the even Fibonacci terms? What about the converse statements? Will they hold? That's how we formulated the assumption regarding the necessary and sufficient conditions for recognizing a Fibonacci number. Finally, at the last stage, we managed to get positive answers to all these questions and even obtain some new interesting relationships and formulas for the Fibonacci numbers.

Chapter 12

Problem 23. Solve the system of equations

$$x^2 - xy = 24,$$
$$y^2 - xy = -8.$$

Solution. Subtracting the second equation from the first gives $x^2 - y^2 = 32$ and adding the equations gives $x^2 - 2xy + y^2 = 16$. Thus, the system can be rewritten as

$$x^2 - y^2 = 32,$$
$$x^2 - 2xy + y^2 = 16.$$

Or equivalently, as

$$(x - y)(x + y) = 32,$$
$$(x - y)^2 = 16.$$

The system now is reduced to the following two systems of equations:

$$(x - y)(x + y) = 32,$$
$$x - y = 4.$$

or

$$(x - y)(x + y) = 32,$$
$$x - y = -4.$$

Solve the first system:

$$x + y = 8,$$
$$x - y = 4.$$

The solutions are $x = 6$, $y = 2$.
Solve the second system:

$$x + y = -8,$$
$$x - y = -4.$$

The solutions are $x = -6$, $y = -2$.

Answer: $(6, 2)$, $(-6, -2)$.

Problem 24. Solve the system of equations

$$\sqrt[3]{x} \cdot \sqrt{y} + \sqrt{x} \cdot \sqrt[3]{y} = 12,$$
$$xy = 64.$$

Solution. Rewrite the first equation as an exponential with fractional powers and then factor it:

$$x^{\frac{1}{6}} y^{\frac{1}{6}} \left(x^{\frac{1}{6}} y^{\frac{1}{3}} + x^{\frac{1}{3}} y^{\frac{1}{6}} \right) = 12.$$

From the second equation, $xy = 64$. This implies that $x^{\frac{1}{6}} y^{\frac{1}{6}} = 2$. Thus, the first equation can be simplified to $x^{\frac{1}{6}} y^{\frac{1}{3}} + x^{\frac{1}{3}} y^{\frac{1}{6}} = 6$, and the system can be rewritten as

$$x^{\frac{1}{6}} y^{\frac{1}{3}} + x^{\frac{1}{3}} y^{\frac{1}{6}} = 6,$$
$$x^{\frac{1}{6}} y^{\frac{1}{6}} = 2.$$

Let $x^{\frac{1}{6}} = u$ and $y^{\frac{1}{6}} = v$. Then

$$uv^2 + u^2 v = 6,$$
$$uv = 2.$$

Let's solve this system.

$$uv(v + u) = 6,$$
$$uv = 2.$$

312 The Equations World

It follows that

$$v + u = 3,$$
$$uv = 2.$$

The solutions are $u_1 = 1$, $v_1 = 2$ or $u_2 = 2$, $v_2 = 1$. Therefore, $x_1 = 1$, $y_1 = 64$ or $x_2 = 64$, $y_2 = 1$.

Answer: $(1, 64)$, $(64, 1)$.

Problem 25. Solve the system of equations

$$x^3 + y^3 = 65,$$
$$x^2y + xy^2 = 20.$$

Solution. Factor the equations

$$(x + y)(x^2 - xy + y^2) = 65,$$
$$xy(x + y) = 20.$$

Make substitutions $x + y = u$, $xy = v$. It follows that

$$x^2 - xy + y^2 = (x + y)^2 - 3xy = u^2 - 3v,$$

and the system can be rewritten as

$$u(u^2 - 3v) = 65,$$
$$uv = 20.$$

Therefore,

$$u^3 - 3uv = 65,$$
$$uv = 20.$$

Substituting $uv = 20$ in the first equation gives

$$u^3 - 60 = 65, \qquad \text{or equivalently,} \qquad u^3 = 125,$$
$$uv = 20. \qquad\qquad\qquad\qquad\qquad uv = 20.$$

Solutions of the last system are $u = 5$, $v = 4$. Thus,

$$x + y = 5,$$
$$xy = 4.$$
$$x_1 = 1, y_1 = 4 \text{ or } x_2 = 4, y_2 = 1.$$

Answer: $(1, 4)$, $(4, 1)$.

Problem 26. Solve the system of equations

$$x^{-1} + y^{-1} = 5,$$
$$x^{-2} + y^{-2} = 13.$$

Solution. Make the substitutions $x^{-1} = u$, $y^{-1} = v$. Then the system can be rewritten as

$$u + v = 5,$$
$$u^2 + v^2 = 13.$$

Express u in terms of v from the first equation and substitute it in the second equation:

$$u = 5 - v,$$
$$(5 - v)^2 + v^2 = 13.$$

The second equation simplifies to $v^2 - 5v + 6 = 0$. Therefore, $v_1 = 2$, $v_2 = 3$, so $u_1 = 3$, $u_2 = 2$. Thus, $x_1 = \frac{1}{3}$, $y_1 = \frac{1}{2}$ or $x_2 = \frac{1}{2}$, $y_2 = \frac{1}{3}$.

Answer: $\left(\frac{1}{3}, \frac{1}{2}\right)$, $\left(\frac{1}{2}, \frac{1}{3}\right)$.

Problem 27. Solve the system of equations

$$x^2 + 2y + \sqrt{x^2 + 2y + 1} = 1,$$
$$2x + y = 2.$$

Solution. Introducing the new variable t as $t = \sqrt{x^2 + 2y + 1}$ ($t \geq 0$) modifies the first equation to $t^2 - 1 + t = 1$ or equivalently, $t^2 + t - 2 = 0$. The solutions of this quadratic equation are $t = 1$, $t = -2$. The negative root $t = -2$ has to be rejected. Therefore, the original system simplifies to

$$x^2 + 2y + 1 = 1,$$
$$2x + y = 2.$$

Solving for y from the second equation and substituting into the first equation gives

$$x^2 + 2(2 - 2x) + 1 = 1,$$
$$y = 2 - 2x.$$

The first equation can be rewritten as $x^2 - 4x + 4 = 0$ or $(x - 2)^2 = 0$, from which $x = 2$. Therefore, $y = 2 - 2x = -2$.

Answer: $(2, -2)$.

Problem 28. Evaluate the expression $xy + 2yz + 3xz$, if x, y, and z are the positive solutions of the system of equations

$$x^2 + xy + \frac{y^2}{3} = 25,$$

$$\frac{y^2}{3} + z^2 = 16,$$

$$z^2 + zx + x^2 = 9.$$

Solution. As in the solution of problem 18 in chapter 12, we will not be solving the system to find the values of the variables, but instead will be making use of the geometrical interpretation of each equation. First, let's recall the *Law of cosines*, which relates the lengths of the sides of a triangle to the cosine of one of its angles. It states that for any triangle with the sides a, b, and c, and angle γ between sides a and b the following equality holds:

$$a^2 + b^2 - 2ab \cdot \cos \gamma = c^2.$$

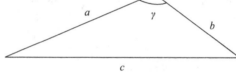

Consider triangle AOC with sides x, $\frac{y}{\sqrt{3}}$, 5, and an angle of $150°$ between the sides x and $\frac{y}{\sqrt{3}}$.

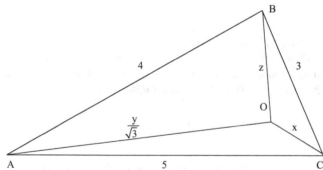

If we rewrite the first equation as $x^2 + \left(\frac{y}{\sqrt{3}}\right)^2 - 2x \cdot \frac{y}{\sqrt{3}} \cdot \underbrace{\cos 150°}_{-\frac{\sqrt{3}}{2}} = 5^2$, then it

represents the application of law of cosines to the triangle. Similarly, if we rewrite the third equation of the system as $z^2 + x^2 - 2zx \cdot \underbrace{\cos 120°}_{-\frac{1}{2}} = 3^2$, then it has to

hold for the triangle BOC with the sides x, z, 3, and angle $120°$ between the sides x and z.

Finally, the second equation of the system we consider as the Pythagorean Theorem applied to the right triangle AOB with the sides $\frac{y}{\sqrt{3}}$, z, and 4. Note that $\angle AOB + \angle BOC + \angle COA = 90° + 120° + 150° = 360°$. Putting these three triangles together so that their equal sides coincide, a new triangle ABC will be formed. Its area equals the sum of the areas of the triangles AOB, BOC, and COA. Recall that the area of the right triangle AOB equals half the product of its legs, $S_{AOB} = \frac{1}{2}AO \cdot OB$, and the area of each of the other two triangles BOC and AOC can be calculated as half the product of two sides by the sine of the angle they form, $S_{BOC} = \frac{1}{2}BO \cdot OC \cdot \sin \angle BOC$, $S_{AOC} = \frac{1}{2}AO \cdot OC \cdot \sin \angle AOC$. So, we obtain $S_{ABC} = S_{AOB} + S_{BOC} + S_{AOC} = \frac{1}{2} \cdot \frac{y}{\sqrt{3}} \cdot z + \frac{1}{2} \cdot zx \cdot \sin 120° +$ $\frac{1}{2} \cdot \frac{y}{\sqrt{3}} \cdot x \cdot \sin 150° = \frac{yz}{2\sqrt{3}} + \frac{\sqrt{3}zx}{4} + \frac{xy}{4\sqrt{3}} = \frac{1}{4\sqrt{3}}(2yz + 3xz + xy)$, from which $2yz + 3xz + xy = 4\sqrt{3}S_{ABC}$. Triangle ABC has sides with lengths 3, 4, and 5. Since $3^3 + 4^2 = 5^2$, then it is the right triangle with legs 3 and 4 and hypotenuse 5. The area of ABC equals half the product of its legs, $S_{ABC} = \frac{1}{2} \cdot 3 \cdot 4 = 6$. Finally, $2yz + 3xz + xy = 4\sqrt{3}S_{ABC} = 4\sqrt{3} \cdot 6 = 24\sqrt{3}$.

Problem 29. Solve the system of equations

$$x + y + z = 2,$$
$$xy + xz + yz = -1,$$
$$xyz = -2.$$

Solution. Consider the cubic equation $t^3 - 2t^2 - t + 2 = 0$, which we can construct using Viète's formulas for the cubic equation. Rewriting the equation by grouping its terms gives

$$t^2(t - 2) - (t - 2) = 0,$$
$$(t - 2)(t^2 - 1) = 0.$$

Thus, $t - 2 = 0$ or $t^2 - 1 = 0$. It follows that $t = 2$, or $t = \pm 1$. Therefore, the solutions of the system are all permutations of -1, 1, and 2.

Answer: $(1, -1, 2)$, $(1, 2, -1)$, $(-1, 1, 2)$, $(-1, 2, 1)$, $(2, -1, 1)$, $(2, 1, -1)$.

References

Jean-Pierre Tignol, *"Galois' Theory of Algebraic Equations"*, World Scientific, 2011

Polya, G., *Mathematical Discovery*, John Wiley & Sons, 1997

Johnson, A., *"Famous Problems and Their Mathematicians"*, Greenwood Village, CO: Teacher's Idea Press, 1999

Mario Livio, *"The Golden Ratio: The Story of Phi, the World's Astonishing Number"*, Broadway Books, 2002

Mario Livio, *"The Equation That Couldn't Be Solved: How Mathematical Genius Discovered the Language of Symmetry"*, Souvenir Press, 2006

М. И. Сканави, *"Сборник конкурсных задач по математике для поступающих во втузы"*, Москва "Высшая Школа" , 1980 (in Russian)

В. Гусев, В. Литвиненко, А.Мордкович, *"Практикум по решению математических задач"*, Москва "Просвещение", 1985 (in Russian)

В.Чистяков, *"Старинные задачи по элементарной математике"*, Минск "Вышэйшая школа", 1978 (in Russian)

Ш. Горделадзе, М. Кухарчук, Ф. Яремчук, *"Збірник Конкурсних Задач з Математики"*, Київ, Вища Школа, 1988 (in Ukrainian)

M. Hazewinkel, *"Trigonometric functions"*, Encyclopedia of Mathematics, Springer, 2001

S. Gindikin, *"Carl Fridrich Gauss"*, *Quantum*, November/December 1999

L. Ryzhkov and Y. Ionin, *"Homogeneous Equations"*, *Quantum*, May/June 1998

А. Панчишкин и Е. Шавгулидзе, *"Тригонометрические функции в задачах"*, Москва "Наука", 1986 (in Russian)

E. Beckenback and R. Bellman, *"An Introduction to Inequalities"*, Washington, D.C. Mathematical Association of America, 1961

C. Boyer, *"A History of Mathematics"* (2nd edition), John Willey & Sons, Inc., 1991

A. Yegorov, "What you add is what you take", *Quantum*, November/December 1994

M. Saul and T. Andreescu, *"Completing the square"*, *Quantum*, November/December 1998

A. Yegorov, "Inequalities become equalities", *Quantum*, March/April 2000

A. Yaglom and I. Yaglom, *"Challenging Mathematical Problems with Elementary Solutions"*, Translated by J. McCawley, San Francisco: Holden-Day, 1957

В. Болтянский, Ю. Сидоров, М. Шабунин, *"Лекции и задачи по элементарной математике"*, Москва, "Наука", 1972 (in Russian)

Nickalls, R. W. D. (July 2006), *"Viéte, Descartes and the cubic equation"*, Mathematical Gazette, *90*: 203–208

I. Bashmakova, G. Smirnova, *"The Beginnings and Evolution of Algebra"*, Mathematical Association of America, 2000

Я. Перельман, *"Занимательная Алгебра. Занимательная Геометрия"*, Москва АСТ, 1999 (in Russian)

Я. Перельман, *"Задачи и Головоломки"*, АСТ Москва, 2008 (in Russian)

Я. Суконник, *"Математические задачи повышеной трудности"*, Киев, "Радянська Школа", 1985 (in Russian)

Н. Васильев, В. Гутенмахер, Ж. Раббот, А. Тоом, *"Заочные Математические Олимпиады"*, Москва, "Наука", 1986 (in Russian)

В. Лейфура, *"Диофантові рівняння"*, *"У Світі Математики"*, Київ, Радянська Школа", 1985 (in Ukrainian)

Yuri, V. Matiyasevich, *Hilbert's Tenth Problem*, MIT Press, Cambridge, Massachusetts, 1993

Nikolai Vorob'ev, *"Fibonacci Numbers"*, Dover Publications Inc., 2011

Петраков И.С., *"Математические кружки в 8-10 классах"*, Москва, "Просвещение", 1987 (in Russian)

Сергев И.Н., *"Зарубежные Математические Олимпиады"*, Москва, "Наука", 1987 (in Russian)

B. Pritsker, *"Geometrical Kaleidoscope"*, Dover Publications, 2017

Index